NEW DEVELOPMENTS IN CELL APOPTOSIS RESEARCH

NEW DEVELOPMENTS IN CELL APOPTOSIS RESEARCH

ALAN J. CORVIN
EDITOR

Nova Science Publishers, Inc.
New York

For permission to use material from this book please contact us:
Telephone 631-231-7269; Fax 631-231-8175
Web Site: http://www.novapublishers.com

Library of Congress Cataloging-in-Publication Data

New developments in cell apoptosis research / Alan J. Corvin (editor).
p. ; cm.
Includes bibliographical references and index.
ISBN 13 978-1-60021-453-0
ISBN 10 1-60021-453-3
1. Apoptosis. I. Corvin, Alan J.
[DNLM: 1. Apoptosis. QU 375 N5314 2006]
QH671.N495 2006
571.9'36--dc22 2006030104

Published by Nova Science Publishers, Inc. ✦ New York

Contents

Preface

Apoptosis is the regulated form of cell death. It is a complex process defined by a set of characteristic morphological and biochemical features that involves the active participation of affected cells in a self-destruction cascade. This programmed cell death plays a critical role in physiological functions such as cell deletion during embryonic development, balancing cell number in continuously renewing tissues and immune system development. Additionally, a dysregulation of apoptosis is underlying in numerous pathological situations such as Parkinson, Alzheimer's disease and cancer. A number of studies have pointed out an association between consumption of fruits and vegetables, and certain beverages such as tea and wine, which are rich in polyphenols, with reduced risk of chronic diseases, including cancer. Apoptosis is also the regulatory mechanism involved in the removal of unnecessary cells during development and in tissue homeostasis in a wide range of organisms from insects to mammals. The aim of this book is to provide new studies in the field of apoptosis research.

Chapter 1 - Contrary to what might be common sense, the ordered development and maintenance of multicellular organisms not only requires adequate nutrient supply and a flourishing growth potential, but also relies on complex mechanisms of cell suicide. In fact, both tissue sculpting during embryogenesis and their homeostatic regulation in adulthood are the outcome of a dynamic competition between cell proliferation and death, and a network of signals has evolved to induce and keep under a tight control the process of cell elimination. In addition, cells damaged by diverse noxious agents must be efficiently cleared away through a regulated death program to avoid lethal dysfunctions. Accordingly, cells have developed a variety of molecular controllers that sense every potentially harmful metabolic change, damage or genetic abnormality. The unifying response to all these unbalancing or stressful conditions is the activation of a form of programmed cell death termed apoptosis. Apoptotic cells utilize a well-defined biochemical machinery to orderly dismantle their architecture, and become endowed with stereotypical morphological changes which ultimately lead to their engulfment by phagocytic cells. Apoptosis signaling can be grossly subdivided in an initiator phase, which depends either on death receptor activation or on intracellular insults, and in an execution phase, whose biochemical hallmark is the activation of proteases termed caspases. In between these two extremes, a sophisticated system composed by molecular checks and balances determines whether a cell is committed to survive or to die. This final output depends on several integrated parameters, such as the transcriptional regulation of apoptosis inducers or brakes and their subcellular distribution and interactions. In recent years, the comprehension of the molecular bases of apoptosis allowed the identification of an articulated network of crosstalks between the death regulators and signaling pathways that coordinate biological processes such as cell cycle and proliferation. Disturbances in the fine tuning of this intricate death/survival equilibrium result either in an uncontrolled growth or in an excessive degeneration of tissues, leading to a variety of diseases, including cancer, neurodegenerative and autoimmunity disease. Further layers of complexity are added by the finding that other forms of cell death, spanning from autophagy to necrosis through a variety of ill-defined intermediate conditions, are relevant in defined physiologic and pathologic settings. A deep understanding of the mechanisms that regulate the apoptotic machinery, its interactions with other transduction pathways and how these orchestrate the fate of the cell is therefore pivotal for the development of drugs that reinstate the appropriate apoptotic response, either where it is lacking, as in cancer, or where it is exacerbated, as in neurodegenerative diseases.

Chapter 2 - Oxysterols are known to be present at high concentrations in atherosclerotic plaques and contribute to the development of atherosclerosis by various mechanisms. It is found that 7-ketocholesterol, one of the oxysterols accumulated in atherosclerotic lesion, induced apoptosis of vascular smooth muscle cells (SMC). This finding indicates that accumulation of 7-ketocholesterol in atherosclerotic lesions may decrease local cell numbers and render atherosclerotic plaques vulnerable. In this paper, we summarize the significance of apoptosis induced by 7-ketocholesterol in the development of vulnerable plaque, mainly according to our published results. In concrete terms, the findings are as followed:

Lipids extracted from fat deposits in human atheromatous lesions induce apoptosis of SMC.

Among the deposited lipids, 7-ketocholesterol is present at high concentration during the early stage of atherosclerosis and contributes to induce the apoptosis.

Apoptosis of SMC induced by 7-ketocholesterol is accelerated when SMC acquire a high proliferative potential.

7-Ketocholesterol increases intracellular Ca^{2+} in human SMC, and as a result, caspase-3 activity is increased. Ca channel blockers inhibit 7-ketocholesterol-induced apoptosis of SMC.

7-ketocholesterol-induced apoptosis of SMC is inhibited by an inhibitor of hydroxymethyl glutaryl CoA (HMG-CoA) reductase.

From these findings, we propose that 7-ketocholesterol contained in plaques is a strong candidate as a cause of vulnerable plaque.

Chapter 3 - Cell death, specifically apoptosis, is a feature of host-parasitoid relationships involving parasitic wasps. Though there are numerous examples of endosymbiotic virus-induced (e.g., polydnaviruses) and venom-mediated apoptosis in host hemocytes, there is very little evidence for this form of cell death being triggered in other host tissues. Is this due to specificity in tissue responsiveness? In terms of timing of events during parasitism (i.e., at the initial stages of parasitism), this is a likely scenario. But throughout the duration of the association between the parasitoid and host, several developmental and physiological changes in host tissues may be regulated by apoptotic signals. There simply is very little evidence to draw any conclusions. Somewhat surprisingly, in most instances when apoptosis has been identified in host tissue (hemocytes), investigations of the pathways involved or specific intracellular changes have apparently not been conducted. Consequently, apoptosis is known to be induced by polydnaviruses, but what relationships exist between this form of cell death and those triggered by wasp venoms are unclear, nor have any attempts been made to compare mechanisms of oncosis evoked by these viruses and parasitoid venoms. What follows is an attempt to show homology in the modes of death evoked by pathogenic insect viruses, endosymbiotic viruses that reside within endoparasitic wasps, and venoms produced by ectoparasitic wasps. As will be discussed, oncosis and apoptosis are intimately linked by common signal transduction pathways and mitochondria. The focus of this paper is to attempt to provide evidence, admittedly at times circumstantial, and speculation on the possible linkage between the modes of action of ectoparasitic wasp venoms and insect viruses.

Chapter 4 - The cellular response to genotoxic stress that damages DNA includes cell cycle arrest, activation of DNA repair, and in the event of irreparable damage, induction of apoptosis. However, the signals that determine cell fate, that is, survival or apoptosis, are largely unknown. The δ isoform of protein kinase C (PKCδ) has been implicated in many important cellular processes, including regulation of apoptotic cell death. The available information supports a model in which certain sensors of DNA lesions activate PKCδ. This activation is triggered in part by tyrosine phosphorylation of PKCδ by c-Abl tyrosine kinase. PKCδ is further proteolytically activated by caspase-3. The cleaved catalytic fragment of PKCδ translocates to the nucleus and induces apoptosis. Importantly, accumulating data have revealed the nuclear targets for PKCδ in the induction of apoptosis. A pro-apoptotic function of activated PKCδ is mediated by at least several downstream effectors known to be associated with the elicitation of apoptosis. Recent findings also demonstrated that PKCδ is involved in cell cycle-specific activation and induction of apoptotic cell death. Moreover, previous studies have shown that PKCδ regulates transcription by phosphorylating various transcription factors, including the p53 tumor suppressor that is critical for cell cycle arrest and apoptosis in response to DNA damage. These findings collectively support a pivotal role for PKCδ in the induction of apoptosis with significant impact. This review is focused on the current views regarding the regulation of cell fate by PKCδ in response to DNA damage.

Chapter 5 - The mammalian heat shock proteins Hsp70, Hsp27 and Hsp90 are known as major molecular chaperones which preserve the native structure of cellular proteins and promote refolding or degradation of stress-damaged protein molecules. For a long time, it was generally accepted that these Hsps are involved in cytoprotection from proteotoxic (protein-damaging) exposures only but not from genotoxic (DNA-damaging) ones. If so, it would be hard to imagine that the Hsps are able somehow to protect mammalian cells against severe ionizing radiation that induces numerous breaks in genomic DNA thereby leading to cell death. Nevertheless, more and more data are published about real contribution of Hsp70, Hsp27 and Hsp90 to protection from the cell-killing effects of irradiation with gamma- or X-rays. Taken together these data allow to conclude that the Hsp-conferred radioprotection is mainly based on an ability of certain Hsps to block the irradiation-responsive apoptotic pathway and/or to stimulate alternative 'cell survival' pathways that interfere with apoptotic signal transduction. The present review affords an intriguing insight into a phenomenon of the Hsp-mediated cellular defense against the radiation-induced apoptosis. Implication of this phenomenon in some radiation-related issues of biology and medicine such as DNA damage/repair, genomic instability, oncogenesis and tumor cell resistance to radiotherapy is also considered herein. Moreover, several prospective approaches are discussed toward pharmacological manipulating the *in vivo* Hsp expression/activity to beneficially influence the outcome of anticancer radiotherapy.

Chapter 6 - There is increasing evidence that RNA products of the mobile genetic elements SINE (such as Alu sequences in primate cells or B1 and B2 elements in rodent cells) can play a considerable role in apoptosis of the cell. All these SINEs are represented in a great number in genomes, possess intragenic promoters and are transcribed by DNA-dependent RNA polymerase III. A series of experiments have shown that diverse stress influences cause different SINE expression activation in the cells. Depending on the severity of a damage caused by the stress, the cell can survive and recover or enter apoptosis. Accumulation of the SINE RNA products in the cell can serve as a pro-apoptotic factor through several mechanisms. (i) It was shown that stress influences not only cause SINE expression activation, but also induce expression of another class of mobile genetic elements – LINEs, coding for reverse transcriptase. Alu RNAs and some other SINE RNA products are the proper templates for reverse transcriptase and can actively replicate through retrotransposition and insert into the active chromatin sites. This in turn can result in disruption of normal gene expression pattern and, as a result, in apoptosis pathway triggering. (ii) In addition, Alu RNA was shown to be able to inhibit translation, which also can induce apoptosis. Alu RNA is highly homologous to 7SL RNA and can bind to SRP (signal recognition particle) proteins possessing affinity to ribosome. It is possible that such Alu-containing SRP bind to ribosomes independent of the presence of N-terminal export signal in nascent polypeptide, and inhibit total translation in the cell. On the other hand, Alu RNAs contain double strand sites and, upon the increase of their level in the cell, can regulate pro-apoptotic protein PKR (dsRNA-activated protein kinase). PKR activation also inhibits total translation. But the specific translation of newly synthesized mRNAs (these can be mRNAs of stress or/and pro-apoptotic proteins) in this case is activated. Considering the possible role of SINE RNA products in apoptosis, of particular interest is the problem of the activation of their transcription by RNA polymerase III during this process. This activation seems to be specific to SINEs, as the transcription of the other studied genes directed by RNA polymerase III is suppressed during apoptosis. The conceivable mechanisms of SINE transcription regulation in apoptosis are considered in this chapter on the basis of literature and the data.

Chapter 7 - Chromosomal DNA dysfunction plays a role in mammalian cell death. Oxidative stress induces chromatin dysfunction such as single strand and double strand DNA fragmentation leading to cell death through apoptosis or necrosis. More than 1 Mb giant DNA fragments, 200-800 kb or 50-300 kb high molecular weight (HMW) DNA fragments and internucleosomal low molecular weight (LMW) DNA fragments are produced not only by oxidative stresses such as ROS, ionizing irradiation and ultraviolet radiation, but also by some chemotherapeutic drugs that induce apoptosis or necrosis in several types of mammalian cells. The mechanismof DNA fragmentation during cell death is poorly understood, although several endonucleases have been implicated in the process. Some nucleases involved in chromosomal DNA fragmentation in apoptosis and necrosis are classified.

Chapter 8 - Growth, differentiation and apoptosis are mutually affected during organogenesis. Elucidating these interactions should be important for understanding the developmental regulation of morphogenesis, homeostasis and organ size control, all of which are commonly found in multicellular organisms. We report here that the overgrowth phenotype by *dco*^{3}, an allele of *discs overgrown* gene which is the *Drosophila* ortholog of mammalian Casein kinase I δ/ε, causes apoptosis during development of the wing imaginal disc. This apoptosis strongly depends on c-Jun N-terminal Kinase (JNK), a stress-activated protein kinase. When apoptosis was blocked by inactivating JNK, the overgrowth phenotype became prominently malignant, indicating that the role of this apoptosis is for elimination of the overgrown tissue. To find the initial event during this apoptosis, we generated a genetic mosaic in the developing wing imaginal disc in which partial cell populations show the overgrown phenotype by *dco*^{3}. Interestingly, JNK activation was induced at the either side of the boundary between the *dco*^{3} mutant and normal cell populations. Furthermore, activation of JNK outside of the *dco*^{3} cell population (that is normal cell population) contributed to the efficient removal of the *dco*^{3} cell population. Although this type of nonautonomous apoptosis has widely been known to be a result of abrupt change in morphogen signal gradient by such as Decapentaplegic (Dpp, a TGF-β superfamily member) and Wingless (Wg, a Wnt family member), the nonautonomous apoptosis by *dco*^{3} was not accompanied by aberration of these morphogen signal levels. In particular, unlike the well-established relationship between Dco and several components of Wg signal pathway, *dco*^{3} did not significantly affect Wg signaling in the wing imaginal disc. These results demonstrated that the nonautonomous apoptosis that begins at the boundary between abnormal and normal cell populations is not only found in the case of inappropriate morphogen signaling but also associated with deregulated tissue growth such as tumorigenesis to restore and assure normal morphogenesis.

Chapter 9 - Recently, increasing interest has been focused on the elucidation of the cellular and molecular aspects of cadmium (Cd)-dependent regulation of gene expression and signal transduction pathways in different model system. From available literature data, it appears that the scenario is multifaceted since Cd evokes in cells a number of responses that involve not only death-signalling- but also cellular protective reactions against the toxicity. Concerning breast cancer, there are few studies on the role played by Cd on ER^{-}(estrogen receptor-negative) human breast cancer cells, that are expected to be insensitive to the metalllo-estrogenic effect exerted by Cd on ER^{+} breast cancer cells. Therefore, we wanted to examine the implications and cellular effects of Cd treatment on ER^{-} MDA-MB231 cells, obtained from mammary adenocarcinoma and endowed with high malignant potential, and, in parallel, on immortalized non-tumoral HB2 breast epithelial cells for comparison. In particular, in the present study we selected the minor concentration of the metal/exposition time capable of inducing an approximately 50% decrease of cell number (IC_{50}), estimated actual cellular uptake of Cd by atomic absorption spectrophotometry, and analyzed the typology of the cytotoxic effect by the evaluation of i) the physical state of the DNA, ii) the membrane localization of phosphatydilserine, and, iii) the expression levels of stress response genes such as those coding for heat shock proteins and metallothioneins, and for factors and enzymes involved in the onset of apoptosis, by conventional and semi-quantitative multiplex PCR. The results indicate that, although both cytotypes do not show some signs of "classical" apoptosis (oligonucleosomal fragmentation of DNA, annexin binding of plasmalemma), a very different scenario emerges from the evaluation of gene expression levels. Treated non-tumoral HB2 cells exhibited a pattern typical of intoxicated cells with up-regulation of *hsp27*, *hsp70* and *hsc70* only, down-regulation or absence of expression of other genes assayed (e.g *caspase-8* and *metallothioneins*), and accumulation of the spliced form of XBP1 mRNA. On the other hand, Cd treatment of neoplastic cells resulted in the up-regulation of some genes (e.g. *hsp27*, *metallothioneins*) that can be interpreted as an attempt to resist to the toxic effect of the metal, but also in the down-regulation of other genes (e.g other *hsps* and *Bcl-2*) that, occurring in parallel with the up-regulation of death-promoting genes, such as those coding for Bad and some caspases, might account for final cell death which shares some aspects with both oxidative stress and apoptosis-like program.

Chapter 10 - Death is an intricate process in all living cellular organisms. Apoptosis, also known as programmed cell death, is a normal physiological process occurring in all eukaryotic cells. This normal cell death process has been known for over half a century. Mechanisms leading to apoptotic cell death have also been studied extensively. Induction of tumour cell apoptosis becomes a new research direction for the treatment of malignant disease. In the past decade, photodynamic therapy (PDT) has been developed as a tumour therapy. The treatment relies on the accumulation of photosensitizers in the tumour mass. Tumour cell death, triggered by irradiating the tumour mass with visible light, is one of the major mechanisms leading to tumour eradication. Generation of singlet oxygen and induction of oxidative stress are the major intracellular events after PDT. In view of the short half-life of reactive oxygen species, the localization of photosensitizers in various subcellular compartments, such as the plasma membrane, mitochondria, lysosomes, Golgi body, and endoplasmic reticulum has been implicated in the induction of various modes of cell death. Photosensitizers targeting mitochondria received particular interest as mitochondria are the key organelles in the mitochondria-mediated apoptotic pathway. Other cellular events, such as activation of transcription factors and intracellular stress signaling pathways and changes of Ca^{2+} and lipid metabolism are also seen in PDT treated cells. This review focuses on the understanding of PDT-induced tumour cell apoptosis.

Chapter 11 - Death is an intricate process in all living cellular organisms. Apoptosis, also known as programmed cell death, is a normal physiological process occurring in all eukaryotic cells. This normal cell death process has been known for over half a century. Mechanisms leading to apoptotic cell death have also been studied extensively. Induction of tumour cell apoptosis becomes a new research direction for the treatment of malignant disease. In the past decade, photodynamic therapy (PDT) has been developed as a tumour therapy. The treatment relies on the accumulation of photosensitizers in the tumour mass. Tumour cell death, triggered by irradiating the tumour mass with visible light, is one of the major mechanisms leading to tumour eradication. Generation of singlet oxygen and induction of oxidative stress are the major intracellular events after PDT. In view of the short half-life of reactive oxygen species, the localization of photosensitizers in various subcellular compartments, such as the plasma membrane, mitochondria, lysosomes, Golgi body, and endoplasmic reticulum has been implicated in the induction of various modes of cell death. Photosensitizers targeting mitochondria received particular interest as mitochondria are the key organelles in the mitochondria-mediated apoptotic pathway. Other cellular events, such as activation of transcription factors and intracellular stress signaling pathways and changes of Ca^{2+} and lipid metabolism are also seen in PDT treated cells. This review focuses on the understanding of PDT-induced tumour cell apoptosis.

In: New Developments in Cell Apoptosis Research
Editor: Alan J. Corvin, pp. 1-52
ISBN: 1-60021-453-3

Chapter I

Apoptosis and Disease: Unbalancing the Survival Equilibrium

Federica Chiara and Andrea Rasola*
Department of Biomedical Sciences, Università di Padova, Padova, Italia

Abstract

Contrary to what might be common sense, the ordered development and maintenance of multicellular organisms not only requires adequate nutrient supply and a flourishing growth potential, but also relies on complex mechanisms of cell suicide. In fact, both tissue sculpting during embryogenesis and their homeostatic regulation in adulthood are the outcome of a dynamic competition between cell proliferation and death, and a network of signals has evolved to induce and keep under a tight control the process of cell elimination. In addition, cells damaged by diverse noxious agents must be efficiently cleared away through a regulated death program to avoid lethal dysfunctions. Accordingly, cells have developed a variety of molecular controllers that sense every potentially harmful metabolic change, damage or genetic abnormality. The unifying response to all these unbalancing or stressful conditions is the activation of a form of programmed cell death termed apoptosis. Apoptotic cells utilize a well-defined biochemical machinery to orderly dismantle their architecture, and become endowed with stereotypical morphological changes which ultimately lead to their engulfment by phagocytic cells. Apoptosis signaling can be grossly subdivided in an initiator phase, which depends either on death receptor activation or on intracellular insults, and in an execution phase, whose biochemical hallmark is the activation of proteases termed caspases. In between these two extremes, a sophisticated system composed by molecular checks and balances determines whether a cell is committed to survive or to die. This

* Correspondence: Department of Biomedical Sciences, Università di Padova, I-35121 Padova, Italia; tel: +39 049 8276062; fax: +39 049 8276361; Email: rasola@bio.unipd.it

final output depends on several integrated parameters, such as the transcriptional regulation of apoptosis inducers or brakes and their subcellular distribution and interactions. In recent years, the comprehension of the molecular bases of apoptosis allowed the identification of an articulated network of crosstalks between the death regulators and signaling pathways that coordinate biological processes such as cell cycle and proliferation. Disturbances in the fine tuning of this intricate death/survival equilibrium result either in an uncontrolled growth or in an excessive degeneration of tissues, leading to a variety of diseases, including cancer, neurodegenerative and autoimmunity disease. Further layers of complexity are added by the finding that other forms of cell death, spanning from autophagy to necrosis through a variety of ill-defined intermediate conditions, are relevant in defined physiologic and pathologic settings. A deep understanding of the mechanisms that regulate the apoptotic machinery, its interactions with other transduction pathways and how these orchestrate the fate of the cell is therefore pivotal for the development of drugs that reinstate the appropriate apoptotic response, either where it is lacking, as in cancer, or where it is exacerbated, as in neurodegenerative diseases.

Introduction

The term apoptosis was proposed in 1972 to discriminate on a morphological basis between programmed cell death (PCD) and another type of cell demise, necrosis (Kerr et al., 1972), but the phenomenon of physiological cell death had been already discovered independently by several investigators in differing tissues and in various organisms (Vogt, 1842, Glucksmann, 1951, Lockshin and Williams, 1965), reviewed in (Hacker and Vaux, 1997). In the last few years, tremendous advances have been made in the comprehension of the apoptotic process. We now know that apoptotic PCD is driven by a genetically determined cell suicide program and is endowed with a highly stereotypical series of morphological and biochemical changes, as displayed by electron microscopic analyses (Figure 1). Cells committed to collapse by apoptotic death show cytoplasmic shrinkage, formation of dark bodies in the cytosol, dynamic blebbing of plasma membrane vesicles (zeiosis) and loss of plasma membrane lipid asymmetry (Zornig et al., 2001). In addition, apoptotic adherent cells round up and ultimately detach from the matrix (Frisch and Screaton, 2001). Organelles within the shrunken apoptotic cytoplasm retain a largely normal appearance, except for some dilation of the endoplasmic reticulum and swelling of the outer mitochondrial membrane (Vander Heiden et al., 1997). However, apoptotic morphological changes are especially evident in the nucleus, where the chromatin condenses and forms globular or crescent-shaped compact aggregates near the nuclear membrane (chromatin margination), which in turn expands and becomes convoluted, whilst the nucleolus undergoes abnormal enlargement and granulation. This chromatin compaction with the eventual rupture of the nuclear envelope is termed karyorrhexis (Leist and Jaattela, 2001; Zornig et al., 2001). Chromatin is also cleaved by different endonucleases that cut DNA initially into 50-300 kb fragments and subsequently into 180 bp fragments, which appear as a typical "ladder" when size fractionated on an agarose gel (Kerr et al., 1972; Wyllie et al., 1980; Brown and Rose,

1992; Cohen et al., 1992; Oberhammer et al., 1993; Walker and Sikorska, 1997). Finally, apoptotic bodies containing various cellular fragments are formed and blebbed away of the dying cell. The PCD process is defined as apoptosis when these features include the activation of cystein proteases termed caspases, which are largely responsible for most of these catabolic changes (Strasser et al., 2000; Zornig et al., 2001).

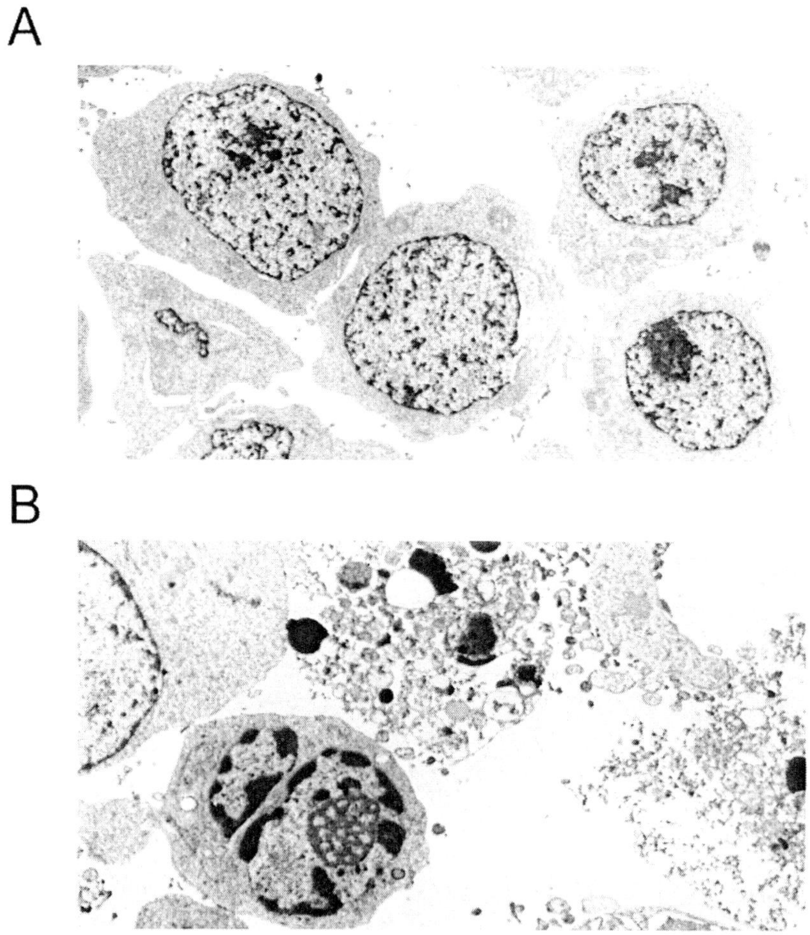

Figure 1. Electron microscopy inspection of lymphoma T cells in healthy conditions (A) or undergoing apoptosis (B) highlights the ultrastructural changes occurring during apoptotic cell death:perinuclear chromatin condensation, karyorrhexis, cytosolic vesicles and zeiosis are clearly visible.

In contrast to this ordered, step-wise cell dismantling, the necrotic cell swells up in an unrestricted way (cellular oedema) following rapid and large changes in critical ions, osmolarity, and pH, and its chromatin takes on the appearance of a lumpy mass in which DNA is non-specifically cleaved. Nuclei eventually disappear to leave a ghost, and neither karyorrhexis nor zeiosis are observed. Membrane integrity is rapidly disrupted and the cell is eventually lysed. Cell debris are dispersed in the environment, where they can induce an inflammatory reaction and even autoimmune reactions. Necrosis normally occurs after cell exposure to a high intensity stress, such as strong detergents or oxidants, ionophores, extreme heat or cold or high concentration of poisons (Leist and Jaattela, 2001; Nicotera et al., 1999). Nonetheless, recent evidence indicates that apoptosis and necrosis can be viewed as far ends of a continuum of death modes, with varying contributions of the cellular machinery. These intermediate or alternative programs of cell demise and their relevance in development, physiology and disease will be discussed below.

Apoptosis in Phylogenesis, Ontogenesis and Physiology

Apoptosis is a biological process of fundamental importance. Its functions include a normal embryogenesis, tissue homeostasis and maintenance of tissue architecture, immune selection through deletion of autoreactive or non-reactive lymphocytes, and elimination of non-functional or damaged cells (Jacobson et al., 1997).

The importance of programmed cell death is highlighted by its evolutionary conservation throughout phylogenesis from prokaryotes to vertebrates, and even in plants. The use of similar mechanisms to induce death across very distant species also argues for a common origin of this biological program. In bacteria, PCD plays an important role in a number of processes, such as sporulation (Ameisen, 1996) or the maintenance of the optimal population density, as it is triggered by toxic factors accumulating with increasing density of the culture (Gordeeva et al., 2004). Some strains of *Escherichia coli* produce an inactive protease, which activates and kills the bacterium after phage infection (Lewis, 2000). Hence, this enzyme shares functional features with the otherwise unrelated caspases, *i.e.* the key PCD effectors in animals. In general, death-inducing bacterial proteins are coded by plasmids and prophages, suggesting that acquisition of these proteins turned out to be highly adaptive following phage infection.

In the most primitive eukaryotes, PCD is mainly triggered by oxidative stress and Ca^{2+}. In dinoflagellates and in chlorophyte algae, typical markers of apoptosis, such as chromatin margination, degradation of the nucleus, DNA fragmentation and protoplast shrinkage are observed, and a cysteine protease inhibitor prevents cell death. In addition, antibodies against mammalian caspases cross-react with alga proteins that specifically induce cell death (Vardi et al., 1999; Segovia et al., 2003). PCD in plants regulates common processes such as morphogenesis of the embryo, organ abscission, formation of vessels, and the response to environmental insults. Animal PCD modulators, such as the Bcl-2 family of pro- and antiapoptotic proteins, or the antiapoptotic IAP proteins, are active on PCD when expressed in plants or yeasts (Uren et al., 1999). Moreover, homologs of genes that control apoptosis in animals are encoded in plant genomes and serve the same function in PCD control. In contrast, there are no evident homologs of caspases in plant or fungal genomes, but caspase-like activities regulate PCD in plants, and a caspase-related family of proteases termed metacaspases was found in plant and fungal genomes; (Uren et al., 2000; Hoeberichts and Woltering, 2003; Sanmartin et al., 2005).

Caspases are instead present in most lower eukaryotes. For instance, in *Leishmania major*, an ancient parasitic unicellular protozoan, caspases induce zeiosis, phosphatidylserine redistribution and mitochondrial and nuclear changes (Arnoult et al., 2002). Remarkably, mitochondrial proteins involved in caspase-independent cell death are conserved from bacteria to mammals (Gordeeva et al., 2004). As in higher eukaryotes mitochondria promote cell death both by activating caspases and by releasing caspase-independent pro-apoptotic proteins, one can envision that during evolution mitochondria-dependent caspase activation added as a further death pathway to increase the complexity of previously instated mechanisms. Accordingly, initiatorcaspase activity and membrane death receptors, which activate caspases following extracellular stimuli, appeared only later in evolution (Kobayashi and Endoh, 2003).

In general, most eukaryotes have inhibitors of apoptosis. An intriguing explanation for this could be that these inhibitors were introduced by viruses and then adapted to the cell requirements. Actually, most viruses possess mechanisms to forestall apoptosis in order to make viral reproduction more efficient by using the host to nurture virus propagation (Roulston et al., 1999; Meier et al., 2000).

Several PCD-like features are also shared with other biological routines in simple eukaryotes, and this is reminiscent of the interplay between apoptosis and other processes in higher vertebrates. Chromatin condensation and cytoplasmic shrinkage occur in the multicellular development of a single celled eukaryote, the slime mold *Dictyostelium discoideum*. In the unicellular eukaryotic ciliates, sexual reproduction requires nuclear degradation under conditions in which the cell continues to live, and involves pyknosis and oligonucleosomal DNA fragmentation (Cornillon et al., 1994; Ameisen, 1996).

The nematode *Caenorhabditis elegans* was instrumental in elucidating apoptotic signalling, in that during ontogeny of the adult worm, exactly 131 of the 1090 somatic cells die by apoptosis (Liu and Hengartner, 1999). The developmental pathway of programmed cell death involves 13 proteins. Four of these, EGL-1, CED-3, CED-4 and CED-9, are required for each cellular demise. This basic cell death machinery is remarkably conserved throughout metazoan evolution. CED-3 encodes a caspase-like cysteine protease, CED-4 (whose mammalian homologue is termed Apaf-1) is a CED-3 adaptor/scaffold molecule (Zou et al., 1997; Wang et al., 2002), whereas the counterparts of CED-9 and EGL-1, which suppress and promote death, respectively, are the mammalian anti-apoptotic (CED-9-like) and proapoptotic (EGL-1-like) Bcl-2 family proteins (Kelekar and Thompson, 1998; Antonsson and Martinou, 2000). In *C. elegans*, CED-4 is stanched by an inhibitory complex with CED-9, and only when EGL-1 disrupts this interaction, CED-4 triggers the lethal proteolytic action of CED-3 that eliminates its 131 cellular victims. These results suggest that the genetic pathway of programmed cell death is evolutionarily conserved from worms to mammals, at least (Yuan, 1996). Consistently, human Bcl-2 was found to suppress *C. elegans* cell death (Vaux et al., 1992).

During insect and amphibian metamorphosis, apoptosis ablates cells that are no longer needed, such as muscles and neurons essential for larval locomotion in insects or the amphibian tadpole tail. In insects, these complex phenomena are largely promoted by the steroid hormone ecdysone, which triggers stage- and tissue-specific onset of apoptosis (Jiang et al., 1997). In the fruit fly *Drosophila melanogaster,* cell death is mainly determined by three pro-apoptotic proteins: RPR, GRIM and HID. Importantly, in *D. melanogaster* apoptosis is not always genetically determined, but can also occur in an adaptive way in response to developmental malfunction or genotoxic stress (Ollmann et al., 2000). RPR, HID and GRIM promote caspase activation and cell death by inhibiting interactions between caspases and the IAP inhibitors, similar to the effect of the unrelated mammalian protein Smac/Diablo.

It must be highlighted how non-autonomous forms of cell death, *i.e.* deaths dependent upon cell-to cell or cell-to-environment interactions, appear relatively late along evolution. In *C. elegans*, developmental apoptosis is observed in an invariant, cell-autonomous pattern (Horvitz et al., 1983; Horvitz, 1999), and this form of cell death can be viewed as one of the possible developmental cell fates, similar to proliferation or differentiation. In contrast, non-autonomous cell death occurs, from *Drosophila* to mammals (Sulston and White, 1980; Richardson and Kumar, 2002), after the integration of a plethora of signals, including extracellular survival factors such as hormones or cytokines, death receptor activation, the proteomic profile of the cell, direct interactions with heterotypic cell neighbours, extracellular matrix or stress stimuli (Barde, 1989; Abrams, 1999; Krammer, 2000; Wajant, 2002; Grossmann, 2002; Puthalakath and Strasser, 2002). Cells are normally embedded within specific body microenvironments, and different cell types require specific combinations of signals, selectively available within discrete loci, to stay alive. Therefore, cell death can be depicted as the default state of all metazoan cells, which must be continuously hindered by environmental survival signals (Raff, 1992; Raff et al., 1993). Anoikis, a form of epithelial or endothelial cell apoptosis induced by inadequate cell-matrix interactions, is a patent example of the importance of such confinement. In fact, cells undergo anoikis because they are deprivedof the survival signals constitutively released from their neighbors and from basal stroma (Grossmann, 2002).

In general terms, apoptosis is a mechanism for adjusting cell populations. During embryogenesis, several structures are removed by apoptosis. The tridimensional organization of the developing organs and tissues is designed by an imprecise and stochastic process based on the production of excess cells. This allows a high degree of flexibility, as primordial scaffolds are finely adapted to define regionalization, pattern formation, limb and appendage configuration, development of axes and location along each axis. Cells of different types are therefore matched to precisely sculpting the developing structures, whereas orphan cells that do not establish the right connections are cleared (Jacobson et al., 1997; Barres and Raff, 1999). In addition, a surplus of cells allow generation of high specificity with low information cost. For instance, the quantity of information needed to specifically connect neurons would be immense, but is mathematically demonstrable that much less (genetic) information is required to produce an excess of neurons with little requirement for specificity. In fact, non-autonomous apoptosis is sufficient to guarantee a correct axonal guidance and the consequent self-assembly of neuron networking in the central nervous system (Hidalgo and ffrench-Constant, 2003), and proper innervation of tissues downstream gradients of specific trophic and survival factors (Barde, 1989; Nijhawan et al., 2000). Misrouted neurons simply fail to receive sufficient stimuli and die.

Similarly, cell attrition is extremely profound in the immune system (Hawkins and Vaux, 1997). Approximately 60 genes are necessary to randomly generate a huge variety of immunoglobulin producing cells, as opposed to the 100,000 or so that would be necessary to code individually for all the antibodies that a mammal is capable of producing. Lymphocyte survival requires the wobbling assembly of a productive immune receptor that provides the signal necessary to suppress apoptosis. In this way, lymphocytes bearing inoperative or self-reactive receptors are deleted from the immune repertoire.

Apoptosis is also widely used to eliminate aged or abnormal and potentially harmful cells, and its rate is maintained at pace with cell proliferation to control organ homeostasis (Raff, 1996). This is particularly evident in regenerating epithelia, for instance in the gut (Gregorieff and Clevers, 2005; Crosnier et al., 2006). The impact of death in this regulatory process is subject to a strict genetic control both at the transcriptional level and via several post-translational mechanisms (Ellis and Horvitz, 1986; Vaux et al., 1988; Twomey and McCarthy, 2005), and anomalies in apoptosis regulation can cause cancer (McDonnell and Korsmeyer, 1991), autoimmunity (Watanabe-Fukunaga et al., 1992), or degenerative disorders (Thompson, 1995).

In conclusion, core elements of the apoptotic machinery are widely conserved across species, but organisms endowed with increased complexity utilize apoptosis in a wider set of situations, such as the maintenance of homeostasis in a changing environment. Thus, they possess additional levels of complexity to connect programmed cell death with a variety of other biological routines, and the number of proteins and molecular pathways that control the precise execution of apoptosis increases along evolution.

Cell Death Machinery

Hardcore Death Players

In multicellular organisms, cell death must be tightly regulated to avoid gross defects or diseases. Apoptogenic proteins must be maintained at very low levels, but cells need to rapidly induce their expression or activity when are stimulated to undergo apoptosis. For this reason, several transcriptional and post-transcriptional checks and a high degree of redundancy regulate the many players involved in apoptosis signaling.

The Executioners: Caspases and Other Proteases

The activation of proteases termed caspases is the biochemical hallmark of apoptosis. The term caspases refers to the observation that they are *c*ysteine proteases that cleave after *asp*artate residues (Thornberry and Lazebnik, 1998). To preclude unwarranted cell death, each is maintained as a zymogen that must be activated by proteolytic cleavage. Caspases spark proteolytic cascades by assembling multimeric complexes on a dedicated adaptor or scaffold protein (Shi, 2001), where initiator enzymes (caspase-2, -8, -9, and -10) cut and activate the zymogenic forms of the executioner caspases (caspase-3, -6, and -7), that in turn target a plethora of proteins ultimately leading to cell dismantling (Creagh et al., 2003). Transactivation of procaspases in amplification cascades must generate sufficient proteolytic activity to overwhelm endogenous caspase inhibitors such as the IAP (inhibitor of apoptosis) proteins (see below).

The extensive prodomains of initiator caspases play important roles in function regulation and as signal integrators. Initiator caspases can be activated by external stimuli following the engagement of death receptors by specific ligands (caspase-8 and caspase-10);

by the assembly of multimeric complexes triggered by the release of mitochondrial proteins (caspase-9); in the nucleus, in response to DNA damage (caspase-2); or in the endoplasmic reticulum (ER), where they mediate apoptosis under ER stress (caspase-12; Locksley et al., 2001; Rao et al., 2002; Creagh et al., 2003; Norbury and Zhivotovsky, 2004).

Substrates of effector caspases can be grouped into different classes according to their functions. Amongst these there are components of the apoptotic machinery with either pro- or anti-apoptotic activity, structural proteins, signal transducers, enzymes involved in DNA and cellular repair and macromolecular synthesis (Strasser et al., 2000). Even though many of these molecules are inactivated by caspaseproteolysis, there are also enzymes that can be directly or indirectly induced. For instance, an endonuclease responsible of both chromatin condensation and DNA fragmentation, caspase activated DNase (CAD), requires cleavage of its inhibitor, ICAD, by caspase-3 for activation (Widlak and Garrard, 2005).

Other proteases can also play a role in PCD, even if this is less understood. Calpain is a Ca^{2+}-dependent cysteine protease activated in both necrosis and apoptosis. Calpain and caspase-3 share many common substrates. Its functions seem particularly relevant in apoptosis induced by ER stress, because of the perturbed Ca^{2+} homeostasis (Leist and Jaattela, 2001). Granzyme B is a serine protease with an important role in cytotoxic T cell-mediated apoptosis, as it can activate various caspases to initiate apoptosis (Fan et al., 2005).

Choosing your Favourite Killer: Extrinsic vs. Intrinsic Death Pathways

Two main signaling cascades trigger apoptosis: the so-called extrinsic apoptotic pathway (Figure 2A), which is engaged by the binding of death receptors of the tumor necrosis factor (TNF)-receptor family (TNF-R1, Fas, DR3/6) to their cognate ligands (TNFα, FasL, TRAIL; Locksley et al., 2001), and the intrinsic apoptotic pathway (Figure 2B), which is promoted by a myriad of stress stimuli (chemotherapeutic drugs, cytokine deprivation, oxidants, osmotic stress, oncogene activation, irradiation, Ca^{2+} overload, etc.) that induce mitochondria permeabilization and release of pro-apoptotic proteins, including cytochrome *c* (cyt *c*; Adams, 2003). In the extrinsic pathway, ligand-receptor binding promotes the formation of a death-inducing signaling complex (DISC) that recruitsinitiator caspase-8 or -10 by interactions with adaptor proteins. These caspases activate themselves and effector caspases by induced proximity, *i.e.* cleavage of the zymogenic forms that is favored by their nearness (Hengartner, 2000). Instead, the central event of the intrinsic pathway is the formation of a wheel-shaped, heptameric, caspase initiator complex, the apoptosome. This happens in the cytosol: the scaffold protein Apaf-1 undergoes a conformational change after binding the cyt *c* released from mitochondria in the presence of ATP/dATP, leading to the further assembly in the complex of the initiator caspase-9 and in turn of effector caspases (Li et al., 1997; Adrain and Martin, 2001; Acehan et al., 2002; Danial and Korsmeyer, 2004). Rather than initiating the caspase cascade, in some cells the apoptosome may primarily boost the total level of caspase activity, and this would be required for the demise of specific cell types where caspase activation needs to be higher to reach a lethal level (Adams, 2003).

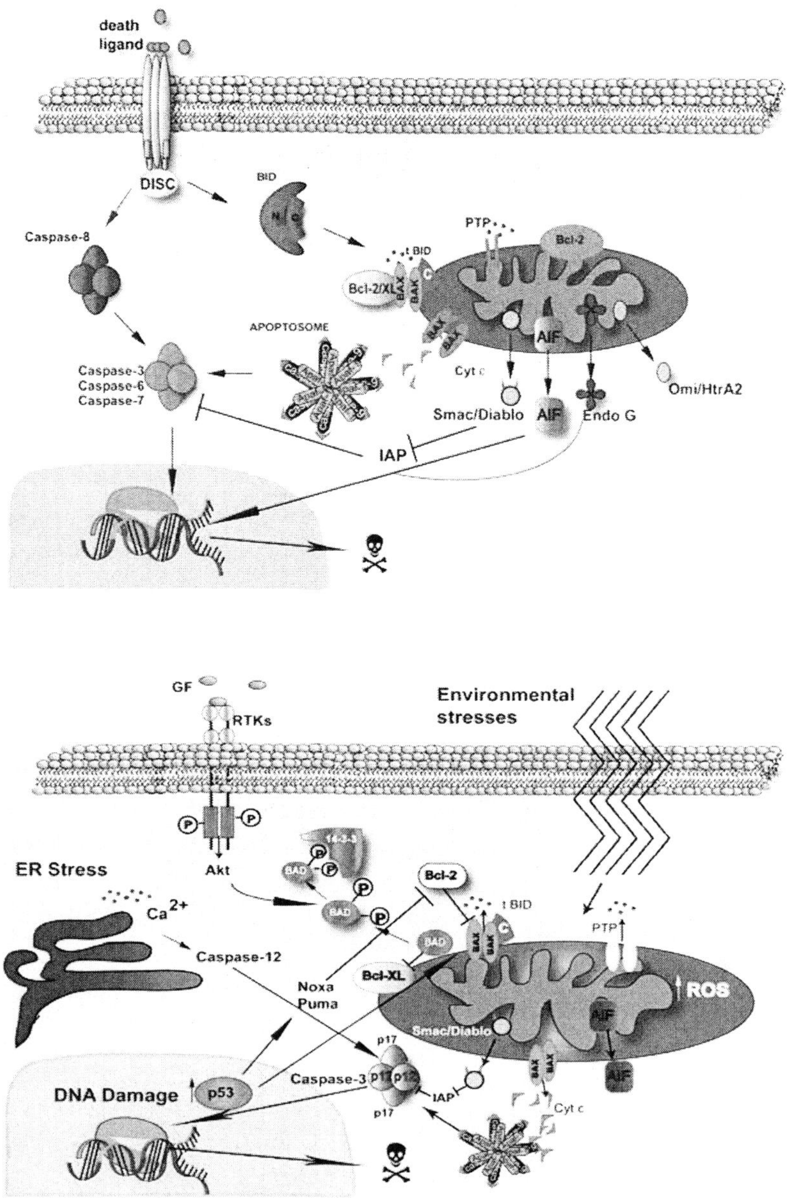

Figure 2. Apoptotic cell death pathways. The extrinsic apoptosis pathway (A) is induced by death receptor ligands (TNF, Trail, FasL etc), which engage their cognate receptors and prompt the formation of the multiprotein death-inducing signaling complex (DISC). Initiator caspases (procaspase-8 or -10) recruited in the DISC activate themselves by induced proximity and trigger a proteolytic cascade leading to effector caspase activation and eventually cell dismantling. Caspase-8 also activates by cleavage the BH-3 only protein Bid, which in turn induces the mitochondrial branch of the cell death pathway. The intrinsic pathway (B) is activated by a variety of stress insults. Upon loss of mitochondrial membrane potential, the outer mitochondrial membrane is permeabilized and leaks out several pro-apoptotic proteins. One of these, cyt*c*, promotes the aggregation of the apoptosome, a multimeric complex that triggers the autocatalitytic activation of the initiator caspase-9 and subsequently of effector caspases. Pro- and anti-apoptotic Bcl-2 homologues regulate the release of apoptogenic proteins from mitochondria, while caspase activity is negatively regulated by the IAPs. Smac/Diablo and Omi/HtrA2 promote caspase activation by counteracting the inhibitory effects of the IAPs, while AIF and EndoG contribute to caspase-independent cell death.

The Arbiters of Survival: Bcl-2 Family Proteins

The master switches that decide whether a cell is committed to survive or must be primed for apoptosis are the Bcl-2 family proteins (Figure 3; Gross et al., 1999; Sharpe et al., 2004). Functionally, they can be subdivided in three main classes. A subset of these molecules (Bcl-2, Bcl-X_L, Bcl-w, A1 and Mcl-1) is characterized by four BH domains (BH1/4) and by a hydrophobic tail that allows membrane insertion. These proteins are localized on the outer mitochondrial membrane (OMM) and on other intracellular membranes and have a central anti-apoptotic function by contributing to the mitochondrial electrochemical homeostasis. Pro-apoptotic Bcl-2 family proteins are subdivided in "multi-BH domain" (Bax, Bak, Bok), which share three BH domains (BH1/3) and a hydrophobic tail, and in "BH3-only", whose sequence homology is restricted to the short BH3 region (Kirkin et al., 2004; Puthalakath and Strasser, 2002). Bax/Bak are the central effectors of apoptosis induction. Once activated, they form oligomers in intracellular membranes and cause OMM permeabilization (Letai et al., 2002, Wei et al., 2001). As a result, mitochondrial depolarize and release apoptogenic proteins. BH3-only proteins are sentinels of stress insults in the cytosol and are recruited by specific cytotoxic signals onto mitochondria (Adams and Cory, 2001), where they act as death inducers.

The network of interactions between the different Bcl-2-like proteins is intricate and not fully clarified. BH3-only proteins probably comprise both sensitisers, which inactivate the anti-apoptotic Bcl-2 proteins, and activators, which directly engage Bax or Bak. Other BH3-only proteins, such as Bim and Puma, have a more general role, as they can antagonize all the anti-apoptotic proteins (Gelinas and White, 2005). In the direct binding model, sensitizers bind only to the pro-survival proteins and liberate the activators that are bound to these. The freed activators become ableto engage Bax/Bak and to induce cell death. In the displacement model, BH3-only proteins simply activate Bax/Bak by displacing them from the Bcl-2 pro-survival proteins that sequester their active forms. Remarkably, in the displacement model death is thedefault, whereas survival is the default in the direct binding model (Willis and Adams, 2005). Whatever the model, apoptosis might require neutralization of two classes of Bcl-2 pro-survival proteins, one comprising Bcl-2, Bcl-X_L and Bcl-w, and the other Mcl-1 and A-1. Accordingly, Bak was found to interact with one member from each of these two inferred pro-survival Bcl-2 classes (Nijhawan et al., 2003; Chen et al., 2005). Targeting only Mcl-1, for instance prompting its degradation following binding with the BH3-only Noxa, failed to kill cells. A second hit was required to unleash another inhibitor, e.g. Bcl-X_L (Willis and Adams, 2005; Chen et al., 2005). Recent data connect Mcl-1 to apoptosis elicited by growth factor withdrawal. In these conditions, Mcl-1 is phosphorylated and targeted for ubiquitinylation and degradation. This would release the BH3-only protein Bim, which in turn can activate Bax/Bak and commit cells to death (Maurer et al., 2006).

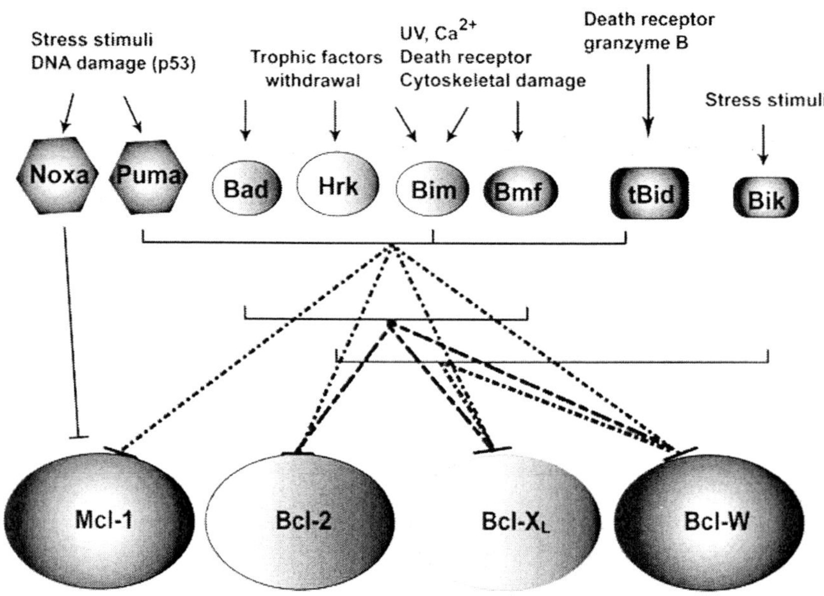

Figure 3. Network of interactions among Bcl-2 family proteins. BH3-only proteins (Noxa, Puma, Bad, Hrk, Bim, Bmf, Bid, Bik) monitor the health status of the cell. When thisis subjected to stress stimuli, BH3-only proteins are activated and initiate apoptosis by binding and neutralizing Bcl-2 pro-survival proteins (Mcl-1, Bcl-2, Bcl-X_L, Bcl-w) *via* their BH3 domain. Bim, Puma and tBid engage all Bcl-2 pro-survival proteins and thus are potent apoptosis inducers, whereas all other BH3-only proteins interact selectively with subsets of the Bcl-2 pro-survival proteins.

Given the importance of the Bcl-2 family proteins in apoptosis regulation, it comes as no surprise the observation that their regulation is extremely complex and occurs in several forms: through transcriptional modulation, phosphorylation, cleavage, N-myristoilation or subcellular relocalization, in order to rapidly respond in a flexible manner to a variety of internal and external stimuli. In addition, each cell type has a different pattern of expression of these proteins. Some of the BH3-only proteins, including Hrk, Noxa, Puma and Bim, are subject to transcriptional control. Hrk expression is induced in the neural tissues upon growth factor withdrawal or treatment with amyloid β protein (Aβ), and its levels peak at the time when these cells are committed to die. Noxa and Puma are induced in response to DNA damage by p53 (see below), whereas Bim expression is controlled by several transduction pathways involving the kinases JNK, ERK and PI3K (Puthalakath and Strasser, 2002). Bim and Bmf are also regulated by sequestration to cytoskeletal structures: the microtubular dynein motor complex and the actin cytoskeleton-based myosin V motor complex, respectively. Ligation of death receptors activates caspase-8, which processes the inactive cytosolic form of Bid into a truncated, active fragment (tBid) that translocates to mitochondria. The BH3-only Bad is inhibited by kinases such as AKT and Erk and could therefore link apoptosis to energy metabolism and survival. Phosphorylation allows Bad sequestration in the cytoplasm by binding to 14-3-3 scaffold proteins, whereas dephosphorylation by calcineurin activates Bad by translocation onto mitochondria. Phosphorylation also regulates Bik by a presently unknown mechanism (Puthalakath and Strasser, 2002).

The equilibrium between cell death and survival not only depends on the regulation of the single Bcl-2-like proteins, but also on the global balance of Bcl-2/Bcl-X_L *vs.* Bax/Bak protein levels.

Some data also raise the possibility that Bcl-2 family members have a function in regulating certain aspects of mitochondrial energetics, such as mitochondrial matrix volume or structure (Scorrano et al., 2002), generation of reactive oxygen species (ROS; Korsmeyer et al., 1995), consumption of exogenously added ATP (Di Lisa et al., 1998), and mitochondrial sensitivity to Ca^{2+} (Murphy et al., 2005). Altogether, these activities of the Bcl-2 family proteins could be functionally intertwined and have a further role in the modulation of cell survival. Multiple roles of other players in the mitochondrial death opera perfectly fit in this picture. For example, in healthy cells cyt *c* is a crucial electron carrier in the respiratory chain: its release has therefore the dual role of promoting apoptosome assembly and of interrupting respiration (Bernardi et al., 1999).

Mitochondria: At the Crossroad between Life and Death

One of the main activities of Bcl-2 family proteins in the regulation of cell survival is the control of outer mitochondrial membrane (OMM) permeability, and massive OMM permeabilization is an irreversible step in death induction. Indeed, mitochondrial damage has deadly implications in multiple cell death paradigms, and mitochondria are integrators of a plethora of death and survival signals (Desagher and Martinou, 2000; Henry-Mowatt et al., 2004).

Several models exist to explain how permeabilization of mitochondrial membranes is mediated during apoptosis. In one model, the OMM might rupture as a result of mitochondrial matrix swelling, which would be caused by opening of a megachannel, called the permeability transition pore (PTP), in the inner mitochondrial membrane (IMM). PTP opening is favoured by Ca^{2+} in the mitochondrial matrix and allows the passage of solutes with molecular masses up to 1500 Da. Multi-BH domain Bcl-2 family proteins could regulate the biophysical properties of the PTP. PTP opening induces a massive mitochondrial depolarization, which in turn uncouples oxidative phosphorylation and leads to matrix swelling, cristae unfolding, rupture of the OMM and eventually release of intermembrane proteins into the cytosol (Bernardi, 1999; Donovan and Cotter, 2004; Forte and Bernardi, 2005). While the molecular composition of the PTP complex remains elusive, its regulation by the matrix cyclophilin D (CyPD), a peptidylprolyl cis-trans isomerase, is now firmly established. Experiments performed on CypD-deficient cells indicate that these are resistant to necrosis induced by oxidative stress and that CypD modulates the sensitivity of the PTP to Ca^{2+} (Basso et al., 2005; Nakagawa et al., 2005; Baines et al., 2005).

A competing model postulates that Bax/Bak oligomers by themselves or following interactions with membrane lipids would decrease OMM stability, leading to pore formation in the outer membrane and allowing efflux of apoptogenic proteins (Gross et al., 1999; Martinou and Green, 2001). Accordingly, in planar lipid bilayers Bax oligomers can form large conductance channels, whereas Bcl-X_L can either form pores or inhibit Bax pore formation (Sharpe et al., 2004). *In vitro*,the Bax and Bak oligomers may coalesce into larger complexes, and can permeabilize mitochondria or protein-free liposomes, allowing passage of cyt *c* (Antonsson and Martinou, 2000) and even of far larger molecules (Kuwana et al.,

2002; Martinou and Green, 2001). Yet another model involves Bcl-2 family proteins cooperating with OMM or IMM proteins, such as the voltage-dependent anion channel (VDAC) or the adenine nucleotide translocator (ANT), to form a cyt *c*-conducting channel (Vander Heiden et al., 2000; Tsujimoto and Shimizu, 2000). Nevertheless, direct evidence for the formation of such pore structures in mitochondria during apoptosis is still missing (Zornig et al., 2001). As VDAC and ANT regulate ATP/ADP exchange between mitochondria and cytosol, their interactions with Bcl-2 family members could fine-tune mechanism for coupling the rate of glycolysis, respiration, ATP production and death/survival equilibrium (Yuan, 2006). A further possibility to explain mitochondrial pore openings would be that alterations in lipidic mixing on the OMM change the bilayer curvature, leading to bilayer discontinuity with increased membrane permeability and release of cyt *c* (Degli Esposti and Dive, 2003).

Mitochondria have been thought of as static structures; however, recent data suggest that they undergo dynamic remodelling characterized by fission and fusion, with important physiological implications (Oakes and Korsmeyer, 2004). For instance, t-Bid has been shown to cause fusion of individual cristae and opening of junctions between cristae and the intermembrane space. This facilitates an efficient cyt *c* release, as about 85% of it resides within the intercristal compartment (Scorrano et al., 2002; Bernardi et al., 2006). In this view, cyt *c* release from mitochondria would be a two-step process: first, cristae remodelling would render the OMM accessible to cyt *c* through the action of proteins involved in mitochondrial dynamics, as OPA1 and PARL (Cipolat et al., 2006, Frezza et al., 2006); second, Bax/Bak-mediated permeabilization of the OMM would induce cyt *c* release in the cytosol. Further in this frame, expression of the hFis1 protein promotes both mitochondrial fragmentation and release of cyt *c* (James et al., 2003).

Opening the Pandora's Box: Mitochondria-released Proteins

Once the electrochemical homeostasis of mitochondria is compromised by the concerted action of pro-apoptotic Bcl-2 family proteins, the cell is irreversibly committed to death. Mitochondria release a number of apoptogenic proteins that underpin both the caspase-dependent and the caspase-independent death programs (Figure 4). These factors include cyt *c* (Yang et al., 1997), the apoptosis inducing factor (AIF; Susin et al., 1996), the high temperature requirement A2 (HtrA2/Omi; Suzuki et al., 2001), the second mitochondria derived activator of caspase/direct IAP binding protein with low pI (SMAC/Diablo; Du et al., 2000) and the endonuclease G (EndoG; Li et al., 2001).

In healthy mitochondria, the flavoprotein AIF acts as a free radical scavenger. When released in response to death stimuli, AIF translocates to the nucleus where induces peripheral chromatin condensation and large-scale DNA fragmentation in a caspase-independent fashion (Ye et al., 2002). Mitochondrial release of AIF has been reportedduring early mammalian development, and AIF was found to be essential for the first wave of PCD that promotes embryonic morphogenesis and cavitation (Penninger and Kroemer, 2003). EndoG is an inner-mitochondrial/matrix-associated protein and its RNAse activity is normally exploited for mitochondrial DNA replication. During cell death EndoG functions as a nuclear endonuclease. Apparently, EndoG activity mainly relies on localisation rather than on relief of inhibition (Li et al., 2001; Saelens et al., 2004).

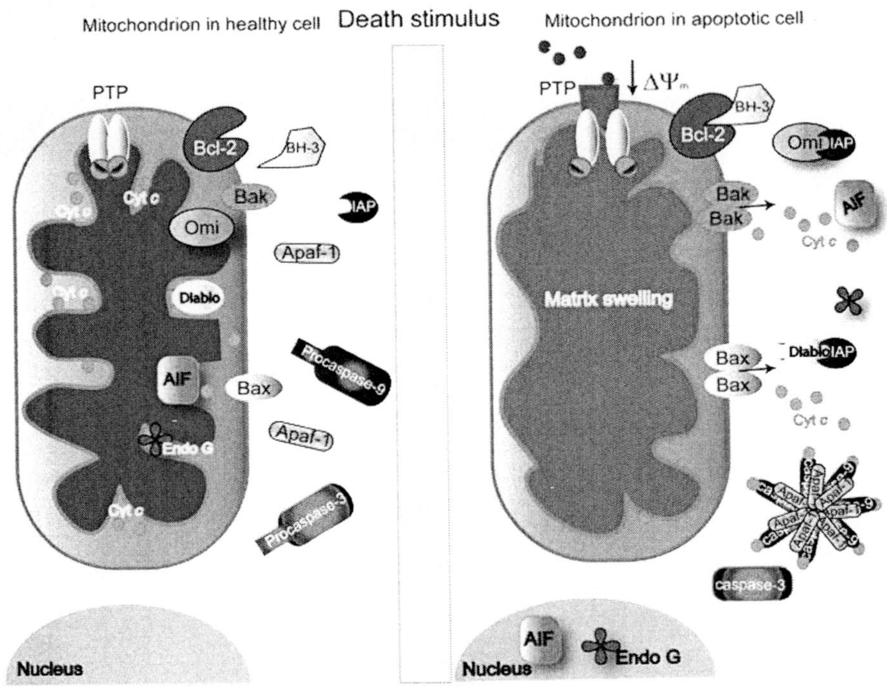

Figure 4. Mitochondrial outer membrane permeabilization in apoptosis. In a healthy cell, caspases arepresent as zymogens and the BH3-only proteins (BH3) are sequestered away from the Bcl-2-like pro-survival proteins (Bcl-2). Apoptotic stimuli elicits the BH3-only protein binding to Bcl-2 on mitochondria and their consequent inhibition. Recruitment of Bax and Bak into oligomers on the mitochondrial outer membrane and/or PTP opening lead to mitochondrial permeabilization. The released cyt *c* induces formation of the heptameric apoptosome constituted by cyt *c*, ATP/dATP, Apaf-1 and procaspase-9, which activates effector caspases as caspase-3. Smac/Diablo and Omi/Htr2 are released from the mitochondrial membrane and inhibit the IAPs in the cytosol, whereas AIF and endonuclease G (endoG) move into the nucleus, where they contribute to DNA degradation.

Bax was reported to induce the release of Smac/Diablo and HtrA2/Omi from isolated mitochondria (Verhagen et al., 2002). Both Smac/Diablo and Omi/HtrA2 block the action of inhibitor of apoptosis (IAPs) proteins. Eight IAPs are known in mammals (XIAP, cIAP1, cIAP2, ML-IAP/LIVIN, ILP2, NIAP, Bruce/Apollan and Survivin). Their main role in apoptosis signalling is the inhibitory binding to different caspase family members (Vaux and Silke, 2005), whose activity is therefore favoured by Smac/Diablo and Omi/HtrA2. In addition, Omi/HtrA2 is a serine protease, and this enzymatic activity is not required to antagonise IAPs. The pro-apoptotic effect linked to Omi/HtrA2 serine protease activity remains poorly understood. The complexity of the interplay between mitochondria-released proteins is underlined by the observations that (i) cyt *c* absence only attenuates apoptosis (Donovan and Cotter, 2004, ii) apoptosis is unaffected by loss of Smac/ Diablo (Okada et al., 2002), and (iii) inhibition of the IAPs does not initiate apoptosis (Verhagen et al., 2000).

p53, the Guardian of the Genome

The p53 tumor suppressor is the key element for the safeguard of a correct genetic information when this is challenged by various genotoxic and non-genotoxic insults.

Functionally, p53is placed in a nodal position, from where it integrates a plethora of different stimuli and links many signaling pathways, to decide whether a genetic damage can be patched up or whether it is so large that the cell must be cleared away by fostering apoptosis (Sengupta and Harris, 2005). Its importance for cellular homeostasis is plastically highlighted by the observation that half of human tumors carry an inactivating p53 mutation. The network of checks keeping p53 under control is huge (Harris and Levine, 2005), but the effect of its activation following an irreparable cellular damage can be simplified in transcription induction and transcription-independent activities. p53 induces the transcription of the cell cycle inhibitor p21, of the pro-apoptotic Bcl-2 family proteins Bax, Bid, PUMA and NOXA, of its inhibitor Mdm2, in a negative feedback loop, of the DNA-damage responsive protein GADD45, and of members of the TNF family of receptors, thus promoting both intrinsic and extrinsic apoptotic pathways (Nakano and Vousden, 2001; Oda et al., 2000; Henry-Mowatt et al., 2004). Regarding transcription-independent p53 effects, it was recently shown that p53 directly induces OMM permeabilization by translocating onto mitochondria before the collapse of the mitochondrial membrane potential and cyt *c* release. Once on mitochondria, p53 physically interacts with Bcl-X_L and Bcl-2 and antagonizes their survival activity. On mitochondria p53 also binds the apoptogenic Bak and liberates it from its inhibitory complex with Mcl-1, whereas a cytosolic interaction between p53 and the pro-apoptotic Bax protein directly activates the latter (Moll et al., 2005). Thus, p53 mutations that occur in tumors may act as 'double hits', eliminating both the transcriptional and the direct mitochondrial functions of p53. Given the complexity of the p53 network it seems plausible that these two groups of functions display a high degree of regulation plasticity under different stress conditions and in different cellular environments (Moll et al., 2005).

Animal Models and Redundancy of Apoptosis Signaling

The physiological role of cell death has been determined in elegant *in vivo* studies, based upon the generation and functional characterization of transgenic gene overexpression and gene deletion murine models. Remarkably, these models have indicated a high degree of functional redundancy in mammalian systems and have hindered definitive conclusions about the role of each component of the apoptosis machinery. Gene-knockout of the *bcl-2* gene results in mice with grey hair, polycystic kidney disease and lymphocytopenia (Nakayama et al., 1994), whereas ablation of the *bcl-x* gene is embryonic lethal, as massive apoptosis occurs in post-mitotic, immature neurons of the brain and of the spinal chord and in the haematopoietic system (Motoyama et al., 1995). Bax deficiency, like over-expression of Bcl-2, facilitates tumorigenesis. In the nervous system Bax is required for cell death following growth factor deprivation (Twomey and McCarthy, 2005). *Apaf-1* null embryos exhibit severe cranio-facial deformation, neural and brain overgrowth and retain inter-digital webs, indicative of insufficient apoptotic cell death (Yoshida et al., 1998). Similarly, animals with mutations in *caspase-3* and *caspase-9* die early in development and show defects in apoptosis in the nervous system resulting in brain overgrowth (Kuida et al., 1998; Kuida et al., 1996; Ranger et al., 2001). *Caspase-8* null mice die *in utero* for defects in heart muscle development and display fewer haematopoietic progenitor cells (Varfolomeev et al., 1998;

Colussi and Kumar, 1999). Interestingly, knock-in mice that express a mutated cyt *c* retaining its electron transfer function, but unable to promote apoptosome assembly, revealed the existence of a cyt *c*- and apoptosome-independent, but Apaf-dependent, mechanism of caspase activation and apoptosis induction in thymocytes (Hao et al., 2005).

This complexity of PCD phenotypes following manipulation of PCD components can only be explained by postulating that all components of this sophisticated mechanism exist in bewildering multiplicity, consistent with the fact that apoptosis is required for tissue repair and remodelling throughout vertebrate life in order to cope with infection and injury. The other side of the coin is the risk of acquiring neoplastic mutations during these regenerative processes. Hence, the presence of effective, overlapping and redundant mechanisms could have been imperative during vertebrate evolution to restrict the clonal autonomy and the probability of tumor development in somatic cells (Meier et al., 2000).

Alternative Pathways to Death

Other pathways and organelles contribute to apoptosis signaling, and apoptosis is not the only type of PCD.

Other Poison Cupboards: The Endoplasmic Reticulum and the Lysosomes

The endoplasmic reticulum (ER), as the mitochondria to which is closely associated, is a meeting point of several pro-and anti-apoptotic mechanisms. The main physiological roles of the ER are the control of protein folding and Ca^{2+} storage (Mattson and Chan, 2003; Boehning et al., 2003). Several ER stresses, such as disturbed glycosylation, misfolded proteins, perturbed Ca^{2+} homeostasis, or glucose deprivation provoke the unfolded protein response, whose persistence induces cell death (Kaufman, 2002). An ER-driven apoptotic cascade involves caspase-12 as initiator, which activates caspase-9 and effector caspases in a cyt *c*-independent way (Rao et al., 2002). In this pathway, the ER is sufficient to sense, initiate and propagate death signaling, but the dismantling of the cell is executed by the cytosolic effector caspases. Significantly, caspase-$12^{-/-}$ mice were found to be resistant to amyloid-β-peptide-induced apoptosis in an Alzheimer's disease model (Nakagawa et al., 2000).

Bcl-2 family members are localized on ER membranes, where they contribute to the modulation of intracellular Ca^{2+} homeostasis in a complex and subtle interplay with mitochondria (Scorrano et al., 2003), even though the exact mechanisms by which these interactions affect apoptosis are still under investigation (Demaurex and Distelhorst, 2003; Annis et al., 2004; Distelhorst and Shore, 2004). An intriguing set of data postulates a role of Bcl-2/Bcl-X_L in controlling the exchange of Ca^{2+} between ER and cytosol, and in turn its mitochondrial uptake that can lead to PTP opening and death. To further complicate the picture, the pro-apoptotic proteins Bak and/or Bax maintain the ER Ca^{2+} concentration in healthy cells, but during apoptosis they promote Ca^{2+} release (Annis et al., 2004). Release of Ca^{2+} from the ER can also initiate cell death by directly activating the Ca^{2+}-dependent protease calpain and/or caspases, by controlling the distribution and activity of Bcl-2s or by modulating the expression of apoptosis regulatory proteins (Mesaeli and Phillipson, 2004). Similar to the mitochondria, the ratio between ER pro- and anti-apoptotic proteins seems to control the cell's fate. Accordingly, the ER stress causes translocation of the pro-apoptotic Bax (Zong et al., 2003) and Bim (Morishima et al., 2004) from cytosol to the ER. The capacity of Bcl-2 to control the ER Ca^{2+} store is decreased by Bcl-2 phosphorylation and correlates with the binding of BH3-only proteins (Bassik et al., 2004) and of the IP3R receptors, a channel family that induces Ca^{2+} efflux from the ER (Chen et al., 2004; Oakes et al., 2005; Walter and Hajnoczky, 2005).

Other apoptotic pathways predict the release of proteases, mainly cathepsins and lipid hydrolases, from lysosomes into the cytosol in response to a wide variety of death stimuli (Jaattela, 2004). The control of lysosomal permeabilization is poorly understood and could impinge on Bcl-2 family proteins. A partial and selective release of proteases would contribute to apoptosis signaling by multiple molecular pathways, including mitochondria permeabilization, whereas a complete lysosomal breakdown would yield necrosis (Guicciardi et al., 2004). The other protein degradation route, the proteasome, is also involved in apoptosis regulation as it modulates the levels of pro- and anti-apoptotic proteins (Adams, 2004).

Necrosis, Autophagy and Senescence

In vivo, PCD is the result of the interwinding of apoptosis and of non-apoptotic death paradigms, and the relative speed of each process determines the final mode of cell demise. Caspase inhibition may unveil or even increase caspase-independent death routines in which cells display non-apoptotic morphological features, as apoptosis-like PCD, necrosis-like PCD or autophagy (Leist and Jaattela, 2001; Abraham and Shaham, 2004). In addition, apoptosis can occur independently of caspase activation, with only partial condensation of chromatin and without internucleosomal DNA fragmentation, but maintaining many other apoptotic features (Kroemer and Martin, 2005). In these cases, several other proteases (cathepsins, calpains, granzyme B) may promote death. These proteases can cleave some of the classic caspase substrates, thus mimicking at least in part the effects of caspase activity (Leist and Jaattela, 2001).

Death receptor can trigger non-apoptotic forms of PCD by activation of serine/threonine kinases, such as RIP, DAPK and JNKs, and/or production of reactive oxygen species (ROS) that elicit OMM permeabilization (Jaattela, 2004). Energy depletion can block the assembly of the apoptosome, switching the mode of cell demise from apoptosis to necrosis (Gramaglia et al., 2004).

Many insults induce apoptosis at lower doses and necrosis at higher doses. Even in response to a certain death-inducing agent, features ofboth apoptosis and necrosis may coexist in the same cell. A form of death-receptor mediated necrosis was recently termed necroptosis (Degterev et al., 2005). After TNFα stimulation, two membrane-associated complexes form: one activates the transcription factor NF-κB and promotes survival, the other induces caspase-dependent apoptosis. In these conditions, caspase inhibition is insufficient to halt cell death, which continues with a necrotic morphology.

Necrosis appears to have evolved as a warning system for events that might compromise the integrity of the organism as a whole. As such, recent data suggest that cells can program necrosis in a number of processes including neurodegenerative diseases, infection, inflammatory diseases, exposures to toxins and cancer. If necrosis is programmed, this implies that it is initiated in response to specific cues, usually when caspases are inhibited, either as a consequence of extracellular signaling or in response to intracellular perturbations. A key feature appears an irreversible bioenergetics damage that results in cell death (Zong and Thompson, 2006). In some cases, caspase-independent necrotic cell death can be forestalled by antioxidants or by eliminating the activity of the protein kinase RIP. TNF may induces necrosis in virally infected cells by initiating specific signal transduction cascades, and viruses harbour proteins that suppress programmed necrosis (Edinger and Thompson, 2004).

Autophagy or type II PCD is an important catabolic mechanism involved in the turnover of intracellular organelles and of large protein complexes through a lysosomal degradation pathway (Levine and Yuan, 2005). The appearance of large autophagic vacuoles, the autophagosomes, is characteristic of this type of PCD. Autophagosomes are membranous vesicular compartments that sequester cytoplasmic components and eventually fuse with the lysosome, leading to protease-mediated degradation of the internal material. Under normal conditions, autophagy displays a survival function by allowing recycling of molecular components to fuel ATP synthesis when cells are under starvation. However, this process can be rapidly upregulated under stress, and in these conditions autophagy might act as a form of cell death. This type of PCDis not characterized by a prompt destruction of DNA. Only when the large majority of the cytoplasm has been destroyed, chromatin condensation becomes apparent, and oligonucleosomal DNA fragmentation occurs. Thus, in some instances there is potential recovery of cells undergoing autophagy, and this is used as a recycling mechanism; however, above a certain stress threshold cells do not halt the process and commit a form of autophagic suicide (Penaloza et al., 2006).

Ceramide, Fas-Associated protein with Death Domain (FADD), Death Associated Protein kinase (DAP), phosphatidylinositol 3-kinase (PI3K) and Bcl-X_L/Bcl-2, which are all involved in apoptosis regulation, have also been shown to modulate autophagy (Tardy et al., 2006). Once death receptors are engaged by their cognate ligands, inhibition of the initiator caspase-8 promotes autophagy along a kinase pathway involving RIP and JNK. In turn, caspase-8 cleaves RIP and prevents it from activating autophagic cell death. The executioners of autophagy are the poorly characterized products of the Atg genes, among which Beclin-1 is involved in the formation of the autophagosome (Baehrecke, 2005).

The boundaries among these different types of cell death are not precisely defined, and several features overlap. The high degree of crosstalk among the different organelles may allow autophagy to convert to apoptosis or *vice versa*. It was recently found that the combined impairment of apoptosis and autophagy promotes necrotic cell death under nutrient depletion conditions. Thus, inhibiting autophagy can restore cell death in apoptosis resistant cells, but this necrosis can associate with inflammation. Thus if one death pathway is blocked, a cell may still die by a second, highly organized biological routine (Degenhardt et al., 2006; Penaloza et al., 2006).

To counter the harmful action of many potential DNA-damaging agents, including reactive oxygen species and telomere shortening beyond a certain limit, a cell can enter a state of irreversible cycle arrest termed senescence. Senescent cells are identified by characteristic morphologic and functional alterations, such as large cell size, flat vacuolated morphology, inability to synthesize DNA, heterochromatin foci in the nucleus, and by the presence of the marker senescence-associated β -galactosidase (Balaban et al., 2005). In human cells, telomeric and nontelomeric signals induce senescence primarily via p53 through one of its downstream effectors, the cell cycle inhibitor p21, or via another cycle inhibitor, p16. Both pathways converge on pRB, the product of the retinoblastoma gene, which is a main cell cycle regulator. Senescence may have evolved to protect organisms from acute signals that may cause unrestricted growth resulting from faulty DNA replication, but also as a survival mechanism that allow cells to overcome the prolonged pressure of multiple insults (Dimri, 2005).

Networks of Cell Death Regulation

The Scenario: Interplays in a Death Drama

Apoptosis pathways can be viewed as rheostats whose setting is determined by the abundance and activity of the different components, thus providing further layers of complexity in PCD signaling. Caspase activation is regulated at several levels. Engagement of death receptors is challenged by decoy receptors (DcR1/3) that compete for ligand binding (Ashkenazi, 2002). Caspases in the DISC can be inhibited by c-FLIP, a protein structurally similar to caspases but devoid of protease activity (Krueger et al., 2001). IAP proteins are regulated at the transcriptional level and inhibited by auto-ubiquitylation or by Smac/DIABLO and Omi/HtrA2 released by damaged mitochondria. Smac/DIABLO displaces the binding of IAPs to caspases; conversely, IAPs target Smac/DIABLO and caspases for proteasomal degradation (Salvesen and Duckett, 2002). Omi/HtrA2 competes with caspases for binding with IAPs, and cleaves bound IAPs (Verhagen et al., 2002). The relative expression levels of Bcl-2 family members, IAPs, c-FLIP and decoys might titrate the response of the cell to apoptosis stimulation.

There is an extensive cross-talk between extrinsic and intrinsic apoptotic pathways (Sprick and Walczak, 2004). The BH3-only protein Bid is activated by the DISC and translocates onto the OMM, where it activates Bax/Bak thus promoting apoptosome aggregation (Yin, 2006), and Bax/Bak double knockout mice resist to death receptor-induced apoptosis (Wei et al., 2001). Several transcription factors, such as p53 and NF-κB, regulate the transcription of a wide array of components of both the extrinsic and the intrinsic apoptotic pathways (Slee et al., 2004; Karin and Lin, 2002). Furthermore, p53 acts in a transcription-independent way by increasing the rate of death receptor transport from the Golgi to the plasma membrane (Bennett et al., 1998), by activating caspase-8 (Ding et al., 2000) and by directly interacting with Bcl-2 family members. p53 activation is pro-apoptotic, whereas NF-κB normally yields a survival signal, but in certain contexts NF-κB can induce genes encoding for death receptor and ligands. In turn, p53 inhibitors and several components of the NF-κB pathway are inactivated by caspases (Jaattela, 2004).

Many other transcription factors, such as oestrogen receptors, retinoic acid receptors and PPARγ, modulate the expression of genes encoding for Bcl-2 family proteins (Reed, 2002) or, as in the case of c-Myc, c-Jun and c-Fos, induce apoptosis when overexpressed (Zornig et al., 2001).

Heat shock proteins (HSPs) are molecular chaperones that mediate major survival functions intervening on different steps of the apoptotic pathways: apoptosome aggregation; AIF inhibition; p53/c-Myc-activity modulation; regulation of the function of Bcl-2 family proteins; maintenance of tyrosine kinase and serine/threonine kinase (e.g. PKB/Akt) activity (Pandey et al., 2000; Bruey et al., 2000; Beere and Green, 2001; Ravagnan et al., 2001).

Ligand-mediated activation of receptor tyrosine kinases (RTKs) transduces signals from the extracellular environment to the cell, ultimately modulating gene transcription. Their activity impinges upon cell survival by regulating the transcription pattern of several apoptosis-related genes, and also by activating a number of anti-apoptotic proteins in a transcription-independent fashion (Pawson, 2002). Among these, the serine-threonine kinase PKB/Akt plays a pivotal role in cell survival. It inactivates the BH3-only apoptogenic protein Bad, caspase-9 and members of the Forkhead and YAP transcription factor families, whose targets are pro-apoptotic proteins such as Bim and FasL. Furthermore, PKB/Akt causes p53 degradation and activates the NF-κB transcription factor. In turn, PKB/Akt is cleaved and inactivated by caspase-3 (Brazil et al., 2004).

Phosphorylation on the mitochondrial outer membrane is also involved in apoptosis regulation, even if by poorly understood mechanisms (Pagliarini and Dixon, 2006).

The Integration of Different Biological Activities: Death in Daily Life

Evolution has led to an extensive integration of mammalian apoptosis machinery with multiple cellular physiological processes, with the result of placing death/survival decisions under the control of a network of internal and external stimuli. At the same time, cell death may be mediated by multiple mechanisms that are part of other cellular processes, whereas portions of cell death machinery may have integrated in other types of signalling cascades.

An example of the complex way by which cells manage largesets of signal to convey them towards a specific outcome is given by the close and multiple links that integrate cell death and proliferation pathways. The same transduction cascades, such as signaling through the mitogen-activated protein kinases (MAPKs), can prompt either proliferation or death. The three main MAPK subfamilies (ERKs, JNKs and p38 MAPKs) modulate gene expression in response to extracellular stimuli (Johnson and Lapadat, 2002). ERKs are mainly involved in cell growth and proliferation and deliver survival signals in the majority of biological conditions, whereas JNKs and p38 MAPKs are mostly co-activated in response to several stress stimuli, but their effects on apoptosis signaling can be diverse, with signal- and cell-specific variations (see below for an example with oncologic relevance). Several growth factor cascades promote progression through the cell cycle by activating of cyclin-dependent kinases (CDKs). p53 activation directly blocks cell cycle progression by upregulating the expression of the CDK inhibitor of p21 (Senderowicz, 2004). Cyclin D3, a positive cell cycle regulator interacts with caspase-2 and stabilize it, suggesting that they act in concert to balance cell division and apoptosis (Mendelsohn et al., 2002). The BH3-only pro-apoptotic protein Bid promotes cell cycle progression and is also involved in the maintenance of genomic stability by engaging at mitosis checkpoint (Yin, 2006).

The DISC component c-FLIP can divert signals from death to proliferation, activating ERKs and the transcription factor NF-κB; Bcl-2 has antiproliferative effects; components of the G1 cell cycle progression machinery (E2F1, cyclin E) induce apoptosis when overexpressed, and, in response to apoptotic stimuli, fully differentiated post-mitotic neurons are prompted to re-express components of the cell-cycle machine, whose inhibition can prevent apoptosis (Budd, 2002; Bonnefoy-Berard et al., 2004).

As a general rule, apoptosis appears as a collateral consequence of a deregulated cell-cycle machinery. In fact, interventions that bypass or disrupt cell-cycle checkpoints correlate more tightly with increased apoptosis than with enhanced proliferation. The proto-oncogene Myc drives the expression of a number of proteins required for cell-cycle progression. In addition to promoting proliferation, Myc could activate several apoptotic programs, depending on the presence or absence of collateral survival signals. For instance, in pancreatic islet β cells, Myc activation was sufficient to boost proliferation. However, this was followed by an apoptotic response resulting in islet degeneration. Inhibition of this latter Myc activity by expression of the anti-apoptotic protein Bcl-X_L promoted an unrestricted β cell expansion and the formation of pancreatic tumors (Pelengaris et al., 2002). Similarly, Bcl-2 gene was able to promotes formation of B cell lymphomas by enhancing lymphocyte survival, but only when coexpressed with Myc. Thus, a selective pressure during lymphomagenesis upregulates both Bcl-2 and c-Myc (McDonnell and Korsmeyer, 1991).

Non-autonomous devices also coordinate proliferation and death. During tissue development, proliferating cells can eliminate by apoptosis adjacent, slower growing neighbours in a process referred to as competition. The converse behaviour, termed compensation, also exists (Abrams and White, 2004).

Apoptosis is strictly connected with energy metabolism. Glucose is a primary trophic caspase inhibitor through activation of the survival kinase PKB/Akt (Gottlob et al., 2001; Rathmell et al., 2003). In turn PKB/Akt promotes recruitment onto mitochondria of the glycolytic enzyme hexokinase, which could promote survival by coupling mitochondrial ATP synthesis to glucose metabolism through association with the channel VDAC that regulates the flux of metabolites across mitochondrial membranes (Majewski et al., 2004; Yuan, 2006).

There are also multiple instances in which components of the apoptotic machinery display functions unrelated to death (Launay et al., 2005). Mammalian caspases play important roles in regulating innate immunity: following activation by caspase-5, caspase-1 processes IL-1β and IL-18, cytokines capable of inducing a multitude of immune responses (Thornberry et al., 1992; Reddy, 2004); caspase-8 and caspase-6 positively regulate T and B cell proliferation, respectively, and are therefore involved in adaptive immune response (Launay et al., 2005). In the nervous system, low-level caspase-3 activation occurs locally in synaptic terminals in response to various stimuli, possibly playing a role in synaptic plasticity (Mattson and Duan, 1999).

Apoptosis in Pathology

Breaking the Brakes: Pathological Apoptosis Reduction

Apoptosis dysregulation is involved in a wide variety of diseases, which include autoimmune diseases, pathogen infections, cancer and degenerative affections (Thompson, 1995).

The Lethal Game: Viruses, Immune System and PCD

The immune response is characterized by rapid cell proliferation in response to pathogens and equally rapid apoptotic death of cells after the pathogen has been eliminated. An abnormal survival of activated lymphocytes, which produce moleculesthat may be toxic for the host, can have dire consequences, eventually resulting in autoimmune attack. Experiments on transgenic mice indicate that overexpression of Bcl-2 or Bim deficiency in B lymphocytes cause a prolonged humoral immune responses and pathological accumulation of plasma cells, which can eventually lead to fatal systemic lupus erythematosus-like autoimmune disease. Mutations in the Fas receptor, one of the death-inducing TNF receptor family members, or in its ligand (FasL), cause lymphadenopathy and autoimmunity in mice and humans (Worth et al., 2006). The autoimmune lymphoproliferative syndrome (ALPS) is caused by germline mutations in the Fas receptor that prompt an inappropriate survival of activated T lymphocytes. Interestingly, this autoimmune disease is tumorigenic, as ALPS patients have an increased incidence of lymphoma, possibly due to the apoptosis-resistant T cells which can sustain further transforming mutations (Johnstone et al., 2002). In addition, defects in physiological clearance of apoptotic cells may produce auto-antibodies to self-antigens, resulting in splenomegaly and glomerulonephritis (Wu et al., 2006). To avoid this, the apoptotic process includes mechanisms that organize disposal of cell corpse by the regulated externalization of the lipid phosphatidylserine, which is confined to the inner leaflet of the plasma membrane in healthy cells (Williamson and Schlegel, 2002). Once on the cell surface, phosphatidylserine is recognized as an "eat me" signal by specific receptors on phagocytic cells that rapidly eliminate the apoptotic cell (Savill and Fadok, 2000; Zwaal et al., 2005), thereby preventing inflammation of the surrounding tissue. Deficiencies in this process in knockout animals lead to disturbances in self-tolerance, and often induce a systemic autoimmune disease (Wu et al., 2006).

Virally infected and tumor cell are targeted by effector lymphocytes, such as CTLs and NK cells, by several means, including death receptor-ligand interactions. In addition, these cells release cytolytic granules that contain apoptogenic enzymes, *e.g.* perforin, a pore-forming protein that facilitates the delivery of other granule components into target cells, and granzyme B, a serine protease that cleaves several proteins, including caspases (Creagh et al., 2003). Therefore, apoptosis is a main defense mechanism against viruses and other intracellular pathogens. In turn, evolution has selected a variety of mechanisms by which these pathogens inhibit the death of host cells, with the aim to promote their replication and/or survival. For example, the adenovirus protein E1A promotes viral replication but also activates the pro-apoptotic protein p53 in the host. To counteract this, adenoviruses produce two molecules that block host apoptosis: E1B55kD, which directly inhibits p53, and

E1B19kD, a Bcl-2 homolog, which inhibits apoptosis signaling further downstream (Berk, 2005). Several other types of viruses have Bcl-2 homologues to prevent host cell apoptosis caused by diverse stress stimuli. Other anti-apoptotic viral molecules include: the cowpox virus-encoded CrmA, a potent caspase inhibitor that not only prevents host cell death, but also the caspase-mediated production of interleukin-1 and interferon-γ, thereby reducing inflammation and subverting the host immune response; the baculovirus protein p35, a wide-spectrum caspase inhibitor; soluble TNF-R homologs that neutralize TNF-R ligands and molecules that mediate loss of TNF-RI and Fas from the cell surface, encoded by some pox viruses and adenoviruses, respectively; viral homologues of cellular FLIP, a molecule involved in the DISC assembly following death receptor engagement, which inhibit recruitment and activation of the initiator caspase-8 (Strasser et al., 2000).

A Life-Threatening Survival: Apoptosis and Cancer Development

In cancer, proper tissue or organ homeostasis is dysregulated by the progressive accumulation of several genetic alterations that imbalances proliferative, growth-inhibitory and apoptotic signals, leading to a net tissue expansion. In parallel, dynamic changes in the surrounding microenvironment are often required to foster the expansion of the tumor, eventually resulting in the acquisition of a malignant phenotype (Hanahan and Weinberg, 2000). These processes are interdependent and must arise together in the same cell to promote tumorigenesis. In addition, for a tumor cell to propagate, it must survive drastic structural and/or metabolic alterations, as well as an extremely stressful microenvironment. Examples are hypoxic conditions or nutrient deprivation in the inner cells of the primary tumor mass, and the host anti-tumour immune response. All these conditions are powerful apoptosis triggers, and apoptosis is therefore used by the organism to eliminate neoplastic cells. Moreover, signals that drive entry into the cell cycle also prime the apoptotic machinery, in a phenomenon called antagonistic pleiotropy (Green and Evan, 2002). Another crucial restriction point for tumour development is constituted by metastasization, the capability of tumour cells to migrate from the primary site of accretion to successfully colonize distant sites, which constitutes the most common cause of cancer death. However, metastasis development is potently counteracted by anoikis and amorphosis, *i.e.* cell death stimulated by the loss of cell-matrix interactions and by the loss of cytoskeletal architecture, respectively. The presence of these multiple safeguard barriers renders both primary tumor development and metastasis formation extremely inefficient and rare processes. As a consequence, for a successful tumor development, neoplastic cells must acquire the fundamental capability to evade apoptosis. Defects in failsafe programs are required to uncouple the growth machinery from checkpoints that sense inappropriate proliferation (Schmitt, 2003). In this model, apoptosis resistance is permissive for tumor development (Pommier et al., 2004). Mutations that confer resistance can affect the core apoptotic machinery or uncouple the damage response from an intact death machinery.

Apoptosis is disabled at distinct stages in tumorigenesis, and tumor cells impair different components of the apoptotic machinery, either increasing the expression of anti-apoptotic proteins or inactivating pro-apoptotic molecules. Nonetheless, the complexity of apoptosis signaling implies that the analysis of individual molecules is not always exhaustive for the comprehension of the tumor phenotype. Levels of apoptosis regulators may also fluctuate with differentiation and cycling of cells, redundant circuitries can compensate for quantitative changes in some subroutines, contributing to the heterogeneity of responses to treatment (Johnstone et al., 2002).

Disruption of the intrinsic apoptotic pathway is extremely frequent in cancer. High levels of anti-apoptotic Bcl-2 family proteins, or an increase in the antiapoptotic to proapoptotic Bcl-2 protein ratio, augment cancer cell malignancy. Overexpression of Bcl-2/X_L or Mcl-1 occurs in more than half of all tumors and contributes to resistance to chemotherapy (Kirkin et al., 2004). Bcl-2/X_L over-expression on its own might not suffice for a full-blown tumorigenesis. In follicular lymphoma, a t(14:18) translocation results in Bcl-2 transcriptional activation and causes an indolent tumor form. To progress towards a more aggressive disease new mutations are required. For instance, if an activating translocation occurs in the *myc* locus, Bcl-2 inhibits c-Myc induced apoptosis, whereas c-Myc overcomes the growth-suppression mediated by Bcl-2 (McDonnell and Korsmeyer, 1991; Green and Evan, 2002). Conversely, proapoptotic Bcl-2 proteins are inactivated in many tumors. Impaired Bax expression has been reported in breast, colon and gastric cancers, hepatocellular carcinoma and other tumors (Ionov et al., 2000). Levels of Bak are reduced in several tumor types, as gastric and colorectal cancers, and Bak down-regulation is required for Ras-induced transformation of intestinal epithelial cells. Bid-deficient mice develop a disease similar to chronic myelomonocytic leukemia (Kirkin et al., 2004).

Caspase mutations are rare in human tumors, but epigeneticinactivations are more frequent. Caspase-8 is silenced by gene methylation in neuroblastomas, causing resistance to chemotherapy. Alternatively spliced isoforms of caspase-2/9 act as dominant negative molecules on apoptosis induction in some tumor cells.Apaf-1 is inactive in chemoresistant metastatic melanoma (Johnstone et al., 2002).

Over-expression of IAPs is commonly observed in malignancies. Down-regulation of XIAP induces apoptosis in chemoresistant cells, and XIAP expression correlates with poor prognosis in acute myelogenous leukemia. Accordingly, a negative regulator of XIAP, XAF1, is reduced in many tumors. c-IAP2 is activated in certain lymphomas. Survivin is overexpressed in the most common human cancers of lung, colon, pancreas, prostate and breast, in non-Hodgkin lymphomas and is associated with poor prognosis (Vaux and Silke, 2005). The heat shock proteins Hsp27, Hsp70 and Hsp90 are over-expressed in various cancers (Parcellier et al., 2003).

The extrinsic apoptotic pathway is also affected in tumorigenesis. Fas is mutated or down-regulated in lymphoid and solid tumors. Tumor cells of varying origin frequently exhibit *de novo* expression of Fas ligand, which, coupled with resistance to Fas-mediated apoptosis, enables many cancers to counterattack against immune system components that express Fas on their surface. In addition, through a poorly understood mechanisms, FasL can increase proliferation of tumor cells (Houston and O'Connell, 2004). Similarly, high c-FLIP expression in many malignant cells is associated with tumor escape from T-cell immunity. Death receptors 4/5 are mutated in metastatic breast cancer, and the decoy receptor DcR3 is amplified in lung and colon tumors (Igney and Krammer, 2002).

Dysregulation of proteins involved in cell-matrix interactions, like the focal-adhesion kinase, and of certain anti-apoptotic molecules, such as Bcl-2 Bim, Bmf and XIAP, induces metastasis formation and anoikis resistance, whereas matrix metalloproteases could interfere with death-receptor-induced cell death by cleaving receptors off the cell surface (Mehlen and Puisieux, 2006).

RTKs become potent oncoproteins and induce cell survival when mutated, overexpressed or structurally altered, and actually they are aberrantly activated in most tumor types. Oncogenic activation of RTKs leads to transcriptional deregulation of several apoptosis-related genes and to PKB/Akt activation. Moreover, the *PKB/Akt* gene is amplified in breast, ovarian and pancreatic cancers; its negative regulator PTEN is functionally silenced in a number of tumors, and its positive regulator PI-3K is constitutively activated by oncogenes as Ras or Bcr-Abl (Zwick et al., 2002; Sulis and Parsons, 2003).

The complexity of the role played by RTKs in oncogenesis is highlighted by the study of the effects of the receptor for the Hepatocyte Growth Factor (HGFR). HGFR elicits a complex morphogenetic program known as invasive growth, by which epithelial cells orchestrate survival, proliferation, and motility. We recently observed these effects of HGFR in a colorectal cancer cell (CRC) model, where HGFR sustains and increases tumour cell invasive properties and survival through a functional crosstalk with Wnt/β catenin signaling, known to be instrumental in CRC development (Rasola et al., 2006). In this model, survival provided by HGF is permissive for CRC progression, as it might sustain tumor capability to bear further mutations and to proceed towards malignancy. It is reasonable to think that HGFR accomplishes these long-term responses integrating a network of transient, epigenetic signals and transcription modulation, which altogether might produce durable regulatory effects in CRC. In fact, we found that HGF-driven survival requires the coordinated engagement of thePI3K/Akt/mTOR/p70S6K and ERK MAPK transduction pathways, and the abrogation of p38 MAPK pro-apoptotic activity (Fassetta et al., 2006). We could measure a similar death-promoting role of p38 MAPK in another tumour model, ovarian carcinoma cells exposed to chemotherapeutics. However, in ovarian cancer cells HGF not only induces survival pathways by simultaneously activating PI3K/Akt and ERK MAPK, but it also strongly enhances the activity of p38 MAPK. The net result is a surprising enhancement of apoptosis (Rasola et al., 2004; Coltella et al., 2006). HGF treatment changes the transcriptional pattern of ovarian cancer cells, and the biological output (death sensitization) is the resultant of the complex modulation of dozens of genes (Olivero et al., 2006). These antinomic effects of HGFR activation on survival in different tumor types could rely upon a differential contribution of each signaling pathway recruited by the receptor throughout carcinoma progression, or upon the integration of concomitant, unrelated metabolic cascades.

MAPK signaling is deregulated in a variety of malignancies. ERKs are hyperactivated in leukemias, epithelial and other tumors. The JNK/c-Jun pathway is oncogenic and promotes cell cycle progression (Wada and Penninger, 2004).

The study of the p53 tumor suppressor gene has further highlighted the importance of apoptosis inhibition during tumorigenesis. Apoptosis is the sole p53 function selected against during c-Myc-mediated transformation, and p53 is mutated in half of human tumors. In several other tumors, mutations are found in p53 upstream regulators involved in DNA damage response or cell-cycle checkpoints (ATM, Chk2, Rb, Mdm2, $p19^{ARF}$), or in downstream effectors, such as certain Bcl-2 family members (Norbury and Zhivotovsky, 2004; Slee et al., 2004). Many oncoproteins that drive cell proliferation, as Myc, E1A, the Ras/Raf pathway, sensitize cells to apoptosis. All these proteins induce the p53 activator

ARF and, to initiate transformation, disruption of the ARF-p53 pathwayis often required (Schmitt, 2003). Remarkably, most mechanisms leading to cell proliferation (deregulated expression of c-Jun, c-Myc or Ras and every pathway that activates the transcription factors of the E2F family) promote apoptosis, and inactivation of the apoptotic machinery is concomitantly required to promote tumorigenesis.

NF-κB is constitutively active in several tumors, and in conjunction with other oncogenic lesions (e.g. Ras overexpression or the Bcr-Abl fusion protein) promotes the developmentof lymphoid-, myeloid- and epithelial-derived cancers (Karin and Lin, 2002).

Alternative death pathways can also be enhanced in transforming conditions: Ras and Src-family kinases can induce caspase-independent PCD (Jaattela, 2004); many lysosomal enzymesare overexpressed in cancer cells, entailing a higher susceptibility to lysosomal-mediated cell death (Guicciardi et al., 2004). These PCD pathways can limit tumor formation, and tumor progression often requires resistance to lysosomal death.

Autophagy has a complex role in tumorigenesis. It can act as a tumor suppressor mechanism at an early stage of tumour development, causing cell death that would both limit unrestricted cancer cell proliferation and maintain a low DNA mutation rate. Accordingly, autophagy is downregulated early in tumorigenesis. Expression of the autophagy-related protein Beclin-1 reduces tumorigenic capacity, and DRAM (damage-regulated autophagy modulator), a p53 target gene encoding a lysosomal autophagic protein, is an effector of p53-mediated death (Tardy et al., 2006; Crighton et al., 2006). However, at advanced stages of tumour development, autophagy may promote tumour progression by protecting cancer cells from insults of the tumor microenvironment, such as low-oxygen and nutrient deprivation, or from chemotherapeutics ('protective autophagy'). Notably, when tumour cells induce protective autophagy, its inhibition could sensitize tumour cells to the treatment by activating apoptosis (Kondo et al., 2005).

Senescence constitutes a potent anticancer mechanism, and the first step in creating an *in vitro* model of human cancer requires abolition of senescence. Many kinds of oncogenic or stressful stimuli and also chemotherapeutics can induce a senescence response under the control of the p53 and RB tumor suppressor proteins when the damage is irreparable. Cells that do not senesce and proliferate despite dysfunctional telomeres or reactive oxygen species production develop chromosomal aberrations, which result in neoplastic transformation, and more than 90% of tumors contain detectable telomerase activity. The senescence responseensures that these cells are permanently withdrawn from the cell cycle, rendering them incapable of forming a tumor. However, the accumulation of nondividing senescent cells may compromise tissue renewal or repair, and among the genes that are upregulated by the senescence response, many encode proteins that can alter the tissue microenvironment. In these cases, paradoxically, senescence may promote tumor progression by secreting certain matrix metalloproteases, growth factors, and cytokines (Campisi, 2005; Dimri, 2005).

Apoptosis Therapy in Cancer: Convincing the Enemy to Commit Suicide

As most mechanisms leading to proliferation normally promote apoptosis, tumorigenesis requires the plugging of mitogenic lesions upon an anti-apoptotic platform that counteracts the incessant apoptotic pressure of hyperproliferative conditions and ectopic growth.

Consequently, cancer cells strongly depend on survival lesions to underpin their capability to accumulate mutations. Shrewdly targeting this survival platform by the addition of an extra apoptotic stimulus can be potently destructive for tumor cells, much less for normal cells that do not boost proliferation balancing on the tightened rope of apoptosis inhibition. Conventional cancer therapy takes advantage from this situation and induces apoptosis by targeting vital processes or structures in tumor cells. Anticancer treatments include DNA-damaging agents, antimetabolites, mitotic inhibitors, nucleotide analogues, inhibitors of topoisomerases and γ-irradiation (Igney and Krammer, 2002). When damage is perceived as irreparable, the tumor cell irreversibly triggers the apoptotic program. Nonetheless, these approaches fail to correct anti-apoptotic lesions specifically carried by tumor cells, since oncoproteins are also part of the normal proliferative machinery: hence, adrawback of traditional cancer therapy is the lack of selectivity towards the tumor. Another major inconvenience is treatment resistance, due to interference in drug-target interaction or to defects in the responses triggered by the drug-induced insult. If the apoptotic pathway inactivated in tumor development is the same as, or overlaps, that leading to cell death by drug, most cancers would be expected to be resistant to drug-induced apoptosis. Therefore, therapeutic interventions that can lower the threshold for apoptosis of tumor cells could become useful approaches to treat cancer, especially when used in combination with other therapeutic modalities (Blagosklonny, 2004).

In addition, the gross genetic instability of most tumor cells renders complicateto address how many survival pathways might need to be targeted by drugs, taking into account that these cells could have an apoptotic machinery partially damaged (Johnstone et al., 2002). Notably, suppression of apoptosis itself has the potential of promoting mutation and genome instability, since apoptosis is a major road for elimination of mutated cells (Green and Evan, 2002). Hence, it is arduous to find singletherapeutics endowed with the capability to eradicate tumor cells. Conversely, a limited number of genetic lesions can be required for the evolutions of many tumors and tumor cell proliferation may remain dependent on the initial oncogene (oncogene addiction; Hingorani and Tuveson, 2003). In addition, cancer cells may paradoxically be more prone to apoptosis than normal cells, as transformation can induce intracellular stress signals in an attempt of the cancerous cell to self-terminate. Exploiting these features can cause tumor cell-restricted apoptosis, whereas generic chemotherapy can select for additional oncogenic mutations. A detailed understanding of the molecular basis of drug action is required to design combined regimens that improve efficacy without impinging on common apoptotic pathways. A map of the network of molecular interactions is mandatory to restore failsafe defects, to avoid overlapping killing mechanisms in combined treatments.

In this frame, the new generation of anticancer strategies include the use of gene therapy, antisense oligonucleotides, RNA interference, small-molecule drugs and peptide-based inhibitors, with the aim to trigger the core components of the cell death machinery. Among these approaches, the most promising are the use of antisense oligonucleotides or small molecule inhibitors directed against Bcl-2 family components (Juin et al., 2004) or the direct activation of the extrinsic apoptotic pathways in cancer. Toxicity has limited the therapeutic use of soluble TNF or FasL, but another ligand of death receptors, TRAIL, has generated tremendous enthusiasm because it selectively induces apoptosis in various malignant cell types, but not in normal cells. However, clinical efficacy of soluble TRAIL might be hampered because ofthe widespread expression of different TRAIL receptors throughout the body (Ashkenazi, 2002; Kelley and Ashkenazi, 2004).

Moreover, other approaches modulate pathways not directly involved in the execution steps of the apoptosis process by targeting a broad array of kinases, phosphatases, transcriptional factors, cell-surface receptors, chaperones and the proteasome. Among these apoptosis regulatory targets, strategies towards the prevention or interception of aberrant RTK activity have a key importance. These include inhibitory monoclonal antibodies such as trastuzumab (which inhibits HER2/Neu), cetuximab and gefitinib (inhibit EGFR), imatinib mesylate (inhibits PDGFR) and bevazucimab (inhibits VEGF), and small inhibitor molecules, such as kinase inhibitors (e.g. Gleevec, Iressa; Krause and Van Etten, 2005). All these strategies affect cell survival by targeting downstream pathways, as PKB/Akt or the MAPK cascades (Zwick et al., 2002; Senderowicz, 2004).

The proteasome inhibitor bortezomib (Velcade) has revealed a differential response in normal and cancer cells, and is now approved for the treatment of multiple myeloma. Other approaches are the development of classes of drugs that activate the apoptotic function of p53, such asthe nutlins (Vassilev, 2005), and interventions on the epigenetic modifications that affect gene expression in cancer, e.g. DNA acetylation or methylation (Yoo and Jones, 2006; Bremer et al., 2006).

Genetic mapping of the tumor and proteomic approaches mayhelp to rapidly identify novel apoptosis-related proteins, thus allowing more integrated and possibly patient-directed therapies. Furthermore, the use of selective apoptosis inducers might not eliminate side effects because of the similarity of genes thatevoke apoptosis in normal and tumor cells. An intriguing possibility to overcome these iatrogenic insults will be the use of survival factors to protect normal cells in cancer patients undergoing cytotoxic therapy. As tumor cells may integrate diverse strategies to overcome death signals, combined approaches are likely to be the more effective to reinstate apoptosis and to avoid cancer relapses.

The Lure for Death: Pathological Apoptosis Enhancement

Rapid progress in understanding the molecular basis of neurodegeneration has been tightly linked with recent discoveries in the field of PCD. Apoptosis is deregulated and increased in a number of neurological disorders, including Alzheimer's disease (AD), Parkinson disease (PD) and Amyotrophic Lateral Sclerosis (ALS). All these diseases share common biochemical features, such as mitochondrial dysfunction, ER stress and protein misfolding, that are central in inducing apoptosis and neurodegeneration (Cotman and Anderson, 1995).

Apoptosis in Neuronal Development and During Age-related Neurodegeneration

Apoptosis is crucial for the normal development of the nervous system to limit the excess of neurons produced, most of which must die to ensure an appropriate synaptic connectivity. This process is essentially controlled by the combination of a limited supply of trophins and the absolute trophic dependence of the developing neurons, which successfully eliminates unnecessary cells by allowing them to die. Immature neurons that do not succeed in accessing enough trophic support are cleared away by an active apoptotic process.

Physiological apoptosis in the developing brain and age-related apoptosis in the adult brain share similar molecular effector mechanisms, but key differences exist in the mechanisms by which apoptosis is triggered. One important change is that the adult cells are not competent to die, and consequently they no longer require trophins to survive.

Key effectors of neuron apoptosis are the proteins of the p53 family, composed by the tumour suppressor p53 and by its two homologs, p63 and p73. These molecules are present in multiple isoforms and display several posttranslational modifications, and this accounts for the complexity of their reciprocal regulatory networks (Melino et al., 2003). All three p53 family members are transcription factors involved in the control of cell cycle arrest and apoptosis. Notably, they have been implicated in neuron death signaling following growth factor withdrawal (mainly nerve growth factor, NGF). Initially, the hypothesis was that the proapoptotic function of p53 was opposed by the prosurvival functions of p63/p73, and the resulting balance would decide the fate of neurons deprived of neurotrophin stimulation (Pozniak et al., 2000). However, several other findings indicate that p53 is not the only proapoptotic effector in neurons: p53 null neurons are only partially protected from NGF withdrawal and, in $p73^{-/-}$ mice, the absence of p53 rescues neurons only in part. Finally, in sympathetic neurons of $p53^{-/-}$ mice, neuronal death is only partially affected. Recent findings strongly support the central role of p63 proapoptotic function in neuronal development (Jacobs et al., 2005). p63 may induce neuronal apoptosis in the absence of p53, whereas p53 requires p63 expression in order to trigger cell death. Bax expression is required forp63-induced neuronal death and can mediate both the extrinsic and the intrinsic apoptotic pathways (Gressner et al., 2005). The MAP kinase Jnk has a central function in Bax- and p63-mediated apoptosis of sympathetic neurons. This signaling cascade, involving JNK, BH3-only Bcl-2 family members, and Bax, is also required for neuronal demise triggered by a variety of pathological conditions, including brain ischemia and β amyloid neurotoxicity.

The expression of Bcl-2 is high in the central nervous system during development and is downregulated after birth, whereas the expression of Bcl-2 in the peripheral nervous system is maintained throughout life. Although the development of the nervous system in Bcl-2-knockout mice is normal, there is a subsequent loss of motor, sensory and sympathetic neurons after birth, suggesting that Bcl-2 is crucial for the maintenance of neuronal survival. Bcl-X_L is expressed in developing brain; but unlike Bcl-2 expression, Bcl-X_L expression continues to increase into adult life (Gonzalez-Garcia et al., 1995; Michaelidis et al., 1996).

Toxic insults resulting from biochemical or genetic accidents might trigger neurodegenerative diseases by co-opting apoptotic signalling pathways, for example through generation of reactive oxygen species (ROS), accumulation of toxic molecules, *i.e.* glutamate, or caspase activation.

ROS are harmful to cells because of their ability to oxidize proteins, lipids and DNA, thus affecting their structure and function. Under physiological conditions, ROS are rapidly cleared in scavenging reactions by antioxidant enzymes such as superoxide dismutase (SOD), catalase, glutathione-reductase, glutathione peroxidase and peroxiredoxins. ROS production inherent to oxidative stress is generally associated with neuronal apoptosis. Dysregulation of

ROS scavenging capacity, due to deficits in glutathione peroxidase or SOD activity, have been implicated in ageing and chronic neurodegeneration, respectively.

In neurons of the central nervous system, excessive glutamate induces increased Ca^{2+} influx, resulting in over-stimulation of numerous Ca^{2+}-regulated pathways. The consequent raise in free radical production and lipid peroxidation causes mitochondrial depolarization, impaired ion channel function and initiation of the apoptotic program (Mattson, 1998; Lu et al., 2002). Dynamic changes in free Ca^{2+} concentration within the ER and alterations in ER Ca^{2+} homeostasis are important in regulating both protein synthesis and post-translational proteins folding. The proper folding of proteins is controlled by several families of chaperones. Disruption of ER Ca^{2+} homeostasis alters the activity of Ca^{2+}-dependent chaperonesresponsible for post-translational protein processing and activates ER stress response and apoptosis (Pahl, 1999; Michalak et al., 2002).

Apoptosis in Neurodegenerative and in Neuromuscular Diseases

Deregulation of PCD is involved in the most diffuse neurodegenerative pathologies, among which are Alzheimer's disease, Parkinson disease and amyotrophic lateral sclerosis.

Alzheimer's disease (AD) is a progressive neurodegenerative disorder that is characterized by two hallmark lesions: extracellular amyloid plaques and neurofibrillary tangles. Amyloid plaques arise from abnormal proteolytic processing of the amyloid precursor protein (APP), an integral membrane protein whose cleavage leads to the formation of the Aβ peptide, the major plaque constituent (Billings et al., 2005). APP is first cleaved by either α- or β-secretase in the extracellular space to yield both a secreted N-terminal fragment and a membrane-bound C-terminal fragment. The latter is then cleaved by γ-secretase to form the amyloid-β (Aβ) peptides (Mattson, 1998). The direct pathologic role of extracellular amyloid plaques has not been assessed yet. Recent evidence suggests that oligomers of Aβ activate caspases through the extrinsic pathway, implicating binding of extracellular Aβ to cell death receptors, while other studies suggest that the intrinsic pathway may be more relevant (Lustbader et al., 2004). In any case, a loss of mitochondrial function is considered central in the development and pathogenesis of AD. Mitochondrial dysfunction can result in excessive production of ROS, opening of the permeability transition pore (PTP; Bernardi et al., 2006), and release of small proteins that trigger the initiation of apoptosis (Saelens et al., 2004). Direct evidence for mitochondrial dysfunction in AD comes from several reports: in isolated mitochondria, Aβ inhibits activities of mitochondrial electron transport chain complexes, β-ketoglutarate dehydrogenase, pyruvate dehydrogenase, and cytochrome c oxidase, and promotes Ca^{2+}-induced PTP opening; Aβ may bind to alcohol dehydrogenase within mitochondria and activates apoptosis through mitochondrial stress (Ojaimi and Byrne, 2001; Byrne, 2002; Lustbader et al., 2004).

ER stress could also be a key event in triggering and mediating neuronal death in AD. In fact, it has been reported that accumulation of Aβ in the ER or in endosomes activates the ER-dependent unfolded protein response. When this stress is prolonged, it may lead to processing of the ER-resident protease procaspase-12 as well as to activation of calpain, caspase-3, caspase-6, and eventually to apoptosis. Aβ has also been shown to directly trigger Ca^{2+} release from the ER, thus inducingneurotoxicity and neuron death (Verkhratsky and Toescu, 2003).

Neurofibrillary tangles, the second hallmarks of AD, are the consequence of microtubule destabilization and aggregation of the microtubule-associated protein Tau. Tau is known for its role in microtubule stabilization, which is important for the generation and maintenance of neurites. In AD, neurons accumulate a heavily phosphorylated form of Tau, which aggregates into paired helical filaments, thus losing its microtubule-binding and stabilizing function (Garcia and Cleveland, 2001). Consequently, the anterograde transport of vesicles necessary for synapse function might be slowed down. Notably, vesicles carrying the amyloid precursor protein (APP) are retained in the cell soma, being more processed with enhanced production of toxic Aβ peptide, leading to Aβ-triggered caspase activation (Xu et al., 1997). It has also been found that early during the disease progression, axonal swellings cause a proximal accumulation of mitochondria. This may contribute to the loss of distal synapses, since dendritic mitochondria have been shown to play an important role in maintaining plasticity of spines and synapses (Liu et al., 2004; Stokin et al., 2005).

Parkinson disease (PD) is the most common neurodegenerative disorder of movement, and is characterized by age-dependent increases in resting tremor, rigidity, bradykinesia, muscular rigidity and gait abnormalities (Moore et al., 2005). Loss of dopaminergic neurons in the substantia nigra of the brain is the main feature in all forms of human parkinsonism, and strong evidences show that the mitochondria-dependent apoptotic pathway is predominantly activated (Tatton et al., 2003), although others could not find apoptotic cells in the substantia nigra of PD patients (Vila and Przedborski, 2003). However, the discovery of Mendelian-inherited genes that cause PD has greatly enhanced our understanding of the biochemical changes that mediate neurodegeneration in PD, and has allowed to establish new links with the mitochondrial physiology. These studies revealed mutations in six different genes: α-synuclein, parkin, DJ-1, PTEN-induced kinase 1 (PINK1), OMI/HTRA2 and leucine-rich repeat kinase 2 (LRRK2; Moore et al., 2005).

Although the function of α-synuclein is still unclear, one of the main pathological hallmarks of PD, intracytoplasmic inclusion bodies known as Lewy bodies, turned out to be composed of α-synuclein1 fibrils. One hypothesis for explaining α-synuclein pathogenicity in the aggregated form is that protofibrils of α-synuclein bind synaptic vesicles and form pores that could lead to permeabilization of the vesicle membranes, thereby releasing excess dopamine into the cytosol. The missense mutation found in the *α-synuclein* gene of PD patients enhances the formation of α-synuclein protofibrils. However, a direct link between α-synuclein and intrinsic apoptosis comes from the observation that mice lacking the gene for the α–synuclein are resistant to the mitochondrial toxin 1-methyl-4-phenyl-1,2,3,6-tetrahydropyridine (MPTP; Dauer et al., 2002), and that overexpression of α-synuclein mutants leads to mitochondrial dysfunction both *in vitro* and *in vivo*. Conversely, α-synuclein-deficient mice are more resistant to the respiratory chain inhibitors MPTP, malonate, and 3- nitroproprionic acid, indicating that α-synuclein may play a role in modulating mitochondrial toxicity and oxidative stress. Transgenic mice overexpressing mutant α-synuclein develop neuronal degeneration and display a decreased complex IV activity in the spinal cord and accumulation of intraneuronal inclusions. These inclusions contain both α-synuclein and degenerating mitochondria (Klivenyi et al., 2006).

LRRK2 is in part localized to mitochondria (West et al., 2005). LRRK2 encodes a complex multi-domain protein whose function is poorly determined. However, preliminary data indicate that LRRK2 encodes a kinase that might be associated with the outer mitochondrial membrane (OMM). Significantly, three PD-associated mutations raise the LRRK2 autophosphorylation and possibly its activity (Zimprich et al., 2004).

Parkin encodes an ubiquitin E3 ligase (Shimura et al., 2000), which is a component of the ubiquitin-proteasome system (UPS) involved in cellular protein degradation. Ubiquitin E3 ligases mediate the specific ubiquitination of substrates, thereby targeting them for degradation by the proteasome. α-synuclein and parkin mutations may cause protein misfolding and UPS inhibition, converging in a unique pathway that leads to dopaminergic degeneration. The mechanism by which damage to the UPS causes mitochondrial injury and apoptosis is still under investigation. Several pro-apoptopic proteins, such as p53, Bcl-2 family members and SMAC/Diablo are normally degraded by the UPS (Jesenberger and Jentsch, 2002). Therefore, the UPS has a crucial role in regulating potentially toxic components of the apoptotic machinery. Proteasome inhibition leads to the accumulation of these proteins, which results in PTP opening, mitochondrial depolarization and apoptosis. Parkin mutations could also be implicated in mitochondrial dysfunction in other ways. For instance, respiratory chain complex I activity is decreased in cells from PD patients when parkin is mutated (Muftuoglu et al., 2004).

In addition, ithas been recently identified the presence of a PINK1-parkin pathway that regulates mitochondrial function in loss-of-function PINK1 mutant flies (Clark et al., 2006; Park et al., 2006). PINK1 mutants are sensitive to the free radical inducer paraquat and to the complex I inhibitor rotenone, suggesting that the absence of PINK1 leads to decreased resistance to reactive oxygen species. PINK1 mutants were also more sensitive to the protein folding inhibitor dithiothreitol and to osmotic stress. PINK1 expression was found to protect cells from apoptosis induced by various agents, whereas PD-inducing PINK1 mutations abrogated this protective effect. At the cellular level, a similar effect (mitochondrial depolarization and subsequent PCD) was shown in cells from OMI/HTRA2 knock-out mice. These mice display parkinsonian phenotypes, and it is therefore tempting to speculate that PINK1 and OMI/HTRA2 share a common pathway in the mitochondrial response to cellular stress and modulation of apoptosis, perhaps with OMI/HTRA2 as a downstream target for PINK1, which initiates apoptosis when levels of oxidative stress become too high (Abou-Sleiman et al., 2006).

Endogenous DJ-1 is partially localized in the mitochondrial matrix and in the intermembrane space (Zhang et al., 2005) and it is thought to have a role in modulating the oxidative stress response. DJ-1 might protect neurons from exposure to oxidative stress, as DJ-1 null embryonic cortical neurons show increased sensitivity to oxidative stress (Moore et al., 2005). This suggests that DJ-1 acts as a free radical scavenger, which prevents the accumulation of free radicals derived from the mitochondrial electron transport system. Moreover, DJ-1 has recently been identified as a regulator of p53 transcriptional activity and might therefore represent a molecular bridge between the UPS and the mitochondrial stress. Intriguingly, DJ-1 has been recently identified as a suppressor of PTEN function, paving the scenario of its control on both PINK1 activity and the survival pathway mediated by PI3K (Kim et al., 2005).

Amyotrophic Lateral Sclerosis

Patients with Amyotrophic Lateral Sclerosis (ALS) suffer from muscle atrophy and fatal paralysis as a result of selective degradation of both upper and lower motor neurons. The 10% of ALS cases are familial (FALS), and a subset of these cases result from mutations in the gene encoding for Cu,Zn-superoxide dismutase (Cu,Zn-SOD1), which protects cells from oxidative damage induced by the superoxide anion (Rosen et al., 1993). Although several studies provide *in vivo* evidences that motor neurons in ALS mice die *via* apoptosis, the molecular determinant linking the SOD1 mutation to the apoptotic cell death are still under investigation. In physiological conditions, Cu,Zn-SOD1 is localized to the mitochondrial intermembrane space and matrix and modulates mitochondrial function (Mattiazzi et al., 2002). Several evidences link mutated Cu,Zn-SOD1 to mitochondria dysfunctions and to apoptosis. In rat neuronal cell lines and primary neurons, expression of mutant Cu,Zn-SOD1 sensitizes to apoptosis (Mena et al., 1997). In addition, preliminary data show that ALS progression is enhanced by caspase-9 activation (Inoue et al., 2003), and that overexpression of the antiapoptotic protein Bcl-2 extended survival of ALS animal models (Kostic et al., 1997). Both wild-type and mutant Cu,Zn-SOD1 have been shown to bind to Bcl-2 *in vitro*, but the mutant enzyme has a higher affinity (Pasinelli et al., 2004). Thus, Bcl-2 may link Cu,Zn-SOD1 to the apoptotic pathway. TheCu,Zn-SOD1 mutant mice display structural abnormalities similar to what observed in patients, *i.e.* the presence of large, membrane-bound vacuoles derived from degenerating mitochondria (Dal Canto and Gurney, 1995; Wong et al., 1995). Investigations on isolated brain and spinal cord mitochondria from mutant Cu,Zn-SOD1 transgenic mice showed a lowering in respiratory chain complex activities, oxygen consumption and ATP synthesis (Jung et al., 2002). All these phenomena were attributed to oxidative damage to mitochondrial lipids and to a defect in proteins of the complex IV of the respiratory chain (Mattiazzi et al., 2002; Kirkinezos et al., 2005). Moreover, directly targeting mutant Cu,Zn-SOD1 to the mitochondrial matrix leads to increased apoptotic cell death (Takeuchi et al., 2002).In mitochondria, mutant Cu,Zn-SOD1 may potentiate damage through aberrant interactions with chaperone proteins (Shinder et al., 2001) and disrupt mitochondrial function by affecting mitochondrial protein import (Liu et al., 2004).

Recent studies have determined that apoptosis significantly contributes to pathology in several neuromuscular diseases by inducing the progressive loss of skeletal muscle fibres. Anoikis, a particular form of apoptosis activated when attachment of cells to the extracellular matrix is disrupted. might occur in neuromuscular diseases that are caused by mutations in extracellular proteins (e.g. collagen-VI deficiency and laminin-α2 deficiency) or integrins (integrin-α7 deficiency) which have key roles in the propagation of pro-apoptotic signals when extracellular components are disrupted (Reddig and Juliano, 2005).

Apoptosis contributes to pathogenesis in a mouse model of human muscular congenital dystrophy type 1A (MDC1A). MDC1A is a severe disease that prevents normal neuromuscular function. The most frequent form of human MDC1A is caused by mutations in the LAMA2 gene, which encodes the α2 subunit of the extracellular protein laminin. Laminin-α2 exerts the essential role of connecting extracellular to intracellular components in a functional pathway. An extracellular binding partner of laminin-α2 is entactin that, in turn, binds to collagens. Inside the cell, laminin-α2 binds to α-dystroglycan, which is functionally connected to dystrophin, and to α7-integrin, which forms a transmembrane complex with integrin-β1. Laminin-α2 chain deficiency leads to disruption of this functional axis and, as a result, to loss of integrity of the muscle membrane. Mutations that delete either α-dystroglycan or α7-integrin produce myopathy (Cote et al., 1999). In addition, disruption of the binding of laminin-α2 to α-dystroglycan leads to abnormal activation of apoptosis in muscle cells through a signalling pathway that involves inhibition of the serine–threonine kinase AKT (Langenbach and Rando, 2002). Therefore, it seems that loss of laminin-α2 binding to α-dystroglycan and to integrin complexes might elicit abnormal activation of apoptosis, thereby causing cell loss and contributing to the pathogenesis of MDC1A. Recent studies have shown that apoptosis in MDC1A involves mainly the mitochondrial pathway. In fact, therapeutic approaches based on the inhibition of mitochondrial pathway improved the growth and increased the lifespan of LAMA2-null mice (Girgenrath et al., 2004). Moreover, the muscle-specific overexpression of the apoptosis inhibitor Bcl-2 reduced the severity of the disease in skeletal muscle and extended the mouse lifespan; accordingly, targeted knockout of the apoptogenic Bax protein inhibits disease progression (Dominov et al., 2005).

The mitochondrial pathway of apoptosis has been also involved in the pathology ofmyopathies caused by collagen-VI deficiency. Collagen-VI is a major extracellular matrix protein formed by three α chains. Different types of mutations in each of the three α chains cause two clinically distinct myopathies: the Bethlem myopathy and the Ullrich myopathy (Lampe and Bushby, 2005). Collagen-VI microfibrils interact with several other components of the basal lamina, including collagen-IV, fibronectin, biglycan, decorin and perlecan (Wiberg et al., 2003). Collagen-VI also interacts with cell transmembrane components such as integrins and NG2 proteoglycan (Tillet et al., 2002), and in this way operates to convey signals from the pericellular to the intracellular space. Mutation of collagen-VI are pathogenic because they disrupt the normal signaling mediated by collagen-VI and, as a consequence, increase muscle fiber apoptosis. Studies of muscle fibers from COL6A1 (the gene encoding for collagen-VIα1) knockout mice have revealed abnormalities in the ultrastructure of the sarcoplasmic reticulum (Irwin et al., 2003). Notably, myofibers show a latent mitochondrial dysfunction, endowed with an increased activity of the F_1F_0-ATPase and a reduced threshold for apoptosis activation. Selective inhibition of F_1F_0-ATPase induced mitochondrial depolarization, Ca^{2+} deregulation, and increased apoptosis through PTP opening. These defects, as well as the increased spontaneous apoptosis, were rescued *in vitro* by growing the cells on collagen-VI (therefore proving that the defect in collagen-VI was causative) and *in vivo* by PTP opening inhibition with cyclosporin treatment (see above), which was followed by amelioration of the contractile strength of the mice. Altogether, these observations suggest that a deregulation of PTP-induced apoptosis in muscle fibres has a main role of in the pathologies cause by collagen-VI deficiency (Rizzuto, 2003).

Conclusion

During the past decade there have been major advances in our understanding of the fundamental mechanism and of the biological meaning of cell death. It is now established that apoptotic and non-apoptoticforms of cell death play an essential role in normal development and functioning of multicellular organisms, and that abnormalities in these processes cause diseases such as cancer, neurodegenerative and neuromuscular disease. A detailed understanding of the molecular mechanisms of PCD has been gained, and the main pathways identified and fully described. Yet, much remains to be done to further extend and integrate knowledge of death pathways and of their crosstalk with the other fundamental biochemical routines of the cell. An open issue is to understand the molecular devices that allow stress stimuli to be transmitted from different subcellular compartments to the mitochondrion, which represents the point of integration of the different signaling cascades. It is therefore of pivotal importance the comprehension of theoperational strategies utilized by the cell to finely tune the balance between survival and death signals, in order to devise measures that limit pathological damage to tissues. Moreover, the identification of critical molecules acting as main switches of the mitochondrial apoptotic pathway has provided rational targets for the development of a new generation of useful therapeutics, endowed with minimal undesirable side effects and maximal efficacy.

References

Abou-Sleiman, P.M., M.M. Muqit, and N.W. Wood. 2006. Expanding insights of mitochondrial dysfunction in Parkinson's disease. *Nat Rev Neurosci.* 7:207-219.

Abraham, M.C., and S. Shaham. 2004. Death without caspases, caspases without death. *Trends Cell Biol.* 14:184-193.

Abrams, J.M. 1999. An emerging blueprint for apoptosis in Drosophila. *Trends Cell Biol.* 9:435-440.

Abrams, J.M., and M.A. White. 2004. Coordination of cell death and the cell cycle: linking proliferation to death through private and communal couplers. *Curr Opin Cell Biol.* 16:634-638.

Acehan, D., X. Jiang, D.G. Morgan, J.E. Heuser, X. Wang, and C.W. Akey. 2002. Three-dimensional structure of the apoptosome: implications for assembly, procaspase-9 binding, and activation. *Mol Cell.* 9:423-432.

Adams, J. 2004. The proteasome: a suitable antineoplastic target. *Nat Rev Cancer.* 4:349-360.

Adams, J.M. 2003. Ways of dying: multiple pathways to apoptosis. *Genes Dev.* 17:2481-2495.

Adams, J.M., and S. Cory. 2001. Life-or-death decisions by the Bcl-2 protein family. *Trends Biochem Sci.* 26:61-66.

Adrain, C., and S.J. Martin. 2001. The mitochondrial apoptosome: a killer unleashed by the cytochrome seas. *Trends Biochem Sci.* 26:390-397.

Ameisen, J.C. 1996. The origin of programmed cell death. *Science.* 272:1278-1279.

Annis, M.G., J.A. Yethon, B. Leber, and D.W. Andrews. 2004. There is more to life and death than mitochondria: Bcl-2 proteins at the endoplasmic reticulum. *Biochim Biophys Acta.* 1644:115-123.

Antonsson, B., and J.C. Martinou. 2000. The Bcl-2 protein family. *Exp Cell Res.* 256:50-57.

Arnoult, D., K. Akarid, A. Grodet, P.X. Petit, J. Estaquier, and J.C. Ameisen. 2002. On the evolution of programmedcell death: apoptosis of the unicellular eukaryote Leishmania major involves cysteine proteinase activation and mitochondrion permeabilization. *Cell Death Differ.* 9:65-81.

Ashkenazi, A. 2002. Targeting death and decoy receptors of the tumour-necrosis factor superfamily. *Nat Rev Cancer.* 2:420-430.

Baehrecke, E.H. 2005. Autophagy: dual roles in life and death? *Nat Rev Mol Cell Biol.* 6:505-510.

Baines, C.P., R.A. Kaiser, N.H. Purcell, N.S. Blair, H. Osinska, M.A. Hambleton, E.W. Brunskill, M.R. Sayen, R.A. Gottlieb, G.W. Dorn, J. Robbins, and J.D. Molkentin. 2005. Loss of cyclophilin D reveals a critical role for mitochondrial permeability transition in cell death. *Nature.* 434:658-662.

Balaban, R.S., S. Nemoto, and T. Finkel. 2005. Mitochondria, oxidants, and aging. *Cell.* 120:483-495.

Barde, Y.A. 1989. Trophic factors and neuronal survival. *Neuron.* 2:1525-1534.

Barres, B.A., and M.C. Raff. 1999. Axonal control of oligodendrocyte development. *J Cell Biol.* 147:1123-1128.

Bassik, M.C., L. Scorrano, S.A. Oakes, T. Pozzan, and S.J. Korsmeyer. 2004. Phosphorylation of BCL-2 regulates ER Ca2+ homeostasis and apoptosis. *Embo J.* 23:1207-1216.

Basso, E., L. Fante, J. Fowlkes, V. Petronilli, M.A. Forte, and P. Bernardi. 2005. Properties of the permeability transition pore in mitochondria devoid of Cyclophilin D. *J Biol Chem.* 280:18558-18561.

Beere, H.M., and D.R. Green. 2001. Stress management - heat shock protein-70 and the regulation of apoptosis. *Trends Cell Biol.* 11:6-10.

Bennett, M., K. Macdonald, S.W. Chan, J.P. Luzio, R. Simari, and P. Weissberg. 1998. Cell surface trafficking of Fas: a rapid mechanism of p53-mediated apoptosis. *Science.* 282:290-293.

Berk, A.J. 2005. Recent lessons in gene expression, cell cycle control, and cell biology from adenovirus. *Oncogene.* 24:7673-7685.

Bernardi, P. 1999. Mitochondrial transport of cations: channels, exchangers, and permeability transition. *Physiol Rev.* 79:1127-1155.

Bernardi, P., A. Krauskopf, E. Basso, V. Petronilli, E. Blachly-Dyson, F. Di Lisa, and M.A. Forte. 2006. The mitochondrial permeability transition from in vitro artifact to disease target. *Febs J.* 273:2077-2099.

Bernardi, P., L. Scorrano, R. Colonna, V. Petronilli, and F. Di Lisa. 1999. Mitochondria and cell death. Mechanistic aspects and methodological issues. *Eur J Biochem.* 264:687-701.

Billings, L.M., S. Oddo, K.N. Green, J.L. McGaugh, and F.M. LaFerla. 2005. Intraneuronal Abeta causes the onset of early Alzheimer's disease-related cognitive deficits in transgenic mice. *Neuron.* 45:675-688.

Blagosklonny, M.V. 2004. Prospective strategies to enforce selectively cell death in cancer cells. *Oncogene.* 23:2967-2975.

Boehning, D., R.L. Patterson, L. Sedaghat, N.O. Glebova, T. Kurosaki, and S.H. Snyder. 2003. Cytochrome c binds to inositol (1,4,5) trisphosphate receptors, amplifying calcium-dependent apoptosis. *Nat Cell Biol.* 5:1051-1061.

Bonnefoy-Berard, N., A. Aouacheria, C. Verschelde, L. Quemeneur, A. Marcais, and J. Marvel. 2004. Control of proliferation by Bcl-2 family members. *Biochim Biophys Acta.* 1644:159-168.

Brazil, D.P., Z.Z. Yang, and B.A. Hemmings. 2004. Advances in protein kinase B signalling: AKTion on multiple fronts. *Trends Biochem Sci.* 29:233-242.

Bremer, E., G. van Dam, B.J. Kroesen, L. de Leij, and W. Helfrich. 2006. Targeted induction of apoptosis for cancer therapy: current progress and prospects. *Trends Mol Med.*

Brown, D.A., and J.K. Rose. 1992. Sorting of GPI-anchored proteins to glycolipid-enriched membrane subdomains during transport to the apical cell surface. *Cell.* 68:533-544.

Bruey, J.M., C. Ducasse, P. Bonniaud, L. Ravagnan, S.A. Susin, C. Diaz-Latoud, S. Gurbuxani, A.P. Arrigo, G. Kroemer, E. Solary, and C. Garrido. 2000. Hsp27 negatively regulates cell death by interacting with cytochrome c. *Nat Cell Biol.* 2:645-652.

Budd, R.C. 2002. Death receptors couple to both cell proliferation and apoptosis. *J Clin Invest.* 109:437-441.

Byrne, E. 2002. Does mitochondrial respiratory chain dysfunction have a role in common neurodegenerative disorders? *J Clin Neurosci.* 9:497-501.

Campisi, J. 2005. Senescent cells, tumor suppression, and organismal aging: good citizens, bad neighbors. *Cell.* 120:513-522.

Chen, L., S.N. Willis, A. Wei, B.J. Smith, J.I. Fletcher, M.G. Hinds, P.M. Colman, C.L. Day, J.M. Adams, and D.C. Huang. 2005. Differential targeting of prosurvival Bcl-2 proteins by their BH3-only ligands allows complementary apoptotic function. *Mol Cell.* 17:393-403.

Chen, R., I. Valencia, F. Zhong, K.S. McColl, H.L. Roderick, M.D. Bootman, M.J. Berridge, S.J. Conway, A.B. Holmes, G.A. Mignery, P. Velez, and C.W. Distelhorst. 2004. Bcl-2 functionally interacts with inositol 1,4,5-trisphosphate receptors to regulate calcium release from the ER in response to inositol 1,4,5-trisphosphate. *J Cell Biol.* 166:193-203.

Cipolat, S., T. Rudka, D. Hartmann, V. Costa, L. Serneels, K. Craessaerts, K. Metzger, C. Frezza, W. Annaert, L. D'Adamio, C. Derks, T. Dejaegere, L. Pellegrini, R. D'Hooge, L. Scorrano, and B. De Strooper. 2006. Mitochondrial rhomboid PARL regulates cytochrome c release during apoptosis via OPA1-dependent cristae remodeling. *Cell.* 126:163-175.

Clark, I.E., M.W. Dodson, C. Jiang, J.H. Cao, J.R. Huh, J.H. Seol, S.J. Yoo, B.A. Hay, and M. Guo. 2006. Drosophila pink1 is required for mitochondrial function and interacts genetically with parkin. *Nature.* 441:1162-1166.

Cohen, J.J., R.C. Duke, V.A. Fadok, and K.S. Sellins. 1992. Apoptosis and programmed cell death in immunity. *Annu Rev Immunol.* 10:267-293.

Coltella, N., A. Rasola, E. Nano, C. Bardella, M. Fassetta, N. Filigheddu, A. Graziani, P.M. Comoglio, and M.F. Di Renzo. 2006. p38 MAPK turns hepatocyte growth factor to a death signal that commits ovarian cancer cells to chemotherapy-induced apoptosis. *Int J Cancer.* 118:2981-2990.

Colussi, P.A., and S. Kumar. 1999. Targeted disruption of caspase genes in mice: what they tell us about the functions of individual caspases in apoptosis. *Immunol Cell Biol.* 77:58-63.

Cornillon, S., C. Foa, J. Davoust, N. Buonavista, J.D. Gross, and P. Golstein. 1994. Programmed cell death in Dictyostelium. *J Cell Sci.* 107 (Pt 10):2691-2704.

Cote, P.D., H. Moukhles, M. Lindenbaum, and S. Carbonetto. 1999. Chimaeric mice deficient in dystroglycans develop muscular dystrophy and have disrupted myoneural synapses. *Nat Genet.* 23:338-342.

Cotman, C.W., and A.J. Anderson. 1995. A potential role for apoptosis in neurodegeneration and Alzheimer's disease. *Mol Neurobiol.* 10:19-45.

Creagh, E.M., H. Conroy, and S.J. Martin. 2003. Caspase-activation pathways in apoptosis and immunity. *Immunol Rev.* 193:10-21.

Crighton, D., S. Wilkinson, J. O'Prey, N. Syed, P. Smith, P.R. Harrison, M. Gasco, O. Garrone, T. Crook, and K.M. Ryan. 2006. DRAM, a p53-induced modulator of autophagy, is critical for apoptosis. *Cell.* 126:121-134.

Crosnier, C., D. Stamataki, and J. Lewis. 2006. Organizing cell renewal in the intestine: stem cells, signals and combinatorial control. *Nat Rev Genet.* 7:349-359.

Dal Canto, M.C., and M.E. Gurney. 1995. Neuropathological changes in two lines of mice carrying a transgene for mutant human Cu,Zn SOD, and in mice overexpressing wild type human SOD: a model of familial amyotrophic lateral sclerosis (FALS). *Brain Res.* 676:25-40.

Danial, N.N., and S.J. Korsmeyer. 2004. Cell death: critical control points. *Cell.* 116:205-219.

Dauer, W., N. Kholodilov, M. Vila, A.C. Trillat,R. Goodchild, K.E. Larsen, R. Staal, K. Tieu, Y. Schmitz, C.A. Yuan, M. Rocha, V. Jackson-Lewis, S. Hersch, D. Sulzer, S. Przedborski, R. Burke, and R. Hen. 2002. Resistance of alpha -synuclein null mice to the parkinsonian neurotoxin MPTP. *Proc Natl Acad Sci U S A.* 99:14524-14529.

Degenhardt, K., R. Mathew, B. Beaudoin, K. Bray, D. Anderson, G. Chen, C. Mukherjee, Y. Shi, C. Gelinas, Y. Fan, D.A. Nelson, S. Jin, and E. White. 2006. Autophagy promotes tumor cell survival and restricts necrosis, inflammation, and tumorigenesis. *Cancer Cell.* 10:51-64.

Degli Esposti, M., and C. Dive. 2003. Mitochondrial membrane permeabilisation by Bax/Bak. *Biochem Biophys Res Commun.* 304:455-461.

Degterev, A., Z. Huang, M. Boyce, Y. Li, P. Jagtap, N. Mizushima, G.D. Cuny, T.J. Mitchison, M.A. Moskowitz, and J. Yuan. 2005. Chemical inhibitor of nonapoptotic cell death with therapeutic potential for ischemic brain injury. *Nat Chem Biol.* 1:112-119.

Demaurex, N., and C. Distelhorst. 2003. Cell biology. Apoptosis--the calcium connection. *Science.* 300:65-67.

Desagher, S., and J.C. Martinou. 2000. Mitochondria as the central control point of apoptosis. *Trends Cell Biol.* 10:369-377.

Di Lisa, F., R. Menabo, M. Canton, and V. Petronilli. 1998. The role of mitochondria in the salvage and the injury of the ischemic myocardium. *Biochim Biophys Acta.* 1366:69-78.

Dimri, G.P. 2005. What has senescence got to do with cancer? *Cancer Cell.* 7:505-512.

Ding, H.F., Y.L. Lin, G. McGill, P. Juo, H. Zhu, J. Blenis, J. Yuan, and D.E. Fisher. 2000. Essential role for caspase-8 in transcription-independent apoptosis triggered by p53. *J Biol Chem.* 275:38905-38911.

Distelhorst, C.W., and G.C. Shore. 2004. Bcl-2 and calcium: controversy beneath the surface. *Oncogene.* 23:2875-2880.

Dominov, J.A., A.J. Kravetz, M. Ardelt, C.A. Kostek, M.L. Beermann, and J.B. Miller. 2005. Muscle-specific BCL2 expression ameliorates muscle disease in laminin {alpha}2-deficient, but not in dystrophin-deficient, mice. *Hum Mol Genet.* 14:1029-1040.

Donovan, M., and T.G. Cotter. 2004. Control of mitochondrial integrity by Bcl-2 family members and caspase-independent cell death. *Biochim Biophys Acta.* 1644:133-147.

Du, C., M. Fang, Y. Li, L. Li, and X. Wang. 2000. Smac, a mitochondrial protein that promotes cytochrome c-dependent caspase activation by eliminating IAP inhibition. *Cell.* 102:33-42.

Edinger, A.L., and C.B. Thompson. 2004. Death by design: apoptosis, necrosis and autophagy. *Curr Opin Cell Biol.* 16:663-669.

Ellis, H.M., and H.R. Horvitz. 1986. Genetic control of programmed cell death in the nematode C. elegans. *Cell.* 44:817-829.

Fan, T.J., L.H. Han, R.S. Cong, and J. Liang. 2005. Caspase family proteases and apoptosis. *Acta Biochim Biophys Sin (Shanghai).* 37:719-727.

Fassetta, M., L. D'Alessandro, N. Coltella, M.F. Di Renzo, and A. Rasola. 2006. Hepatocyte growth factor installs a survival platform for colorectal cancer cell invasive growth and overcomes p38 MAPK-mediated apoptosis.*Cell Signal.*

Forte, M., and P. Bernardi. 2005. Genetic dissection of the permeability transition pore. *J Bioenerg Biomembr.* 37:121-128.

Frezza, C., S. Cipolat, O. Martins de Brito, M. Micaroni, G.V. Beznoussenko, T. Rudka, D. Bartoli, R.S. Polishuck, N.N. Danial, B. De Strooper, and L. Scorrano. 2006. OPA1 controls apoptotic cristae remodeling independently from mitochondrial fusion. *Cell.* 126:177-189.

Frisch, S.M., and R.A. Screaton. 2001. Anoikis mechanisms. *Curr Opin Cell Biol.* 13:555-562.

Garcia, M.L., and D.W. Cleveland. 2001. Going new places using an old MAP: tau, microtubules and human neurodegenerative disease. *Curr Opin Cell Biol.* 13:41-48.

Gelinas, C., and E. White. 2005. BH3-only proteins in control: specificity regulates MCL-1 and BAK-mediated apoptosis. *Genes Dev.* 19:1263-1268.

Girgenrath, M., J.A. Dominov, C.A. Kostek, and J.B. Miller. 2004. Inhibition of apoptosis improves outcome in a model of congenital muscular dystrophy. *J Clin Invest.* 114:1635-1639.

Glucksmann, A. 1951. Cell deaths in normal vertebrate ontogeny. *Biol Rev Camb Philos Soc.* 26:59-86.

Gonzalez-Garcia, M., I. Garcia, L. Ding, S. O'Shea, L.H. Boise, C.B. Thompson, and G. Nunez. 1995. bcl-x is expressed in embryonic and postnatal neural tissues and functions to prevent neuronal cell death. *Proc Natl Acad Sci U S A.* 92:4304-4308.

Gordeeva, A.V., Y.A. Labas, and R.A. Zvyagilskaya. 2004. Apoptosis in unicellular organisms: mechanisms and evolution. *Biochemistry (Mosc).* 69:1055-1066.

Gottlob, K., N. Majewski, S. Kennedy, E. Kandel, R.B. Robey, and N. Hay. 2001. Inhibition of early apoptotic events by Akt/PKB is dependent on the first committed step of glycolysis and mitochondrial hexokinase. *Genes Dev.* 15:1406-1418.

Gramaglia, D., A. Gentile, M. Battaglia, L. Ranzato, V. Petronilli, M. Fassetta, P. Bernardi, and A. Rasola. 2004. Apoptosis to necrosis switching downstream of apoptosome formation requires inhibition of both glycolysis and oxidative phosphorylation in a BCL-X(L)- and PKB/AKT-independent fashion. *Cell Death Differ.* 11:342-353.

Green, D.R., and G.I. Evan. 2002. A matter of life and death. *Cancer Cell.* 1:19-30.

Gregorieff, A., and H. Clevers. 2005. Wnt signaling in the intestinal epithelium: from endoderm to cancer. *Genes Dev.* 19:877-890.

Gressner, O., T. Schilling, K. Lorenz, E. Schulze Schleithoff, A. Koch, H. Schulze-Bergkamen, A. Maria Lena, E. Candi, A. Terrinoni, M. Valeria Catani, M. Oren, G. Melino, P.H. Krammer, W. Stremmel, and M. Muller. 2005. TAp63alpha induces apoptosis by activating signaling via death receptors and mitochondria. *Embo J.* 24:2458-2471.

Gross, A., J.M. McDonnell, and S.J. Korsmeyer. 1999. BCL-2 family members and the mitochondria in apoptosis. *Genes Dev.* 13:1899-1911.

Grossmann, J. 2002. Molecular mechanisms of "detachment-induced apoptosis--Anoikis". *Apoptosis.* 7:247-260.

Guicciardi, M.E., M. Leist, and G.J. Gores. 2004. Lysosomes in cell death. *Oncogene.* 23:2881-2890.

Hacker, G., and D.L. Vaux. 1997. A chronology of cell death. *Apoptosis.* 2:247-256.

Hanahan, D., and R.A. Weinberg. 2000. The hallmarks of cancer. *Cell.* 100:57-70.

Hao, Z., G.S. Duncan, C.C. Chang, A. Elia, M. Fang, A. Wakeham, H. Okada, T. Calzascia, Y. Jang, A. You-Ten, W.C. Yeh, P. Ohashi, X. Wang, and T.W. Mak. 2005. Specific ablation of the apoptotic functions of cytochrome C reveals a differential requirement for cytochrome C and Apaf-1 in apoptosis. *Cell.* 121:579-591.

Harris, S.L., and A.J. Levine. 2005. The p53 pathway: positive and negative feedback loops. *Oncogene.* 24:2899-2908.

Hawkins, C.J., and D.L. Vaux. 1997. The role of the Bcl-2 family of apoptosis regulatory proteins in the immune system. *Semin Immunol.* 9:25-33.

Hengartner, M.O. 2000. The biochemistry of apoptosis. *Nature.* 407:770-776.

Henry-Mowatt, J., C. Dive, J.C. Martinou, and D. James. 2004. Role of mitochondrial membrane permeabilization in apoptosis and cancer. *Oncogene.* 23:2850-2860.

Hidalgo, A., and C. ffrench-Constant. 2003. The control of cell number during central nervous system development in flies and mice. *Mech Dev.* 120:1311-1325.

Hingorani, S.R., and D.A. Tuveson. 2003. Targeting oncogene dependence and resistance. *Cancer Cell.* 3:414-417.

Hoeberichts, F.A., and E.J. Woltering. 2003. Multiple mediators of plant programmed cell death: interplay of conserved cell death mechanisms and plant-specific regulators. *Bioessays.* 25:47-57.

Horvitz, H.R. 1999. Genetic control of programmed cell death in the nematode Caenorhabditis elegans. *Cancer Res.* 59:1701s-1706s.

Horvitz, H.R., P.W. Sternberg, I.S. Greenwald, W. Fixsen, and H.M. Ellis. 1983. Mutations that affect neural cell lineages and cell fates during the development of the nematode Caenorhabditis elegans. *Cold Spring Harb Symp Quant Biol.* 48 Pt 2:453-463.

Houston, A., and J. O'Connell. 2004. The Fas signalling pathway and its role in the pathogenesis of cancer. *Curr Opin Pharmacol.* 4:321-326.

Igney, F.H., and P.H. Krammer. 2002. Death and anti-death: tumour resistance to apoptosis. *Nat Rev Cancer.* 2:277-288.

Inoue, H., K. Tsukita, T. Iwasato, Y. Suzuki, M. Tomioka, M. Tateno, M. Nagao, A. Kawata, T.C. Saido, M. Miura, H. Misawa, S. Itohara, and R. Takahashi. 2003. The crucial role of caspase-9 in the disease progression of a transgenic ALS mouse model. *Embo J.* 22:6665-6674.

Ionov, Y., H. Yamamoto, S. Krajewski, J.C. Reed, and M. Perucho. 2000. Mutational inactivation of the proapoptotic gene BAX confers selective advantage during tumor clonal evolution. *Proc Natl Acad Sci U S A.* 97:10872-10877.

Irwin, W.A., N. Bergamin, P. Sabatelli, C. Reggiani, A. Megighian, L. Merlini, P. Braghetta, M. Columbaro, D. Volpin, G.M. Bressan, P. Bernardi, and P. Bonaldo. 2003. Mitochondrial dysfunction and apoptosis in myopathic mice with collagen VI deficiency. *Nat Genet.* 35:367-371.

Jaattela, M. 2004. Multiple cell death pathways as regulators of tumour initiation and progression. *Oncogene.* 23:2746-2756.

Jacobs, W.B., G. Govoni, D. Ho, J.K. Atwal, F. Barnabe-Heider, W.M. Keyes, A.A. Mills, F.D. Miller, and D.R. Kaplan. 2005. p63 is an essential proapoptotic protein during neural development. *Neuron.* 48:743-756.

Jacobson, M.D., M. Weil, and M.C. Raff. 1997. Programmed cell death in animal development. *Cell.* 88:347-354.

James, D.I., P.A. Parone, Y. Mattenberger, and J.C. Martinou. 2003. hFis1, a novel component of the mammalian mitochondrial fission machinery. *J Biol Chem.* 278:36373-36379.

Jesenberger, V., and S. Jentsch. 2002. Deadly encounter: ubiquitin meets apoptosis. *Nat Rev Mol Cell Biol.* 3:112-121.

Jiang, C., E.H. Baehrecke, and C.S. Thummel. 1997. Steroid regulated programmed cell death during Drosophila metamorphosis. *Development.* 124:4673-4683.

Johnson, G.L., and R. Lapadat. 2002. Mitogen-activated protein kinase pathways mediated by ERK, JNK, and p38 protein kinases. *Science.* 298:1911-1912.

Johnstone, R.W., A.A. Ruefli, and S.W. Lowe. 2002. Apoptosis: a link between cancer genetics and chemotherapy. *Cell.* 108:153-164.

Juin, P., O. Geneste, E. Raimbaud, and J.A. Hickman. 2004. Shooting at survivors: Bcl-2 family members as drug targets for cancer. *Biochim Biophys Acta.* 1644:251-260.

Jung, C., C.M. Higgins, and Z. Xu. 2002. Mitochondrial electron transport chain complex dysfunction in a transgenic mouse model for amyotrophic lateral sclerosis. *J Neurochem.* 83:535-545.

Karin, M., and A. Lin. 2002. NF-kappaB at the crossroads of life and death. *Nat Immunol.* 3:221-227.

Kaufman, R.J. 2002. Orchestrating the unfolded protein response in health and disease. *J Clin Invest.* 110:1389-1398.

Kelekar, A., and C.B. Thompson. 1998. Bcl-2-family proteins: the role of the BH3 domain in apoptosis. *Trends Cell Biol.* 8:324-330.

Kelley, S.K., and A. Ashkenazi. 2004. Targeting death receptors in cancer with Apo2L/TRAIL. *Curr Opin Pharmacol.* 4:333-339.

Kerr, J.F., A.H. Wyllie, and A.R. Currie. 1972. Apoptosis: a basic biological phenomenon with wide-ranging implications in tissue kinetics. *Br J Cancer.* 26:239-257.

Kim, R.H., M. Peters, Y. Jang, W. Shi, M. Pintilie, G.C. Fletcher, C. DeLuca, J. Liepa, L. Zhou, B. Snow, R.C. Binari, A.S. Manoukian, M.R. Bray, F.F. Liu, M.S. Tsao, and T.W. Mak. 2005. DJ-1, a novel regulator of the tumor suppressor PTEN. *Cancer Cell.* 7:263-273.

Kirkin, V., S. Joos, and M. Zornig. 2004. The role of Bcl-2 family members in tumorigenesis. *Biochim Biophys Acta.* 1644:229-249.

Kirkinezos, I.G., S.R. Bacman, D. Hernandez, J. Oca-Cossio, L.J. Arias, M.A. Perez-Pinzon, W.G. Bradley, and C.T. Moraes. 2005. Cytochrome c association with the inner mitochondrial membrane is impaired in the CNS of G93A-SOD1 mice. *J Neurosci.* 25:164-172.

Klivenyi, P., D. Siwek, G. Gardian, L. Yang, A. Starkov, C. Cleren, R.J. Ferrante, N.W. Kowall, A. Abeliovich, and M.F. Beal. 2006. Mice lacking alpha-synuclein are resistant to mitochondrial toxins. *Neurobiol Dis.* 21:541-548.

Kobayashi, T., and H. Endoh. 2003. Caspase-like activity in programmed nuclear death during conjugation of Tetrahymena thermophila. *Cell Death Differ.* 10:634-640.

Kondo, Y., T. Kanzawa, R. Sawaya, and S. Kondo. 2005. The role of autophagy in cancer development and response to therapy. *Nat Rev Cancer.* 5:726-734.

Korsmeyer, S.J., X.M. Yin, Z.N. Oltvai, D.J. Veis-Novack, and G.P. Linette. 1995. Reactive oxygen species and the regulation of cell death by the Bcl-2 gene family. *Biochim Biophys Acta.* 1271:63-66.

Kostic, V., V. Jackson-Lewis, F. de Bilbao, M. Dubois-Dauphin, and S. Przedborski. 1997. Bcl-2: prolonging life in a transgenic mouse model of familial amyotrophic lateral sclerosis. *Science.* 277:559-562.

Krammer, P.H. 2000. CD95's deadly mission in the immune system. *Nature.* 407:789-795.

Krause, D.S., and R.A. Van Etten. 2005. Tyrosine kinases as targets for cancer therapy. *N Engl J Med.* 353:172-187.

Kroemer, G., and S.J. Martin. 2005. Caspase-independent cell death. *Nat Med.* 11:725-730.

Krueger, A., S. Baumann, P.H. Krammer, and S. Kirchhoff. 2001. FLICE-inhibitory proteins: regulators of death receptor-mediated apoptosis. *Mol Cell Biol.* 21:8247-8254.

Kuida, K., T.F. Haydar, C.Y. Kuan, Y. Gu, C. Taya, H. Karasuyama, M.S. Su, P. Rakic, and R.A. Flavell. 1998. Reduced apoptosis and cytochrome c-mediated caspase activation in mice lacking caspase 9. *Cell.* 94:325-337.

Kuida, K., T.S. Zheng, S. Na, C. Kuan, D. Yang, H. Karasuyama, P. Rakic, and R.A. Flavell. 1996. Decreased apoptosis in the brain and premature lethality in CPP32-deficient mice. *Nature.* 384:368-372.

Kuwana, T., M.R. Mackey, G. Perkins, M.H. Ellisman, M. Latterich, R. Schneiter, D.R. Green, and D.D. Newmeyer. 2002. Bid, Bax, and lipids cooperate to form supramolecular openings in the outer mitochondrial membrane. *Cell.* 111:331-342.

Lampe, A.K., and K.M. Bushby. 2005. Collagen VI related muscle disorders. *J Med Genet.* 42:673-685.

Langenbach, K.J., and T.A. Rando. 2002. Inhibition of dystroglycan binding to laminin disrupts the PI3K/AKT pathway and survival signaling in muscle cells. *Muscle Nerve.* 26:644-653.

Launay, S., O. Hermine, M. Fontenay, G. Kroemer, E. Solary, and C. Garrido. 2005. Vital functions for lethal caspases. *Oncogene.* 24:5137-5148.

Leist, M., and M. Jaattela. 2001. Four deaths and a funeral: from caspases to alternative mechanisms. *Nat Rev Mol Cell Biol.* 2:589-598.

Letai, A., M.C. Bassik, L.D. Walensky, M.D. Sorcinelli, S. Weiler, and S.J. Korsmeyer. 2002. Distinct BH3 domains either sensitize or activate mitochondrial apoptosis, serving as prototype cancer therapeutics. *Cancer Cell.* 2:183-192.

Levine, B., and J. Yuan. 2005. Autophagy in cell death: an innocent convict? *J Clin Invest.* 115:2679-2688.

Lewis, K. 2000. Programmed death in bacteria. *Microbiol Mol Biol Rev.* 64:503-514.

Li, L.Y., X. Luo, and X. Wang. 2001. Endonuclease G is an apoptotic DNase when released from mitochondria. *Nature.* 412:95-99.

Li, P., D. Nijhawan, I. Budihardjo, S.M. Srinivasula, M. Ahmad, E.S. Alnemri, and X. Wang. 1997. Cytochrome c and dATP-dependent formation of Apaf-1/caspase-9 complex initiates an apoptotic protease cascade. *Cell.* 91:479-489.

Liu, J., C. Lillo, P.A. Jonsson, C. Vande Velde, C.M. Ward, T.M. Miller, J.R. Subramaniam, J.D. Rothstein, S. Marklund, P.M. Andersen, T. Brannstrom, O. Gredal, P.C. Wong, D.S. Williams, and D.W. Cleveland. 2004. Toxicity of familial ALS-linked SOD1 mutants from selective recruitmentto spinal mitochondria. *Neuron.* 43:5-17.

Liu, Q.A., and M.O. Hengartner. 1999. The molecular mechanism of programmed cell death in C. elegans. *Ann N Y Acad Sci.* 887:92-104.

Lockshin, R., and C. Williams. 1965. Programmed cell death. II. Endocrine potentiation of the breakdown of the intersegmantal muscles of silkworms. *J Insect Physiol.* 11:803-809.

Locksley, R.M., N. Killeen, and M.J. Lenardo. 2001. The TNF and TNF receptor superfamilies: integrating mammalian biology. *Cell.* 104:487-501.

Lu, C., S.L. Chan, W. Fu, and M.P. Mattson. 2002. The lipid peroxidation product 4-hydroxynonenal facilitates opening of voltage-dependent Ca2+ channels in neurons by increasing protein tyrosine phosphorylation. *J Biol Chem.* 277:24368-24375.

Lustbader, J.W., M. Cirilli, C. Lin, H.W. Xu, K. Takuma, N. Wang, C. Caspersen, X. Chen, S. Pollak, M. Chaney, F. Trinchese, S. Liu, F. Gunn-Moore, L.F. Lue, D.G. Walker, P. Kuppusamy, Z.L. Zewier, O. Arancio, D. Stern, S.S. Yan, and H. Wu. 2004. ABAD directly links Abeta to mitochondrial toxicity in Alzheimer's disease. *Science.* 304:448-452.

Majewski, N., V. Nogueira, P. Bhaskar, P.E. Coy, J.E. Skeen, K. Gottlob, N.S. Chandel, C.B. Thompson, R.B. Robey, and N. Hay. 2004. Hexokinase-mitochondria interaction mediated by Akt is required to inhibit apoptosis in the presence or absence of Bax and Bak. *Mol Cell.* 16:819-830.

Martinou, J.C., and D.R. Green. 2001. Breaking the mitochondrial barrier. *Nat Rev Mol Cell Biol.* 2:63-67.

Mattiazzi, M., M. D'Aurelio, C.D. Gajewski, K. Martushova, M. Kiaei, M.F. Beal, and G. Manfredi. 2002. Mutated human SOD1 causes dysfunction of oxidative phosphorylation in mitochondria of transgenic mice. *J Biol Chem.* 277:29626-29633.

Mattson, M.P. 1998. Modification of ion homeostasis by lipid peroxidation: roles in neuronal degeneration and adaptive plasticity. *Trends Neurosci.* 21:53-57.

Mattson, M.P., and S.L. Chan. 2003. Calcium orchestrates apoptosis. *Nat Cell Biol.* 5:1041-1043.

Mattson, M.P., and W. Duan. 1999. "Apoptotic" biochemical cascades in synaptic compartments: roles in adaptive plasticity and neurodegenerative disorders. *J Neurosci Res.* 58:152-166.

Maurer, U., C. Charvet, A.S. Wagman, E. Dejardin, and D.R. Green. 2006. Glycogen synthase kinase-3 regulates mitochondrial outer membrane permeabilization and apoptosis by destabilization of MCL-1. *Mol Cell.* 21:749-760.

McDonnell, T.J., and S.J. Korsmeyer. 1991. Progression from lymphoid hyperplasia to high-grade malignant lymphoma in mice transgenic for the t(14; 18). *Nature.* 349:254-256.

Mehlen, P., and A. Puisieux. 2006. Metastasis: a question of life or death. *Nat Rev Cancer.* 6:449-458.

Meier, P., A. Finch, and G. Evan. 2000. Apoptosis in development. *Nature.* 407:796-801.

Melino, G., X. Lu, M. Gasco, T. Crook, and R.A. Knight. 2003. Functional regulation of p73 and p63: development and cancer. *Trends Biochem Sci.* 28:663-670.

Mena, M.A., U. Khan, D.M. Togasaki, D. Sulzer, C.J. Epstein, and S. Przedborski. 1997. Effects of wild-type and mutated copper/zinc superoxide dismutase on neuronal survival and L-DOPA-induced toxicity in postnatal midbrain culture. *J Neurochem.* 69:21-33.

Mendelsohn, A.R., J.D. Hamer, Z.B. Wang, and R. Brent. 2002. Cyclin D3 activates Caspase 2, connecting cell proliferation with cell death. *Proc Natl Acad Sci U S A.* 99:6871-6876.

Mesaeli, N., and C. Phillipson. 2004. Impaired p53 expression, function, and nuclear localization in calreticulin-deficient cells. *Mol Biol Cell.* 15:1862-1870.

Michaelidis, T.M., M. Sendtner, J.D. Cooper, M.S. Airaksinen, B. Holtmann, M. Meyer, and H. Thoenen. 1996. Inactivation of bcl-2 results in progressive degeneration of motoneurons, sympathetic and sensory neurons during early postnatal development. *Neuron.* 17:75-89.

Michalak, M., J.M. Robert Parker, and M. Opas. 2002. Ca2+ signaling and calcium binding chaperones of the endoplasmic reticulum. *Cell Calcium.* 32:269-278.

Moll, U.M., S. Wolff, D. Speidel, and W. Deppert. 2005. Transcription-independent pro-apoptotic functions of p53. *Curr Opin Cell Biol.* 17:631-636.

Moore, D.J., A.B. West, V.L. Dawson, and T.M. Dawson. 2005. Molecular pathophysiology of Parkinson's disease. *Annu Rev Neurosci.* 28:57-87.

Morishima, N., K. Nakanishi, K. Tsuchiya, T. Shibata, and E. Seiwa. 2004. Translocation of Bim to the endoplasmic reticulum (ER) mediates ER stress signaling for activation of caspase-12 during ER stress-induced apoptosis. *J Biol Chem.* 279:50375-50381.

Motoyama, N., F. Wang, K.A. Roth, H. Sawa, K. Nakayama, I. Negishi, S. Senju, Q. Zhang, S. Fujii, and et al. 1995. Massive cell death of immature hematopoietic cells and neurons in Bcl-x-deficient mice. *Science.* 267:1506-1510.

Muftuoglu, M., B. Elibol, O. Dalmizrak, A. Ercan, G. Kulaksiz, H. Ogus, T. Dalkara, and N. Ozer. 2004. Mitochondrial complex I and IV activities in leukocytes from patients with parkin mutations. *Mov Disord.* 19:544-548.

Murphy, E., K. Imahashi, and C. Steenbergen. 2005. Bcl-2 regulation of mitochondrial energetics. *Trends Cardiovasc Med.* 15:283-290.

Nakagawa, T., S. Shimizu, T. Watanabe, O. Yamaguchi, K. Otsu, H. Yamagata, H. Inohara, T. Kubo, and Y. Tsujimoto. 2005. Cyclophilin D-dependent mitochondrial permeability transition regulates some necrotic but not apoptotic cell death. *Nature.* 434:652-658.

Nakagawa, T., H. Zhu, N. Morishima, E. Li, J. Xu, B.A. Yankner, and J. Yuan. 2000. Caspase-12 mediates endoplasmic-reticulum-specific apoptosis and cytotoxicity by amyloid-beta. *Nature.* 403:98-103.

Nakano, K., and K.H. Vousden. 2001. PUMA, a novel proapoptotic gene, is induced by p53. *Mol Cell.* 7:683-694.

Nakayama, K., I. Negishi, K. Kuida, H. Sawa, and D.Y. Loh. 1994. Targeted disruption of Bcl-2 alpha beta in mice: occurrence of gray hair, polycystic kidney disease, and lymphocytopenia. *Proc Natl Acad Sci U S A.* 91:3700-3704.

Nicotera, P., M. Leist, and L. Manzo. 1999. Neuronal cell death: a demise with different shapes. *Trends Pharmacol Sci.* 20:46-51.

Nijhawan, D., M. Fang, E. Traer, Q. Zhong, W. Gao, F. Du, and X. Wang. 2003. Elimination of Mcl-1 is required for the initiation of apoptosis following ultraviolet irradiation. *Genes Dev.* 17:1475-1486.

Nijhawan, D., N. Honarpour, and X. Wang. 2000. Apoptosis in neural development and disease. *Annu Rev Neurosci.* 23:73-87.

Norbury, C.J., and B. Zhivotovsky. 2004. DNA damage-induced apoptosis. *Oncogene.* 23:2797-2808.

Oakes, S.A., and S.J. Korsmeyer. 2004. Untangling the web: mitochondrial fission and apoptosis. *Dev Cell.* 7:460-462.

Oakes, S.A., L. Scorrano, J.T. Opferman, M.C. Bassik, M. Nishino, T. Pozzan, and S.J. Korsmeyer. 2005. Proapoptotic BAX and BAK regulate the type 1 inositol trisphosphate receptor and calcium leak from the endoplasmic reticulum. *Proc Natl Acad Sci U S A.* 102:105-110.

Oberhammer, F., J.W. Wilson, C. Dive, I.D. Morris, J.A. Hickman, A.E. Wakeling, P.R. Walker, and M. Sikorska. 1993. Apoptotic death in epithelial cells: cleavage of DNA to 300 and/or 50 kb fragments prior to or in the absence of internucleosomal fragmentation. *Embo J.* 12:3679-3684.

Oda, K., H. Arakawa, T. Tanaka, K. Matsuda, C. Tanikawa, T. Mori, H. Nishimori, K. Tamai, T. Tokino, Y. Nakamura, and Y. Taya. 2000. p53AIP1, a potential mediator of p53-dependent apoptosis, and its regulation by Ser-46-phosphorylated p53. *Cell.* 102:849-862.

Ojaimi, J., and E. Byrne. 2001. Mitochondrial function and alzheimer's disease. *Biol Signals Recept.* 10:254-262.

Okada, H., W.K. Suh, J. Jin, M. Woo, C. Du, A. Elia, G.S. Duncan, A. Wakeham, A. Itie, S.W. Lowe, X. Wang, and T.W. Mak. 2002. Generation and characterization of Smac/DIABLO-deficient mice. *Mol Cell Biol.* 22:3509-3517.

Olivero, M., T. Ruggiero, S. Saviozzi, A. Rasola, N. Coltella, S. Crispi, F. Di Cunto, R. Calogero, and M.F. Di Renzo. 2006. Genes regulated by hepatocyte growth factor as targets to sensitize ovarian cancer cells to cisplatin. *Mol Cancer Ther.* 5:1126-1135.

Ollmann, M., L.M. Young, C.J. Di Como, F. Karim, M. Belvin, S. Robertson, K. Whittaker, M. Demsky, W.W. Fisher, A. Buchman, G. Duyk, L. Friedman, C. Prives, and C. Kopczynski. 2000. Drosophila p53 is a structural and functional homolog of the tumor suppressor p53. *Cell.* 101:91-101.

Pagliarini, D.J., and J.E. Dixon. 2006. Mitochondrial modulation: reversible phosphorylation takes center stage? *Trends Biochem Sci.* 31:26-34.

Pahl, H.L. 1999. Signal transduction from the endoplasmic reticulum to the cell nucleus. *Physiol Rev.* 79:683-701.

Pandey, P., A. Saleh, A. Nakazawa, S. Kumar, S.M. Srinivasula, V. Kumar, R. Weichselbaum, C. Nalin, E.S. Alnemri, D. Kufe, and S. Kharbanda. 2000. Negative regulation of cytochrome c-mediated oligomerization of Apaf-1 and activation of procaspase-9 by heat shock protein 90. *Embo J.* 19:4310-4322.

Parcellier, A., S. Gurbuxani, E. Schmitt, E. Solary, and C. Garrido. 2003. Heat shock proteins, cellular chaperones that modulate mitochondrial cell death pathways. *Biochem Biophys Res Commun.* 304:505-512.

Park, J., S.B. Lee, S. Lee, Y. Kim, S. Song, S. Kim, E. Bae, J. Kim, M. Shong, J.M. Kim, and J. Chung. 2006. Mitochondrial dysfunction in Drosophila PINK1 mutants is complemented by parkin. *Nature.* 441:1157-1161.

Pasinelli, P., M.E. Belford, N. Lennon, B.J. Bacskai, B.T. Hyman, D. Trotti, and R.H. Brown, Jr. 2004. Amyotrophic lateral sclerosis-associated SOD1 mutant proteins bind and aggregate with Bcl-2 in spinal cord mitochondria. *Neuron.* 43:19-30.

Pawson, T. 2002. Regulation and targets of receptor tyrosine kinases. *Eur J Cancer.* 38 Suppl 5:S3-10.

Pelengaris, S., M. Khan, and G.I. Evan. 2002. Suppression of Myc-induced apoptosis in beta cells exposes multiple oncogenic properties of Myc and triggers carcinogenic progression. *Cell.* 109:321-334.

Penaloza, C., L. Lin, R.A. Lockshin, and Z. Zakeri. 2006. Cell death in development: shaping the embryo. *Histochem Cell Biol.*

Penninger, J.M., and G. Kroemer. 2003. Mitochondria, AIF and caspases--rivaling for cell death execution. *Nat Cell Biol.* 5:97-99.

Pommier, Y., O. Sordet, S. Antony, R.L. Hayward, and K.W. Kohn. 2004. Apoptosis defects and chemotherapy resistance: molecular interaction maps and networks. *Oncogene.* 23:2934-2949.

Pozniak, C.D., S. Radinovic, A. Yang, F. McKeon, D.R. Kaplan, and F.D. Miller. 2000. An anti-apoptotic role for the p53 family member, p73, during developmental neuron death. *Science.* 289:304-306.

Puthalakath, H., and A. Strasser. 2002. Keeping killers on a tight leash: transcriptional and post-translational control of the pro-apoptotic activity of BH3-only proteins. *Cell Death Differ.* 9:505-512.

Raff, M.C. 1992. Social controls on cell survival and cell death. *Nature.* 356:397-400.

Raff, M.C. 1996. Size control: the regulation of cell numbers in animal development. *Cell.* 86:173-175.

Raff, M.C., B.A. Barres, J.F. Burne, H.S. Coles, Y. Ishizaki, and M.D. Jacobson. 1993. Programmed cell death and the control of cell survival: lessons from the nervous system. *Science.* 262:695-700.

Ranger, A.M., B.A. Malynn, and S.J. Korsmeyer. 2001. Mouse models of cell death. *Nat Genet.* 28:113-118.

Rao, R.V., S. Castro-Obregon, H. Frankowski, M. Schuler, V. Stoka, G. del Rio, D.E. Bredesen, and H.M. Ellerby. 2002. Coupling endoplasmic reticulum stress to the cell death program. An Apaf-1-independent intrinsic pathway. *J Biol Chem.* 277:21836-21842.

Rasola, A., S. Anguissola, N. Ferrero, D. Gramaglia, A. Maffe, P. Maggiora, P.M. Comoglio, and M.F. Di Renzo. 2004. Hepatocyte growth factor sensitizes human ovarian carcinoma cell lines to paclitaxel and cisplatin. *Cancer Res.* 64:1744-1750.

Rasola, A., M. Fassetta, F. De Bacco, L. D'Alessandro, D. Gramaglia, M.F. Di Renzo, and P. Comoglio. 2006. A positive feedback loop between Hepatocyte Growth Factor Receptor and -catenin sustains colorectal cancer cell invasive growth. *Oncogene.* in press.

Rathmell, J.C., C.J. Fox, D.R. Plas, P.S. Hammerman, R.M. Cinalli, and C.B. Thompson. 2003. Akt-directed glucose metabolism can prevent Bax conformation change and promote growth factor-independent survival. *Mol Cell Biol.* 23:7315-7328.

Ravagnan, L., S. Gurbuxani, S.A. Susin, C. Maisse, E. Daugas, N. Zamzami, T. Mak, M. Jaattela, J.M. Penninger, C. Garrido, and G. Kroemer. 2001. Heat-shock protein 70 antagonizes apoptosis-inducing factor. *Nat Cell Biol.* 3:839-843.

Reddig, P.J., and R.L. Juliano. 2005. Clinging to life: cell to matrix adhesion and cell survival. *Cancer Metastasis Rev.* 24:425-439.

Reddy, P. 2004. Interleukin-18: recent advances. *Curr Opin Hematol.* 11:405-410.

Reed, J.C. 2002. Apoptosis-based therapies. *Nat Rev Drug Discov.* 1:111-121.

Richardson, H., and S. Kumar. 2002. Death to flies: Drosophila as a model system to study programmed cell death. *J Immunol Methods.* 265:21-38.

Rizzuto, R. 2003. The collagen-mitochondria connection. *Nat Genet.* 35:300-301.

Rosen, D.R., T. Siddique, D. Patterson, D.A. Figlewicz, P. Sapp, A. Hentati, D. Donaldson, J. Goto, J.P. O'Regan, H.X. Deng, and et al. 1993. Mutations in Cu/Zn superoxide dismutase gene are associated with familial amyotrophic lateral sclerosis. *Nature.* 362:59-62.

Roulston, A., R.C. Marcellus, and P.E. Branton. 1999. Viruses and apoptosis. *Annu Rev Microbiol.* 53:577-628.

Saelens, X., N. Festjens, L.V. Walle, M. van Gurp, G. van Loo, and P. Vandenabeele. 2004. Toxic proteins released from mitochondria in cell death. *Oncogene.* 23:2861-2874.

Salvesen, G.S., and C.S. Duckett. 2002. IAP proteins: blocking the road to death's door. *Nat Rev Mol Cell Biol.* 3:401-410.

Sanmartin, M., L. Jaroszewski, N.V. Raikhel, and E. Rojo. 2005. Caspases. Regulating death since the origin of life. *Plant Physiol.* 137:841-847.

Savill, J., and V. Fadok. 2000. Corpse clearance defines the meaning of cell death. *Nature.* 407:784-788.

Schmitt, C.A. 2003. Senescence, apoptosis and therapy--cutting the lifelines of cancer. *Nat Rev Cancer.* 3:286-295.

Scorrano, L., M. Ashiya, K. Buttle, S. Weiler, S.A. Oakes, C.A. Mannella, and S.J. Korsmeyer. 2002. A distinct pathway remodels mitochondrial cristae and mobilizes cytochrome c during apoptosis. *Dev Cell.* 2:55-67.

Scorrano, L., S.A. Oakes, J.T. Opferman, E.H. Cheng, M.D. Sorcinelli, T. Pozzan, and S.J. Korsmeyer. 2003. BAX and BAK regulation of endoplasmic reticulum Ca2+: a control point for apoptosis. *Science.* 300:135-139.

Segovia, M., L. Haramaty, J.A. Berges, and P.G. Falkowski. 2003. Cell death in the unicellular chlorophyte Dunaliella tertiolecta. A hypothesis on the evolution of apoptosis in higher plants and metazoans. *Plant Physiol.* 132:99-105.

Senderowicz, A.M. 2004. Targeting cell cycle and apoptosis for the treatment of human malignancies. *Curr Opin Cell Biol.* 16:670-678.

Sengupta, S., and C.C. Harris. 2005. p53: traffic cop at the crossroads of DNA repair and recombination. *Nat Rev Mol Cell Biol.* 6:44-55.

Sharpe, J.C., D. Arnoult, and R.J. Youle. 2004. Control of mitochondrial permeability by Bcl-2 family members. *Biochim Biophys Acta.* 1644:107-113.

Shi, Y. 2001. A structural view of mitochondria-mediated apoptosis. *Nat Struct Biol.* 8:394-401.

Shimura, H., N. Hattori, S. Kubo, Y. Mizuno, S. Asakawa, S. Minoshima, N. Shimizu, K. Iwai, T. Chiba, K. Tanaka, and T. Suzuki. 2000. Familial Parkinson disease gene product, parkin, is a ubiquitin-protein ligase. *Nat Genet.* 25:302-305.

Shinder, G.A., M.C. Lacourse, S. Minotti, and H.D. Durham. 2001. Mutant Cu/Zn-superoxide dismutase proteins have altered solubility and interact with heat shock/stress proteins in models of amyotrophic lateral sclerosis. *J Biol Chem.* 276:12791-12796.

Slee, E.A., D.J. O'Connor, and X. Lu. 2004. To die or not to die: how does p53 decide? *Oncogene.* 23:2809-2818.

Sprick, M.R., and H. Walczak. 2004. The interplay between the Bcl-2 family and death receptor-mediated apoptosis. *Biochim Biophys Acta.* 1644:125-132.

Stokin, G.B., C. Lillo, T.L. Falzone, R.G. Brusch, E. Rockenstein, S.L. Mount, R. Raman, P. Davies, E. Masliah, D.S. Williams, and L.S. Goldstein. 2005. Axonopathy and transport deficits early in the pathogenesis of Alzheimer's disease. *Science.* 307:1282-1288.

Strasser, A., L. O'Connor, and V.M. Dixit. 2000. Apoptosis signaling. *Annu Rev Biochem.* 69:217-245.

Sulis, M.L., and R. Parsons. 2003. PTEN: from pathology to biology. *Trends Cell Biol.* 13:478-483.

Sulston, J.E., and J.G. White. 1980. Regulation and cell autonomy during postembryonic development of Caenorhabditis elegans. *Dev Biol.* 78:577-597.

Susin, S.A., N. Zamzami, M. Castedo, T. Hirsch, P. Marchetti, A. Macho, E. Daugas, M. Geuskens, and G. Kroemer. 1996. Bcl-2 inhibits the mitochondrial release of an apoptogenic protease. *J Exp Med.* 184:1331-1341.

Suzuki, Y., Y. Imai, H. Nakayama, K. Takahashi, K. Takio, and R. Takahashi. 2001. A serine protease, HtrA2, is released from the mitochondria and interacts with XIAP, inducing cell death. *Mol Cell.* 8:613-621.

Takeuchi, H., Y. Kobayashi, S. Ishigaki, M. Doyu, and G. Sobue. 2002. Mitochondrial localization of mutant superoxide dismutase 1 triggers caspase-dependent cell death in a cellular model of familial amyotrophic lateral sclerosis. *J Biol Chem.* 277:50966-50972.

Tardy, C., P. Codogno, H. Autefage, T. Levade, and N. Andrieu-Abadie. 2006. Lysosomes and lysosomal proteins in cancer cell death (new players of an old struggle). *Biochim Biophys Acta.* 1765:101-125.

Tatton, W.G., R. Chalmers-Redman, D. Brown, and N. Tatton. 2003. Apoptosis in Parkinson's disease: signals for neuronal degradation. *Ann Neurol.* 53 Suppl 3:S61-70; discussion S70-62.

Thompson, C.B. 1995. Apoptosis in the pathogenesis and treatment of disease. *Science.* 267:1456-1462.

Thornberry, N.A., H.G.Bull, J.R. Calaycay, K.T. Chapman, A.D. Howard, M.J. Kostura, D.K. Miller, S.M. Molineaux, J.R. Weidner, J. Aunins, and et al. 1992. A novel heterodimeric cysteine protease is required for interleukin-1 beta processing in monocytes. *Nature.* 356:768-774.

Thornberry, N.A., and Y. Lazebnik. 1998. Caspases: enemies within. *Science.* 281:1312-1316.

Tillet, E., B. Gential, R. Garrone, and W.B. Stallcup. 2002. NG2 proteoglycan mediates beta1 integrin-independent cell adhesion and spreading on collagen VI. *J Cell Biochem.* 86:726-736.

Tsujimoto, Y., and S. Shimizu. 2000. VDAC regulation by the Bcl-2 family of proteins. *Cell Death Differ.* 7:1174-1181.

Twomey, C., and J.V. McCarthy. 2005. Pathways of apoptosis and importance in development. *J Cell Mol Med.* 9:345-359.

Uren, A.G., T. Beilharz, M.J. O'Connell, S.J. Bugg, R. van Driel, D.L. Vaux, and T. Lithgow. 1999. Role for yeast inhibitor of apoptosis (IAP)-like proteins in cell division. *Proc Natl Acad Sci U S A.* 96:10170-10175.

Uren, A.G., K. O'Rourke, L.A. Aravind, M.T. Pisabarro, S. Seshagiri, E.V. Koonin, and V.M. Dixit. 2000. Identification of paracaspases and metacaspases: two ancient families of caspase-like proteins, one of which plays a key role in MALT lymphoma. *Mol Cell.* 6:961-967.

Vander Heiden, M.G., N.S. Chandel, X.X. Li, P.T. Schumacker, M. Colombini, and C.B. Thompson. 2000. Outer mitochondrial membrane permeability can regulate coupled respiration and cell survival. *Proc Natl Acad Sci U S A.* 97:4666-4671.

Vander Heiden, M.G., N.S. Chandel, E.K. Williamson, P.T. Schumacker, and C.B. Thompson. 1997. Bcl-xL regulates the membrane potential and volume homeostasis of mitochondria. *Cell.* 91:627-637.

Vardi, A., I. Berman-Frank, T. Rozenberg, O. Hadas, A. Kaplan, and A. Levine. 1999. Programmed cell death of the dinoflagellate Peridinium gatunense is mediated by CO(2) limitation and oxidative stress. *Curr Biol.* 9:1061-1064.

Varfolomeev, E.E., M. Schuchmann, V. Luria, N. Chiannilkulchai, J.S. Beckmann, I.L. Mett, D. Rebrikov, V.M. Brodianski, O.C. Kemper, O. Kollet, T. Lapidot, D. Soffer, T. Sobe, K.B. Avraham, T. Goncharov, H. Holtmann, P. Lonai, and D. Wallach. 1998. Targeted disruption of the mouse Caspase 8 gene ablates cell death induction by the TNF receptors, Fas/Apo1, and DR3 and is lethal prenatally. *Immunity.* 9:267-276.

Vassilev, L.T. 2005. p53 Activation by small molecules: application in oncology. *J Med Chem.* 48:4491-4499.

Vaux, D.L., S. Cory, and J.M. Adams. 1988. Bcl-2 gene promotes haemopoietic cell survival and cooperates with c-myc to immortalize pre-B cells. *Nature.* 335:440-442.

Vaux, D.L., and J. Silke. 2005. IAPs, RINGs and ubiquitylation. *Nat Rev Mol Cell Biol.* 6:287-297.

Vaux, D.L., I.L. Weissman, and S.K. Kim. 1992. Prevention of programmed cell death in Caenorhabditis elegans by human bcl-2. *Science.* 258:1955-1957.

Verhagen, A.M., P.G. Ekert, M. Pakusch, J. Silke, L.M. Connolly, G.E. Reid, R.L. Moritz, R.J. Simpson, and D.L. Vaux. 2000. Identification of DIABLO, a mammalian protein that promotes apoptosis by binding to and antagonizing IAP proteins. *Cell.* 102:43-53.

Verhagen, A.M., J. Silke, P.G. Ekert, M. Pakusch, H. Kaufmann, L.M. Connolly, C.L. Day, A. Tikoo, R. Burke, C. Wrobel, R.L. Moritz, R.J. Simpson, and D.L. Vaux. 2002. HtrA2 promotes cell death through its serine protease activity and its ability to antagonize inhibitor of apoptosis proteins. *J Biol Chem.* 277:445-454.

Verkhratsky, A., and E.C. Toescu. 2003. Endoplasmic reticulum Ca(2+) homeostasis and neuronal death. *J Cell Mol Med.* 7:351-361.

Vila, M., and S. Przedborski. 2003. Targeting programmed cell death in neurodegenerative diseases. *Nat Rev Neurosci.* 4:365-375.

Vogt, C. 1842. Untersuchungen uber die Entwicklungsgeschichte der Geburtshelferkrote. *Jent und Gassman, Solothurn.* 130.

Wada, T., and J.M. Penninger. 2004. Mitogen-activated protein kinases in apoptosis regulation. *Oncogene.* 23:2838-2849.

Wajant, H. 2002. The Fas signaling pathway: more than a paradigm. *Science.* 296:1635-1636.

Walker, P.R., and M. Sikorska. 1997. New aspects of the mechanism of DNA fragmentation in apoptosis. *Biochem Cell Biol.* 75:287-299.

Walter, L., and G. Hajnoczky. 2005. Mitochondria and endoplasmic reticulum: the lethal interorganelle cross-talk. *J Bioenerg Biomembr.* 37:191-206.

Wang, X., C. Yang, J. Chai, Y. Shi, and D. Xue. 2002. Mechanisms of AIF-mediated apoptotic DNA degradation in Caenorhabditis elegans. *Science.* 298:1587-1592.

Watanabe-Fukunaga, R., C.I. Brannan, N.G. Copeland, N.A. Jenkins, and S. Nagata. 1992. Lymphoproliferation disorder in mice explained by defects in Fas antigen that mediates apoptosis. *Nature.* 356:314-317.

Wei, M.C., W.X. Zong, E.H. Cheng, T. Lindsten, V. Panoutsakopoulou, A.J. Ross, K.A. Roth, G.R. MacGregor, C.B. Thompson, and S.J. Korsmeyer. 2001. Proapoptotic BAX and BAK: a requisite gateway to mitochondrial dysfunction and death. *Science.* 292:727-730.

West, A.B., D.J. Moore, S. Biskup, A. Bugayenko, W.W. Smith, C.A. Ross, V.L. Dawson, and T.M. Dawson. 2005. Parkinson's disease-associated mutations in leucine-rich repeat kinase 2 augment kinase activity. *Proc Natl Acad Sci U S A.* 102:16842-16847.

Wiberg, C., A.R. Klatt, R. Wagener, M. Paulsson, J.F. Bateman, D. Heinegard, and M. Morgelin. 2003. Complexes of matrilin-1 and biglycan or decorin connect collagen VI microfibrils to both collagen II and aggrecan. *J Biol Chem.* 278:37698-37704.

Widlak, P., and W.T. Garrard. 2005. Discovery, regulation, and action of the major apoptotic nucleases DFF40/CAD and endonuclease G. *J Cell Biochem.* 94:1078-1087.

Williamson, P., and R.A. Schlegel. 2002. Transbilayer phospholipid movement and the clearance of apoptotic cells. *Biochim Biophys Acta.* 1585:53-63.

Willis, S.N., and J.M. Adams. 2005. Life in the balance: how BH3-only proteins induce apoptosis. *Curr Opin Cell Biol.* 17:617-625.

Wong, P.C., C.A. Pardo, D.R. Borchelt, M.K. Lee, N.G. Copeland, N.A. Jenkins, S.S. Sisodia, D.W. Cleveland, and D.L. Price. 1995. An adverse property of a familial ALS-linked SOD1 mutation causes motor neuron disease characterized by vacuolar degeneration of mitochondria. *Neuron.* 14:1105-1116.

Worth, A., A.J. Thrasher, and H.B. Gaspar. 2006. Autoimmune lymphoproliferative syndrome: molecular basis of disease and clinical phenotype. *Br J Haematol.* 133:124-140.

Wu, Y., N. Tibrewal, and R.B. Birge. 2006. Phosphatidylserine recognition by phagocytes: a view to a kill. *Trends Cell Biol.* 16:189-197.

Wyllie, A.H., J.F. Kerr, and A.R. Currie. 1980. Cell death: the significance of apoptosis. *Int Rev Cytol.* 68:251-306.

Xu, H., D. Sweeney, R. Wang, G. Thinakaran, A.C. Lo, S.S. Sisodia, P. Greengard, and S. Gandy. 1997. Generation of Alzheimer beta-amyloid protein in the trans-Golgi network in the apparent absence of vesicle formation. *Proc Natl Acad Sci U S A.* 94:3748-3752.

Yang, J., X. Liu, K. Bhalla, C.N. Kim, A.M. Ibrado, J. Cai, T.I. Peng, D.P. Jones, and X. Wang. 1997. Prevention of apoptosis by Bcl-2: release of cytochrome c from mitochondria blocked. *Science.* 275:1129-1132.

Ye, H., C. Cande, N.C. Stephanou, S. Jiang, S. Gurbuxani, N. Larochette, E. Daugas, C. Garrido, G. Kroemer, and H. Wu. 2002. DNA binding is required for the apoptogenic action of apoptosis inducing factor. *Nat Struct Biol.* 9:680-684.

Yin, X.M. 2006. Bid, a BH3-only multi-functional molecule, is at the cross road of life and death. *Gene.* 369:7-19.

Yoo, C.B., and P.A. Jones. 2006. Epigenetic therapy of cancer: past, present and future. *Nat Rev Drug Discov.* 5:37-50.

Yoshida, H., Y.Y. Kong, R. Yoshida, A.J. Elia, A. Hakem, R. Hakem, J.M. Penninger, and T.W. Mak. 1998. Apaf1 is required for mitochondrial pathways of apoptosis and brain development. *Cell.* 94:739-750.

Yuan, J. 1996. Evolutionary conservation of a genetic pathway of programmed cell death. *J Cell Biochem.* 60:4-11.

Yuan, J. 2006. Divergence from a dedicated cellular suicide mechanism: exploring the evolution of cell death. *Mol Cell.* 23:1-12.

Zhang, L., M. Shimoji, B. Thomas, D.J. Moore, S.W. Yu, N.I. Marupudi, R. Torp, I.A. Torgner, O.P. Ottersen, T.M. Dawson, and V.L. Dawson. 2005. Mitochondrial localization of the Parkinson's disease related protein DJ-1: implications for pathogenesis. *Hum Mol Genet.* 14:2063-2073.

Zimprich, A., S. Biskup, P. Leitner, P. Lichtner, M. Farrer, S. Lincoln, J. Kachergus, M. Hulihan, R.J. Uitti, D.B. Calne, A.J. Stoessl, R.F. Pfeiffer, N. Patenge, I.C. Carbajal, P. Vieregge, F. Asmus, B. Muller-Myhsok, D.W. Dickson, T. Meitinger, T.M. Strom, Z.K. Wszolek, and T. Gasser. 2004. Mutations in LRRK2 cause autosomal-dominant parkinsonism with pleomorphic pathology. *Neuron.* 44:601-607.

Zong, W.X., C. Li, G. Hatzivassiliou, T. Lindsten, Q.C. Yu, J. Yuan, and C.B. Thompson. 2003. Bax and Bak can localize to the endoplasmic reticulum to initiate apoptosis. *J Cell Biol.* 162:59-69.

Zong, W.X., and C.B. Thompson. 2006. Necrotic death as a cell fate. *Genes Dev.* 20:1-15.

Zornig, M., A. Hueber, W. Baum, and G. Evan. 2001. Apoptosis regulators and their role in tumorigenesis. *Biochim Biophys Acta.* 1551:F1-37.

Zou, H., W.J. Henzel, X. Liu, A. Lutschg, and X. Wang. 1997. Apaf-1, a human protein homologous to C. elegans CED-4, participates in cytochrome c-dependent activation of caspase-3. *Cell.* 90:405-413.

Zwaal, R.F., P. Comfurius, and E.M. Bevers. 2005. Surface exposure of phosphatidylserine in pathological cells. *Cell Mol Life Sci.* 62:971-988.

Zwick, E., J. Bange, and A. Ullrich. 2002. Receptor tyrosine kinases as targets for anticancer drugs. *Trends Mol Med.* 8:17-23.

In: New Developments in Cell Apoptosis Research
Editor: Alan J. Corvin, pp. 53-68
ISBN: 1-60021-453-3

Chapter II

Significance of Vascular Smooth Muscle Cell Apoptosis Induced by 7-Ketocholesterol in the Development of Vulnerable Plaque

Yoh Miyashita*, Atsuhito Saiki, Kei Endoh, Masahiro Ohira, Nobukiyo Koide, Tomokazu Oyama and Kohji Shirai,
Department of Internal Medicine, Sakura Medical Center, School of Medicine, Toho University, 564-1 Shimoshizu, Sakura-City, Chiba, Japan

Abstract

Oxysterols are known to be present at high concentrations in atherosclerotic plaques and contribute to the development of atherosclerosis by various mechanisms. We found that 7-ketocholesterol, one of the oxysterols accumulated in atherosclerotic lesion, induced apoptosis of vascular smooth muscle cells (SMC). This finding indicates that accumulation of 7-ketocholesterol in atherosclerotic lesions may decrease local cell numbers and render atherosclerotic plaques vulnerable. In this paper, we summarize the significance of apoptosis induced by 7-ketocholesterol in the development of vulnerable plaque, mainly according to our published results. In concrete terms, our findings are as followed.

1) Lipids extracted from fat deposits in human atheromatous lesions induce apoptosis of SMC.

* Address correspondence to: Yoh Miyashita, MD, Sakura Medical Center, Toho University School of Medicine, 564, Shimoshizu, Sakura-Shi, Chiba Prefecture, 285-8741, Japan; Tel. 81-43-462-8811; Fax. 81-43-489-9770; E- mail; mumon@sf6.so-net.ne.jp

2) Among the deposited lipids, 7-ketocholesterol is present at high concentration during the early stage of atherosclerosis and contributes to induce the apoptosis.
3) Apoptosis of SMC induced by 7-ketocholesterol is accelerated when SMC acquire a high proliferative potential.
4) 7-Ketocholesterol increases intracellular Ca^{2+} in human SMC, and as a result, caspase-3 activity is increased. Ca channel blockers inhibit 7-ketocholesterol-induced apoptosis of SMC.
5) 7-ketocholesterol-induced apoptosis of SMC is inhibited by an inhibitor of hydroxymethyl glutaryl CoA (HMG-CoA) reductase.

From these findings, we propose that 7-ketocholesterol contained in plaques is a strong candidate as a cause of vulnerable plaque.

Introduction

Myocardial infarction occurs not necessary due to narrowing of the vessel lumen by intimal thickening of the artery, but mainly due to rupture of the capsule of the plaque [1]. The mechanism of this plaque rupture is not fully understood. Recent studies indicate a decrease in number of vascular smooth muscle cells (SMC) in arteries with advanced atherosclerosis [2, 3], and implicate apoptosis as responsible for cell death in atherosclerotic lesion [4]. Various factors existing in atherosclerotic lesions might induce apoptosis of SMC; for example, lipids and cytokines such as TNF-α. Among the lipids, oxysterols are reported to be present at high concentrations in atherosclerotic plaques compared to other tissues [5], and that they contribute to the development of atherosclerosis [6] by various mechanisms including cytotoxic or apoptotic potential for endothelial cells [7, 8, 9], impairment of vascular endothelial barrier function [10, 11], inhibition of endothelial NO release [12] and artery relaxation [13]. We have reported that 7-ketocholesterol (7KCHO), an oxysterol, induces apoptosis of human SMC [14]. This finding suggests that accumulation of 7KCHO in atherosclerotic lesions may render atherosclerotic plaques unstable by decreasing the number of SMC locally in the plaque area. In this paper, we summarize our published findings, and review the significance of 7KCHO-induced SMC apoptosis as a cause of vulnerable plaque. Furthermore, we also discuss the mechanisms of and the effects of some drugs on 7KCHO-induced apoptosis.

Lipids Deposited in Human Atheromatous Lesion Induce Apoptosis of SMC

Coronary angioscopy in acute myocardial infarction frequently reveals disrupted yellow lesions [15, 16]. The yellow color is due to light reflecting from the lipid-rich yellow core through a thin fibrous cap [16]. Postmortem studies have demonstrated that these lesions contain few cells and the major components are a thin collagenous cap with an underlying lipid-rich core [17]. When we observed these yellow lesions by transmission electron microscopy, we found the presence of apoptotic SMC (Figure 1). Apoptosis of SMC has already been demonstrated in atherosclerotic lesions and implicated to play a role in plaque rupture [18, 19]. From these observations, we hypothesized that lipids accumulated in atherosclerotic lesion may be related to the formation of vulnerable plaques by inducing SMC apoptosis. Therefore, we extracted the lipids from atherosclerotic lesion obtained at postmortem examinations (crude lipid extract), and examined the effect of the lipids deposited in plaque on SMC apoptosis [20].

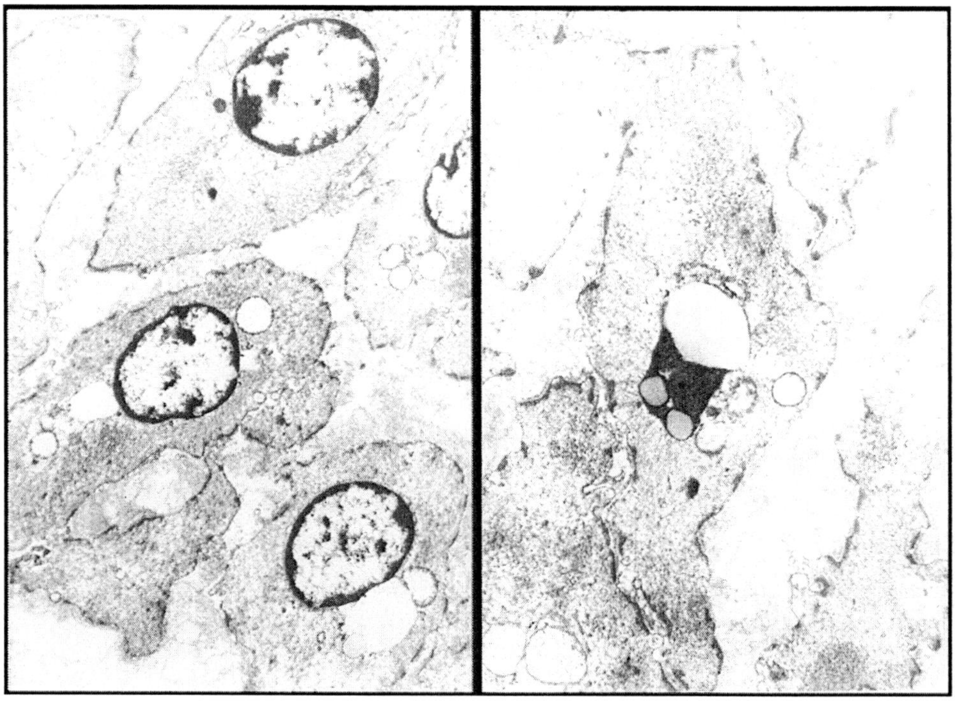

Figure 1. Transmission electron micrograph of smooth muscle cells in fatty streak. Micrographs show normal (A) and apoptotic SMC (B). (Reference 20)

1) Among Deposited Lipids, Oxysterols Have Apoptosis-inducing Potential for SMC

The crude lipid extract was fractionated into oxysterol-rich (Fr. 1), free cholesterol-rich (Fr. 2), and cholesterol ester-rich (Fr. 3) fractions using a cartridge of Sep-Pak ODS. SMC were incubated with crude lipid extract or each lipid fraction, then DNA was extracted from the SMC and subjected to agarose electrophoresis. The DNA from smooth muscle cells incubated with crude lipid extract showed a ladder pattern with a monomer size of 180 base pairs. A 180 base-pair monomer ladder pattern was also observed by the addition of Fr. 1, but no ladder pattern was found by the addition of Fr. 2 or 3 (Figure 2). DNA nicks were also detected by the TUNEL method. SMC incubated with Fr.1 showed positive TUNEL staining, whereas cells incubated with Fr.2 or 3 showed negative staining (data not shown). Furthermore, the DNA content of SMC was measured by flow cytometry. Cells having DNA content lower than that of G_0/G_1 cells (hypodiploidy) were considered apoptotic (Figure 3, region marked with asterisk). The percentage of apoptotic SMC increased significantly by the addition of crude lipid extract or Fr. 1 compared to control, but did not change significantly by the addition of Fr. 2 or 3.

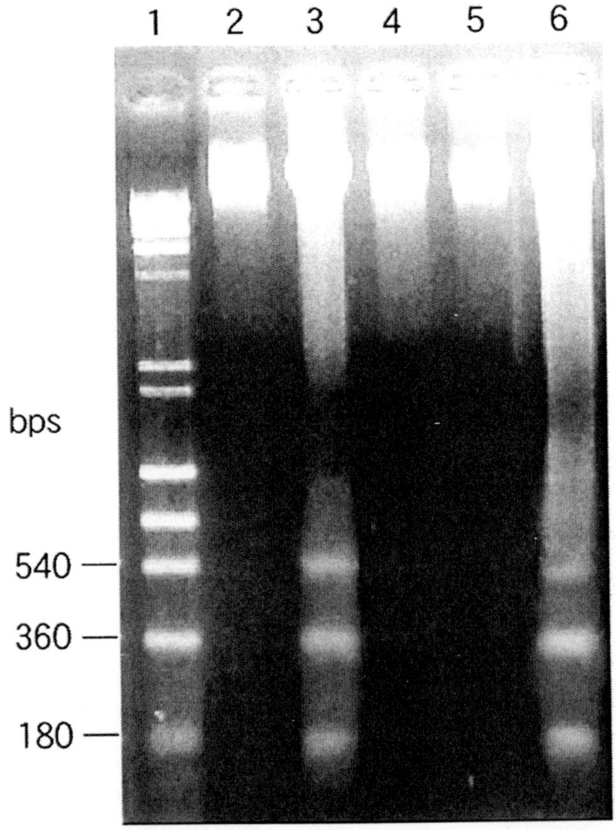

Figure 2. Agarose electrophoresis of DNA from SMC incubated with crude lipid extract, three lipid fractions, or control. Lane 1 shows molecular weight marker. A DNA ladder pattern is observed with the crude lipid extract (lane 6) or the oxysterol-rich fraction (lane 3).However, no ladder pattern is seen with the free cholesterol-rich fraction (lane 4), the cholesteryl ester-rich fraction (lane 5), and control (lane 2). (Reference 20)

To analyze the dose-dependent effect of each lipid fraction in inducing apoptosis, fragmented DNA assay was performed by a quantitative sandwich enzyme immunoassay (ELISA) [14]. The amount of fragmented DNA was significantly higher with the addition of crude lipid extract or Fr. 1 compared to control or Fr. 2 or 3, at both concentrations of 50 and 100 μg/mL of each fraction (Figure 4). Furthermore, crude lipid extract or Fr.1 increased the amount of fragmented DNA in a dose-dependent manner. No significant difference was observed between the amounts of fragmented DNA with the addition of control compared with Fr. 2 or Fr. 3.

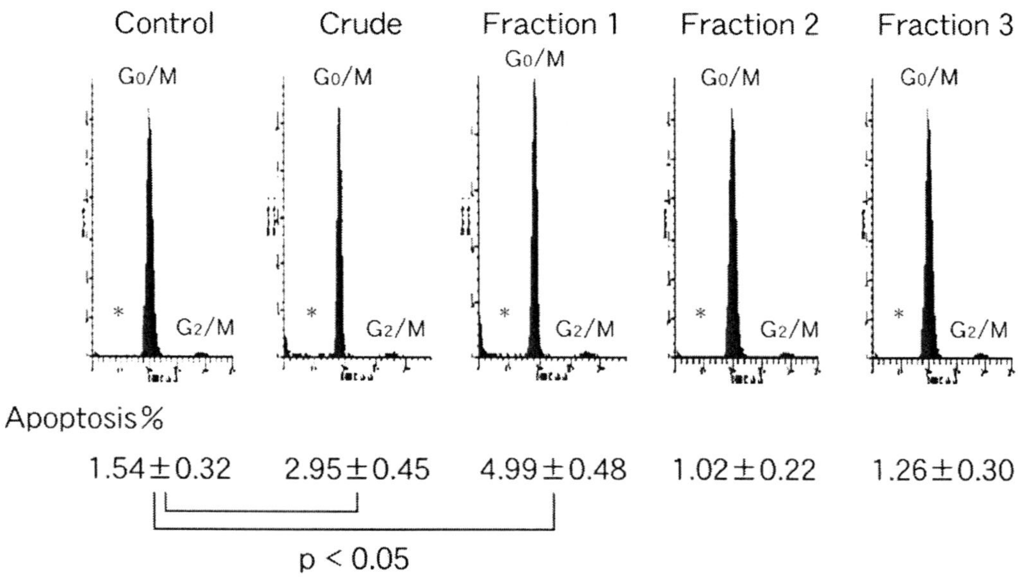

Figure 3. Quantitative analysis of apoptotic cells by flow cytometry. Apoptotic cells were expressed as percent apoptotic cells. Graphs show typical flow cytometric images of SMC treated with crude lipid extract, each lipid fraction or control. Asterisk indicates the apoptotic cell population. Results are mean ± SD. (Reference 20)

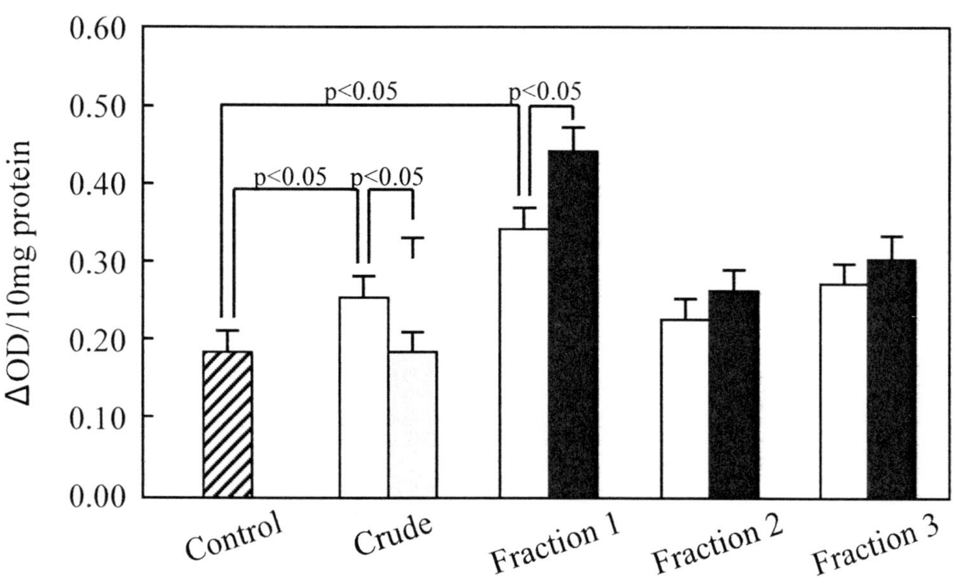

Figure 4. The effect of deposited lipids on fragmented DNA of SMC. Crude or each of three lipid fractions was added at 50 μg/ml (open bar) or 100 μg/ml (solid bar). Results are mean ± SD. (Reference 20)

These findings imply that lipids accumulated in atherosclerotic lesions could reduce local cell numbers through SMC apoptosis rendering atherosclerotic plaques vulnerable, and that oxysterols are the most important lipid component responsible for inducing apoptosis, because among the three fractions, only the oxysterol-rich fraction induced apoptosis of

SMC. Actually, reduction of lipids deposited in atherosclerotic lesion has been reported to stabilize atheromatous plaque [15].

2) Contents of Oxysterols in Lipids Extracted from Atheromatous Lesion

Previous report demonstrates that 27-hydroxycholesterol (27OH-chol) is the major oxysterol in human atheromatous lesions, followed by 7KCHO [5]. To clarify which oxysterol is important for inducing SMC apoptosis, we measured the contents of the two major oxysterols, 27OH-chol and 7KCHO, contained in atheromatous lesion by high-pressure liquid chromatography (HPLC). Atheromatous lesions were classified into four grades by the degree of atherosclerosis and the contents of 27OH-chol and 7KCHO were measured. The contents of both oxysterols were significantly higher in fatty streak than in normal tissue (Figure 5). Furthermore, the oxysterol contents were significantly higher in fibrous plaque and complicated lesion than in fatty streak. The 7KCHO content was significantly higher than 27OH-chol in fatty streak, while the reverse was observed in fibrous plaque and complicated lesion. Theses results indicate that 7KCHO is the predominant oxysterol present in the lesion during the early stage of atherosclerosis.

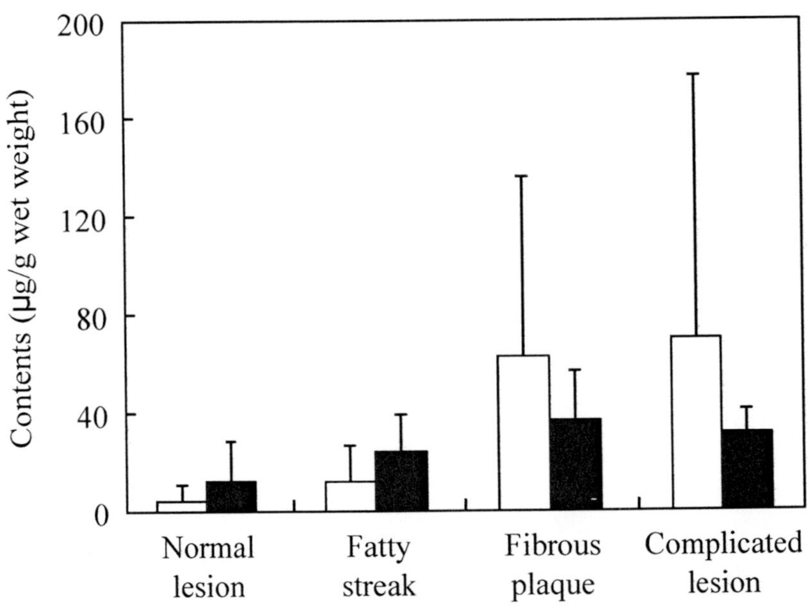

Figure 5. Contents of 7KCHO and 27OH-chol in atheromatous lesions. Three human aortas were divided into normal lesions, fatty streaks, fibrous plaques and complicated lesions, and contents of 7KCHO (solid bar) and 27OH-chol (open bar) in each portion were measured by HPLC. Results are mean ± SD. *: $p<0.05$, **: $p<0.01$ vs. normal lesion. (Reference 20)

The 7KCHO content is approximately 25 μg/g wet weight of fatty streak, and this content almost matches the *in vitro* apoptosis-inducing level, because our previous data indicated that 25 μg/mL of 7-KCHO induced SMC apoptosis *in vitro* [14]. It is well known that plaque rupture depends not only on the severity of atherosclerosis but also on the plaque properties such as yellow plaque [1]. Our results suggest that if 7KCHO exhibits the potential to induce SMC apoptosis in atherosclerotic lesion, a high 7KCHO content may accelerate the change of white plaque to yellow plaque, and finally to plaque rupture even in the early stage of atherosclerosis. Therefore, we propose that plaque rupture may depend not only on the severity of atherosclerosis but also on the amount of 7KCHO deposited in the atherosclerotic lesion. The deposition of 7KCHO in atherosclerotic lesion could explain the observation that myocardial infarction occurs not necessary due to narrowing of the vascular lumen by intimal thickening of the artery.

Apoptosis of SMC Induced by 7KCHO is Accelerated When SMC Acquire a High Proliferative Potential

Diabetes mellitus is a very important risk factor for myocardial infarction, through accelerating atherosclerosis of coronary artery [21, 22]. The major mechanisms reported include accelerated glycation [23] and coagulation [24], increased protein kinase C activity [25], and increased oxidative stress in the body [26]. However, these factors cannot fully explain the mechanism of plaque rupture. Therefore, we examined the difference in sensitivity to 7KCHO between SMC obtained from the Otsuka Long-Evans Tokushima fatty (OLETF) rats, which is a model of type 2 diabetes mellitus [27], and the Long-Evans Tokushima Otsuka (LETO) rats, which can be used as age-matched control for OLETF rat [28].

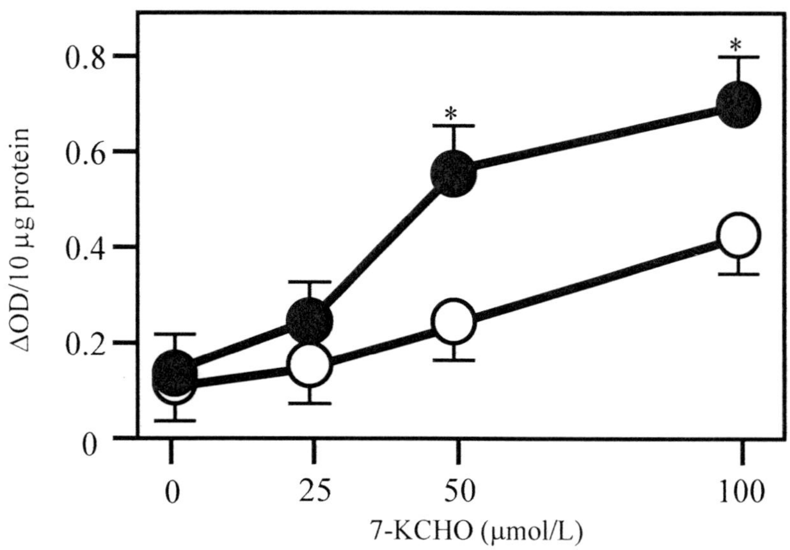

Figure 6. Fragmentation of DNA induced by 7KCHO in OLETF-derived SMC (closed circles) and LETO-derived SMC (open circles). Results are mean ± SEM. * $p<0.05$ vs. LETO. (Reference 28)

7KCHO-induced fragmented DNA was assayed by ELISA [14] to examine whether there is a difference in sensitivity to 7KCHO between OLETF-derived SMC and LETO-derived SMC. By the addition of 7KCHO, the amount of fragmented DNA increased depending on the dose of 7KCHO in both OLETF- and LETO-derived SMC (Figure 6). However, the increase in fragmented DNA was significantly higher in OLETF-derived SMC compared to LETO-derived SMC at 7KCHO concentrations of 50 and 100 μmol/L (Figure 6). These results suggest that OLETF-derived SMC are more susceptible to 7KCHO-induced apoptosis than LETO-derived SMC.

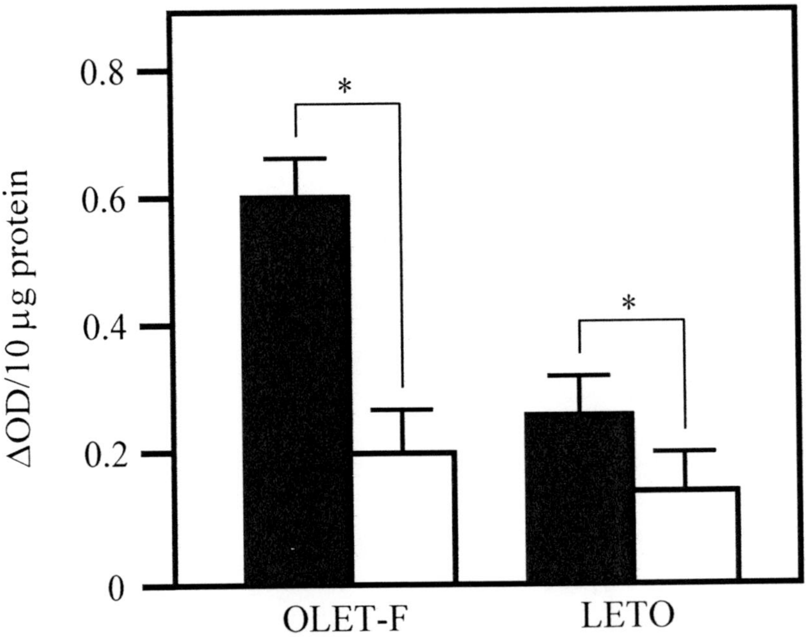

Figure 7. Effect of 7-ketocholesterol on fragmented DNA of SMC in resting phase. Cells were incubated with DMEM containing 10% FBS (solid bar) or without FBS (open bar). Then 7KCHO was added at a final concentration of 50 μmol/L, and incubated for 3 days. Results are mean ± SEM. * $p<0.05$ vs. LETO. (Reference 28)

Augmented SMC proliferation has been proposed as a mechanism of increased atherogenesis in diabetes, based on experimental evidence obtained using OLETF rats compared with LETO rats [29]. Therefore, we hypothesized that the high sensitivity of OLETF-derived SMC to 7KCHO may be due to its high proliferative potential. Then, to clarify whether the proliferative potential of SMC affects 7KCHO-induced apoptosis, we assayed 7KCHO-induced fragmented DNA of SMC in the resting and proliferative phases. First, the effect on SMC in resting phase was examined by assaying 7KCHO-induced fragmented DNA in SMC incubated without fetal bovine serum (FBS). When incubated with 10% FBS, the amount of fragmented DNA was significantly higher in OLETF-derived SMC than in LETO-derived SMC. When incubated without FBS, however, the amounts of fragmented DNA decreased significantly in both OLETF- and LETO-derived SMC, reaching the same magnitude with no significant difference between OLETF- and LETO-derived SMC (Figure 7). Next, to elucidate the effect of increased proliferation of SMC on the susceptibility to 7KCHO-induced apoptosis, we used PDGF-BB to stimulate the proliferation of LETO-derived SMC. The amount of 7KCHO-induced DNA fragmentation was assayed after adding PDGF-BB to LETO-derived SMC. The amount of fragmented DNA in LETO-derived SMC increased significantly when incubated with PDGF-BB compared to without

PDGF-BB (Figure 8). The effect of PDGF-BB was dose-dependent. No significant increase of fragmented DNA was observed by the addition of PDGF-BB without 7KCHO (Figure 8). These results suggest that SMC with an enhanced proliferative potential show increased sensitivity to 7KCHO and hence augmented apoptosis.

Our data suggest that under diabetic condition, SMC are more susceptible to 7KCHO-induced apoptosis, and one of the mechanisms may be due to a high proliferative potential of the cells induced by diabetic condition. Diabetes mellitus is known to be associated with a high frequency of atherosclerosis, especially in the coronary artery. As for the mechanism of the high atherosclerosis rate in diabetic condition, augmented SMC proliferation cannot fully explain the mechanism of plaque rupture. We propose that in atherosclerotic lesions under diabetic condition, essentially proliferation of SMC occurs but apoptosis of these SMC with high proliferative potential is accelerated with accumulation of 7KCHO.

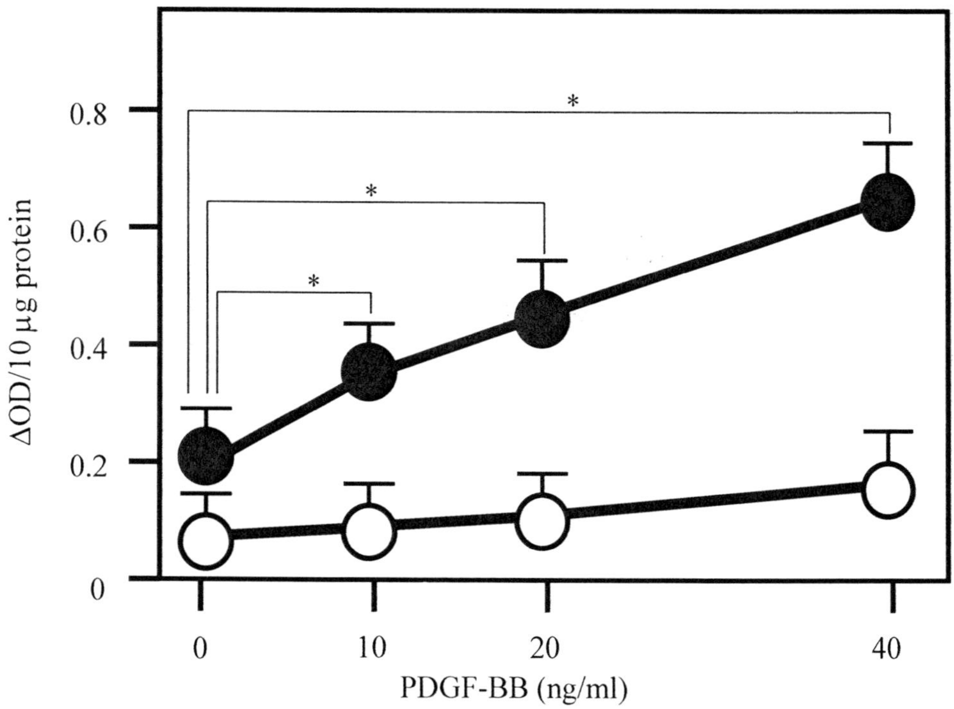

Figure 8. Effect of PDGF-BB on fragmentation of DNA of LETO-derived SMC. Cells were incubated with 7KCHO (closed circle) or without 7-ketocholesterol (open circle). After incubation for 24 hours, PDGF-BB at varying concentrations was added to the cells. 7KCHO was added at a final concentration of 50 μmol/L. Results are mean ± SEM. * $p<0.05$ (Reference 28)

7KCHO Enhances Caspase-3 Activity in SMC by Promoting Ca^{2+} Influx into SMC through Ca-channel

Regarding the mechanism by which apoptosis occurs, activation of caspases in cells plays an important role [30, 31]. Caspases are known to be key proteolytic mediators for the initiation and execution of apoptotic response [32, 33]. These enzymes need Ca^{2+} for activation [34]. Hence, we examined whether 7KCHO promotes Ca^{2+} influx and caspase-3 activity in human SMC. Incubation of SMC with 7KCHO induced Ca^{2+} influx into SMC, and the amount of Ca^{2+} influx induced by 50 μmol/L of 7KCHO was higher than that by 25 μmol/L of 7KCHO (Figure 9A, Metabolism, in press). However, no Ca^{2+} influx was observed by the addition of cholesterol (Figure 9A, Metabolism, in press). The activity of caspase-3 in SMC incubated with 7KCHO was significantly higher compared to that in SMC with cholesterol, and the increase was dependent on the concentration of 7KCHO (Figure 9B, Metabolism, in press). These findings indicate that 7KCHO has the potential of inducing Ca^{2+} influx into cells, and this potential could be one of the important mechanisms of 7KCHO-induced apoptosis.

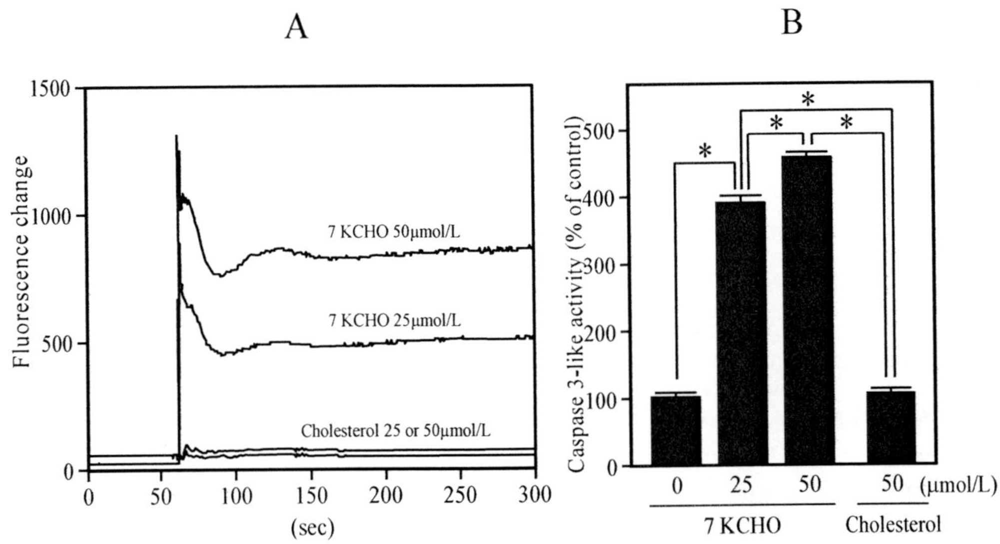

Figure 9. Effects of 7KCHO on Ca2+ influx (A) and the activity of caspase-3 (B). Values represent mean ± SEM of two experiments performed in triplicate. *$P < 0.01$.

Furthermore, to clarify whether 7KCHO-induced Ca^{2+} influx takes place through the Ca channel, Ca^{2+} influx induced by 7KCHO was observed when a Ca channel blocker (niphedipine) was added. In the presence of niphedipine, Ca^{2+} influx induced by 7KCHO was suppressed (Figure 10A, Metabolism, in press). The activity of caspase-3 and the amount of fragmented DNA induced by 7KCHO also decreased significantly in a niphedipine dose-dependent manner (Figure 10B and C, Metabolism, in press). From these results, we conclude that 7KCHO induces Ca^{2+} influx into SMC through the Ca channel, and that niphedipine suppresses SMC apoptosis induced by 7KCHO.

Several studies have shown that the Ca channel blockers have an anti-atherosclerotic action not only in animals but also in humans [35, 36]. This action is speculated to be a pleiotropic effect beyond the blood pressure lowering effect, although some controversies remain. For example, Ca channel blockers have been shown to inhibit smooth muscle cell proliferation [37]. Also Ca channel blockers decrease very low-density lipoprotein secretion by hepatocytes [35, 38]. We propose that the inhibitory effect of Ca channel blockers on 7KCHO-induced apoptosis may stabilize plaques by preventing the decrease of cell number in atherosclerotic lesions, and this probably accounts for the anti-atherosclerotic action as one of the pleiotropic effects recognized in Ca channel blockers.

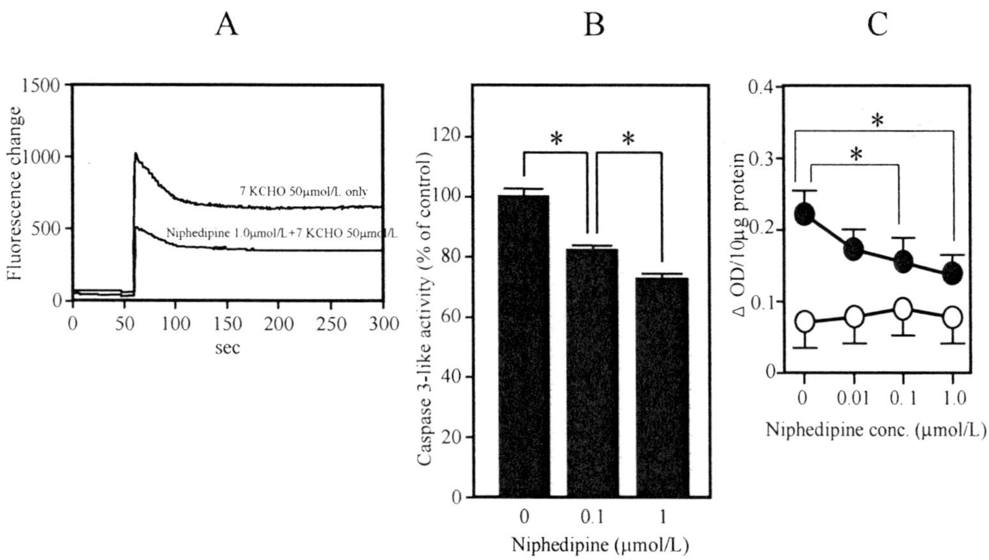

Figure 10. Inhibitory effect of niphedipine on Ca2+ influx (A), activity of caspase-3 (B) and fragmented DNA (C) induced by 7KCHO. Caspase-3 activity (B) was measured by the addition of 7KCHO at 50 μmol/L. In graph C, closed circle shows cells incubated with 7KCHO at 50 μmol/L, and open circle shows cells without 7KCHO. Results are the mean ± SEM. *P < 0.01.

7KCHO-Induced Apoptosis of SMC is Reduced by HMG-CoA Reductase Inhibitor, Pravastatin

3-Hydroxy-3-methylglutaryl coenzyme A reductase inhibitors (statins) are a class of drug with proven effect of lowering low-density lipoprotein cholesterol (LDL-C). Numerous studies have reported that the frequency of coronary heart disease is significantly decreased by statin administration [39, 40, 41]. For example, in the West of Scotland Coronary Prevention Study (WOSCOPS), pravastatin administration for 5 years reduced the number of definite nonfatal case of myocardial infarction by 33%, as compared with placebo [39]. In additional to the LDL-C lowering effect, other pleiotropic effects of statins have also been demonstrated [42, 43, 44, 45]. We examined the effect of pravastatin on 7KCHO-induced SMC apoptosis. In the presence of pravastatin (10^{-5} to 10^{-7}), the amount of 7KCHO-induced DNA fragmentation decreased significantly (Figure 11) [46]. The decrease was both dose and time--dependent. These results suggest that pravastatin suppresses the decrease in cell numbers in atherosclerosis, which leads to stabilization of the plaques. We speculate that this inhibitory effect of pravastatin may contribute to the decrease in frequency of atherosclerotic disease by pravastatin administration [39].

Generally, HMG CoA reductase inhibits cell proliferation [42]. We already mentioned above that SMC having a high proliferative potential are more susceptible to 7KCHO-induced apoptosis. Hence, the mechanism by which pravastatin inhibits 7KCHO-induced apoptosis may be due to the inhibition of cell proliferation.

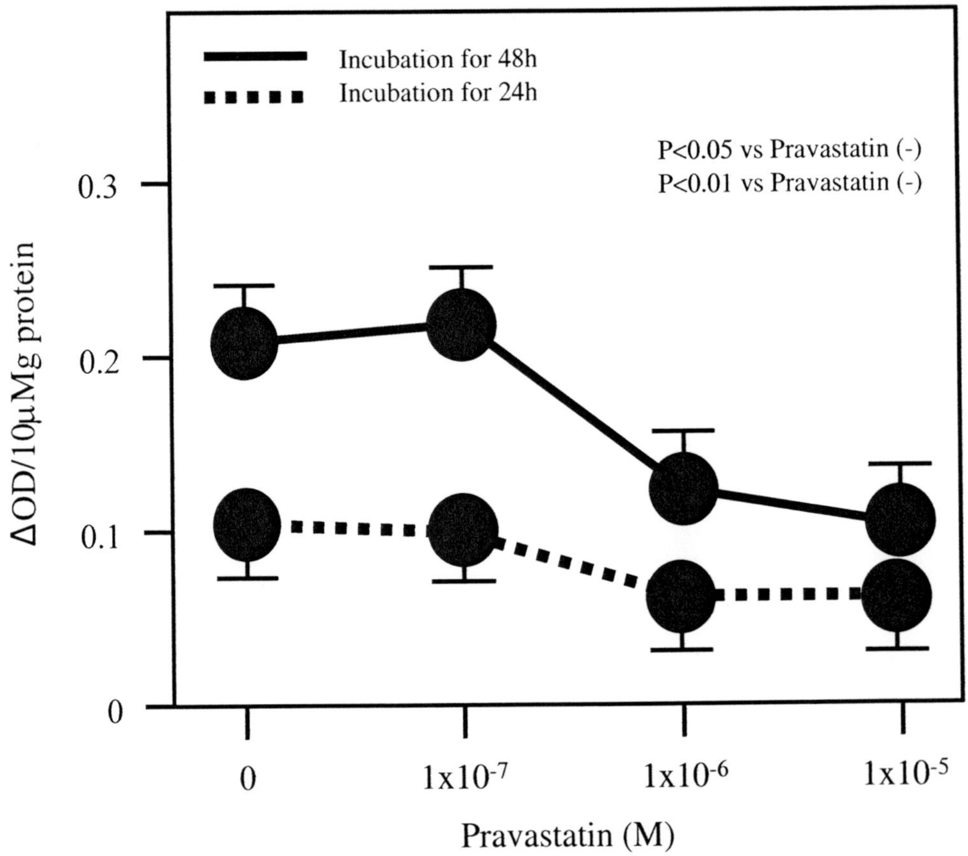

Figure 11. Effect of pravastatin on 7KCHO-induced fragmentation of DNA. SMC was incubated with 7KCHO at 50 μmol/L. Results are the mean ± SD. (Reference 46)

Conclusion

Enhanced oxidative stress may be closely related to the pathogenesis of atherosclerosis [47, 48]. Oxidative stress increases various oxidative products. 7KCHO is probably better defined as an oxidative product that acts directly on atherosclerotic lesion and contributes to the development of atherosclerosis, rather than a marker of oxidative stress. We have already reported that 7KCHO inhibits SMC migration [49] and enhances the activity of MMP-9 by promoting pro-MMP-9 secretion from macrophage [50]. In addition to apoptosis, the inhibition of SMC migration could further promote the decrease of cell number in plaque, while enhancement of MMP-9 activity could dissolve the extracellular matrix (ECM) and further destabilize the plaque. Thus various lines of evidence indicate that 7KCHO containing in plaque is a strong candidate as an important cause of vulnerable plaque. We hope that elucidation of the relationship between 7KCHO and SMC apoptosis will contribute to a decrease in incidence of atherosclerotic disease.

References

[1] Fuster, V; Stein, B; Ambrose, JA; Badimon, L; Badimon, JJ; Chesebro, JH. (1990) Atherosclerotic plaque rupture and thrombosis. *Circulation*, 82 (suppl):II-47-II-59.

[2] Bennett, MR; Evan, GI; Schwartz, SM. (1995) Apoptosis of human vascular smooth muscle cells derived from normal vessels and coronary atherosclerotic plaques. *J Clin Invest*, 95, 2266-2274.

[3] Geng, Y-J; Wu, Q; Muszynski, M; Hansson, GK; Libby, P. (1995) Apoptosis of vascular smooth muscle cells induced by in vitro stimulation with interferon-γ, tumor necrosis factor-α, and interleukin-1β. *Arterioscler Thromb*, 7, 19-27.

[4] Kockx, MM; Knaapen, MW. (2000) The role of apoptosis in vascular disease. *J Pathol*, 190, 267-280.

[5] Brown, AJ; Jessup, W. (1998) Oxysterols and atherosclerosis. *Atherosclerosis*, 142, 1-28.

[6] Smith, LL. (1996) Review of progress in sterol oxidations: 1987-1995. *Lipids*, 31, 453-487.

[7] Sevanian, A; Berliner, J; Peterson, AR. (1991) Uptake, metabolism and cytotoxicity of isomeric cholesterol-5, 6-epoxides in rabbit aortic endothelial cells. *J Lipids Res*, 32, 147-155.

[8] Sevanian, A; Hodis, HN; Hwang, J; Peterson, H. (1995) Involvement of cholesterol oxides in the atherogenic action of oxidized LDL. In: Bellomo G, Finardi G, Maggi E, Rice-Evans C, eds. Free Radicals, Lipoprotein Oxidation and Atherosclerosis. 1st ed. London, England: *Eichelieu Press*, 139-161.

[9] Palladini, G; Finardi, G; Bellomo, G. (1996) Disruption of actin microfilament organization by cholesterol oxides in 73/73 endothelial cells. *Exp Cell Res*, 223, 72-82.

[10] Boissonneault, GA; Hennig, B; Wang, Y; Ouyang, CM; Krahulik, L; Cunnup, L; Oeltgen, PR. (1991) Effect of oxysterol-enriched low-density lipoproteins. *Ann Nutr Metab*, 35, 226-232.

[11] Boissonneault, GA; Hennig, B; Ouyang, CM. (1991) Oxysterols, cholesterol biosynthesis, and vascular endothelial cell monolayer barrier function. *Proc Soc Exp Biol Med*, 196, 338-343.

[12] Deckert, V; Brunet, A; Lantoine, F; Lizard, G; Brussel, EM; Monier, S; Lagrost, L; David-Dufilho, M; Gamber, P; Devynck, M-A. (1998) Inhibition by cholesterol oxides of NO release from human vascular endothelial cells. *Arterioscler Thromb Vasc Bio*, 18, 1054-1060.

[13] Deckert, V; Persegol, L; Viens, L; Lizard, G; Athias, A; Lallemant, C; Gambert, P; Lagrost, L. (1997) Inhibition of arterial relaxation among components of human oxidized low-density lipoproteins: cholesterol derivatives oxidized in position 7 are potent inhibitors of endothelium-dependent relaxation. *Circulation*, 95, 723-731.

[14] Miyashita, Y; Shirai, K; Ito, Y; Watanabe, H; Urano, Y; Murano, T; Tomioka, H. (1997) Cytotoxicity of some oxysterols on human vascular smooth muscle cells was mediated by apoptosis. *J Atheroscler Thromb*, 4, 73-78.

[15] Shiomi, M; Yamada, S; Ito, T. (2005) Atheroma stabilizing effects of simvastatin due to depression of macrophages or lipid accumulation in the atheromatous plaques of coronary plaque-prone WHHL rabbits. *Atherosclerosis*, 178, 287-294.

[16] Miyamoto, A; Prieto, AR; Friedl, SE; Lin, FC; Muller, JE; Nesto, RW; Abela, GS 2nd. (2004) Atheromatous plaque cap thickness can be determined by quantitative color analysis during angioscopy: implications for identifying the vulnerable plaque. *Clin Cardiol*, 27, 9-15.

[17] Davies, MJ. (1996) Stability and instability: two faces of coronary atherosclerosis. *Circulation*, 94, 2013-2020.

[18] Geng, YJ; Libby, P. (1995) Evidence for apoptosis in advanced human atheroma: colocalization with interleukin-1ß-converting enzyme. *Am J Pathol*, 147, 251-266.

[19] Han, DK; Haudenschild, CC; Hong, MK; Tinkle, BT; Leon, MB; Liau, G. (1995) Evidence for apoptosis in human atherogenesis and in a rat vascular injury model. *Am J Pathol*, 147, 267-277.

[20] Masaki Ohtsuka, Yoh Miyashita, Kohji Shirai; Lipids Deposited in Human Atheromatous Lesions Induce Apoptosis of Human Vascular Smooth Muscle Cells. *J Atheroscler Thromb*, 13, 256-262.

[21] Kannnel, WB; McGee, DL. (1979) Diabetes and cardiovascular disease, the Framingham study. *JAMA*, 241, 2035-2038.

[22] Stamler, J; Vaccaro, O; Neaton, JD; Wentworth, D. (1993) Diabetes, other risk factors, and 12-yr cardiovascular mortality for men screened in the multiple risk factor intervention trial. *Diabetes Care*, 16, 434-444.

[23] Sakata, N; Meng, J; Takebayashi, S. (2000) Effect of advanced glycation end products on the proliferation and fibronectin production of smooth muscle cells, *J Atheroscler Thromb*, 3, 169-176.

[24] Moons, AH; Levi, M; Peters, RJ. (2002) Tissue factor and coronary artery disease. *Cardiovasc Res*, 53, 313-325.

[25] Ways, DK; Sheetz, MJ. (2000) The role of protein kinase C in the development of the complications of diabetes. *Vitam Horm*, 60, 149-193.

[26] Mashima, R; Witting, PK; Stocker, R. (2001) Oxidants and antioxidants in atherosclerosis. *Corr Opin Lipidol*, 4, 411-418.

[27] Kawano, K; Hirashima, T; Mori, S; Saitoh, Y; Kurosumi, M; Natori, T. (1992) Spontaneously long-term hyperglycemic rat with diabetic complications. Otsuka Long-Evans Tokushima Fatty (OLETF) strain. *Diabetes*, 11, 1422-1428.

[28] Miyashita, Y; Koide, N; Oyama, T; Itoh, Y; Shirai, K. (2005) Apoptosis Induced by 7-ketocholesterol is Enhanced in Smooth Muscle Cells Derived from OLETF Rats. *J Atheroscler Thromb*, 12, 92-97.

[29] Tamura, K; Kanzaki, T; Tashiro, J; Yokote, K; Mori, S; Ueda, S; Saito, Y; Morisaki, N. (2000) Increased atherogenesis in Otuka Long-Evans Tokushima fatty rats before the onset of diabetes mellitus: association with overexpression of PDGF beta-receptors in aortic smooth muscle cells. *Atherosclerosis*, 149, 351-358.

[30] Dimmeler, S; Haendeler, J; Nehls, M; Zeiher, AM. (1997) Suppression of apoptosis by nitric oxide via inhibition of interleukin-1β-converting enzyme (ICE) -like and cysteine protease protein (CPP) -32-like proteases. *J Exp Med*, 185, 601-607.

[31] Schönbeck, U; Herzberg, M; Petersen, A; Wohlenberg, C; Gerdes, J; Flad, HD; Loppnow, H. (1997) Human vascular smooth muscle cells express interleukin-1β-converting enzyme (ICE) -like, but inhibit processing of the interleukin-1□precursor by ICE. *J Exp Med*, 185, 1287-1294.

[32] Riedl, SJ; Renatus, M; Snipas, SJ; Salvesen, GS. (2001) Mechanism-based inactivation of caspases by the apoptotic suppressor p35. *Biochemistry*, 40, 13274-13280.

[33] Liang, Y; Yan, C; Schor, NF. (2001) Apoptosis in the absence of caspase 3. *Oncogene*, 20, 6570-6578.

[34] McConkey, DJ; Orrenius, S. (1996) The role of calcium in the regulation of apoptosis. *J Leukoc Biol*, 59, 775-783.

[35] Henry, PD. (1990) Calcium antagonist and atherosclerosis. *Atherosclerosis*, 10, 963-965.

[36] Ram, CV. (1999) Antiatherosclerotic effects of calcium-channel blockers: therapeutic implications. *Cardiology*, 10, 77-83.

[37] Fang, LH; Zhang, YH; Ma, JJ; Du, GH; Ku, BS; Tao, HY; Yun, YP; Kim, TJ. (2004) Inhibitory effects of tetrandrine on the serum- and platelet- derived growth factor-BB-induced proliferation of cell cycle progression DNA synthesis, ERK1/2 activation and c-fos expression. *Atherosclerosis*, 174, 215-223.

[38] Pitt, B. (1997) Diversity of calcium antagonist. *Clin Therapeut*, 19, 3-17.

[39] Shepherd, J; Cobbe, SM; Ford, I; Isles, CG; Lorimer, AR; Macfarlane, PW; Mckillop, JH; Packard, CJ. (1995) Prevention of coronary heart disease with pravastatin in men with hypercholesterolemia. *N Engl J Med*, 333, 1301-1307.

[40] Pitt, B; Waters, D; Brown, WV; van Boven, AJ; Schwartz, L; Title, LN. (1999) Eisenberg D, Shurzinske L, and MacCormick LS: Aggressive lipid-lowering therapy compared with angioplasty in stable coronary artery disease. *N Engl J Med*, 341, 70-76.

[41] Byington, RP; Furberg, CD; Crouse, JR third; Espeland, MA; Bond, MG. (1995) Pravastatin, Lipids, and atherosclerosis in the carotid arteries (PLAC-II). *Am J Cardiol*, 76, 54C-59C.

[42] Negre-Aminou, P; van Vliet, AK; van Erck, M; van Thiel, GC; van Leeuwen, RE; Cohen, LH. (1997) Inhibition of proliferation of human smooth muscle cells by various HMG-CoA reductase inhibitor; comparison with other human cell types. *Biochim Biophys Acta*, 21, 259-268.

[43] Malyszko, J; Urano, T; Knofler, R; Takada, Y; Takada, A. (1996) Effects of short-term treatment with mevalotin on platelet aggregation, fibrinolysis, peripheral serotonergic system and serum lipids in Japanese monkeys. *Throm Res*, 81, 397-402.

[44] Hoffman, R; Brook, GJ; Aviram, M. (1992) Hypolipidemic drugs reduce lipoprotein susceptibility to undergo lipid peroxidation: in vitro and ex vivo studies. *Atherosclerosis*, 93, 105-113.

[45] Katznelson, H; Wang, XM; Chia, D; Ozawa, M; Zhong, HP; Hirata, M; Terasaki, PI; Kobashigawa, JA. (1998) The inhibitory effects of pravastatin on natural killer cell activity in vivo and on cytotoxic T lymphocyte activity in vitro. *J Heart Lung Transplant*, 17, 335-340.

[46] Miyashita, Y; Ozaki, H; Koide, N; Otsuka, M; Oyama, T; Itoh, Y; Matsuzaka, T; Shirai, K. (2002) Oxysterol–induced apoptosis of vascular smooth muscle cell is reduced by HMG-CoA reductase inhibitor, pravastatin. *J Atheroscler Thromb*, 9, 65-71.

[47] Lyons, TJ. (1991) Oxidized LDL: a role in the pathogenesis of atherosclerosis in diabetes? *Diabetic Med*, 8, 411-419.

[48] Giugliano, D; Ceriello, A; Paolisso, G. (1996) Oxidative stress and diabetic vascular complications. *Diabetes Care*, 19, 257-267.

[49] Oyama, T; Miyashita, Y; Kinoshita, K; Watanabe, H; Shirai, K; Yagima, T. (2001) Effect of deposited lipids in atheromatous lesions on the migration of vascular smooth muscle cells. *J Atheroscler Thromb*, 9, 308-314.

[50] Ozaki, H; Miyashita, Y; Watanabe, H; Shirai, H. (2005) Enhancement of MMP-9 activity in THP-1 cells by 7-ketocholesterol and its suppression by HMG-CoA reductase inhibitor, fluvastatin. *J Atheroscler Thromb*, 12, 109-113.

In: New Developments in Cell Apoptosis Research
Editor: Alan J. Corvin, pp. 69-96
ISBN: 1-60021-453-3

Chapter III

Cell Death in the Host-Parasitoid Relationship

David B. Rivers[1]*, Ekrem Ergin[2] and Fevzi Uçkan[2]
[1]Department of Biology, Loyola College in Maryland, Baltimore, Maryland, USA
[2]Department of Biology, Faculty of Science-Literature, Kocaeli University, Îzmit &#811, Kocaeli, Turkey

Abstract

Cell death, specifically apoptosis, is a feature of host-parasitoid relationships involving parasitic wasps. Though there are numerous examples of endosymbiotic virus-induced (e.g., polydnaviruses) and venom-mediated apoptosis in host hemocytes, there is very little evidence for this form of cell death being triggered in other host tissues. Is this due to specificity in tissue responsiveness? In terms of timing of events during parasitism (i.e., at the initial stages of parasitism), this is a likely scenario. But throughout the duration of the association between the parasitoid and host, several developmental and physiological changes in host tissues may be regulated by apoptotic signals. There simply is very little evidence to draw any conclusions. Somewhat surprisingly, in most instances when apoptosis has been identified in host tissue (hemocytes), investigations of the pathways involved or specific intracellular changes have apparently not been conducted. Consequently, apoptosis is known to be induced by polydnaviruses, but what relationships exist between this form of cell death and those triggered by wasp venoms are unclear, nor have any attempts been made to compare mechanisms of oncosis evoked by these viruses and parasitoid venoms. What follows is an attempt to show homology in the modes of death evoked by pathogenic insect viruses, endosymbiotic viruses that reside within endoparasitic wasps, and venoms produced by ectoparasitic wasps. As will

* To whom correspondence should be addressed: Dr. David B. Rivers. Department of Biology, Loyola College in Maryland, 4501 North Charles Street, Baltimore, MD USA 21210; Tel (410) 617-2057; Fax (410) 617-5682; Drivers@loyola.edu

be discussed, oncosis and apoptosis are intimately linked by common signal transduction pathways and mitochondria. The focus of this paper is to attempt to provide evidence, admittedly at times circumstantial, and speculation on the possible linkage between the modes of action of ectoparasitic wasp venoms and insect viruses.

Keywords: ectoparasitoid, wasp venom, oncosis, apoptosis, polydnavirus, baculovirus

Introduction

Parasitic Hymenoptera have evolved unique adaptations to cope with their specific lifestyles of being either endoparasitic or ectoparasitic during the egg and larval stages (reviewed by Jervis and Copeland; 1996; Quicke, 1997). Perhaps the most striking difference between these two lifestyles is associated with oviposition: though both endo- and ectoparasitic species inject maternally derived substances that modify and/or regulate the host, the degree of host regulation varies widely among species and between life history strategies (idiobiont vs. koinobiont; ecto vs. endoparasitic) (Gauld and Hanson, 1995; Godfray, 2000). Traditionally, the widely held view has been that such differences generally reflect the more complex existence of endoparasitism (Vinson and Iwantsch, 1980; Thompson, 1993), a situation that requires breaching of the host's physicochemical barriers, egg deposition into a hostile host environment, and larval survival under constantly changing conditions (at least for some endoparasitoids) (Strand and Pech, 1995; Beckage and Gelman, 2004). These parasites are more often thought of as koinobionts; that is, parasitoids whose hosts continue to feed and potentially grow following the initial stage of parasitism (Askew and Shaw, 1986). As well, endoparasitic species tend to have a higher degree of physiological adaptation to the host-parasite association than ectoparasitoids (Quicke, 1997). Not surprisingly, then, the consensus theory in phylogeny is that parasitoid life history evolution has proceeded stepwise from ectoparasitism toward endoparasitism (Whitfield, 1998; Strand, 2000). Obviously, such views place ectoparasitism at the opposing end of the complete spectrum between idiobiont (parasitoids that typically prevent continued host feeding and development) and koinobiont life history strategies.

This theory has become the accepted dogma in parasitoid biology largely because it has gone unchallenged. The overwhelming majority of work on the evolution and biology of host-parasitoid relationships has been conducted using endoparasitic wasp models. Comparatively little is known about ectoparasitoids, and the information that is available, which can yield insight into the complexity of host-ectoparasitoid associations, has been obtained from only a handful of species. Based on this limited data, and also on speculation and circumstantial evidence, we challenge the notion that ectoparastism necessarily be viewed as less complex or adaptive than the endoparasitic condition.

Perhaps the most compelling evidence to challenge current evolutionary models comes from wasp secretions. More specifically, the secretions injected by adult females to suppress, modify or regulate the host condition. The ultimate purpose of such wasp manipulation is to maximize a female parasitoid's reproductive fitness by attempting to ensure progeny survival (Godfray, 1994; Rivers et al., 1998). An initial inspection of endo- and ectoparasitic species reveals that both types of parasitoids use maternally derived secretions [polydnaviruses (PDVs), virus-like particles (VLPs), ovarian fluids, teratocytes (derived from injected eggs), and venoms] for host regulation, although the composition and origins of these substances are very different (Moreau and Guillot, 2005, Quicke, 1997). This can clearly be seen with many endoparasitic species that transmit symbiotic viruses (polydnaviruses, entomopoxvirus) or VLPs in their venom or calyx fluid (Asgari, 2006, Federici and Bigot, 2003, Labrosse et al., 2005) into hosts: Physiological and immunological changes in the host are directly or indirectly due to the action of viral products. In contrast, nearly all ectoparasitic wasps and some endoparasitoids rely on either paralytic or non-paralytic proteinaceous venoms for induction of host alterations (Nakamatsu and Tanaka, 2003; Parkinson et al., 2003; Rivers et al., 2005, Uçkan et al., 2004, Zhang et al., 2005). In fact, no conclusive evidence has been found that shows ectoparasitic wasps inject any factors other than venom into their hosts.

Despite these differences, parasitic wasp venoms and viruses found in endoparasitic species display tremendous homology in terms of initial target sites (hemocytes), suppression of host feeding, arrestment of host development for the purpose of maximizing progeny production, and manipulation of cellular death pathways (apoptotic and oncotic) in host tissues. The modes of action of these venoms are consistent with mechanisms used by several groups of animal viruses. These viral groups vary in their association with insects: Some are symbiotic within endoparasitic wasps (polydnaviruses and entomopoxvirus), others are pathogenic toward an array of insect species (baculoviruses and ascoviruses), and one group is not found in insects at all (rotovirus). Perhaps the most common feature shared in the action of wasp secretions and viral products is the induction, or apparently so, of cell death in select tissues of the insect host. Apoptosis and/or oncosis appear to be necessary means to manipulate the host, whether to ensure successful development (e.g. arrestment of development, suppression of host immune responses, reduction or redirection of host metabolism, release of host lipids for larval feeding) of parasitoid larvae (Asgari, 2006, Nakamatsu and Tanaka, 2003, Zhang et al., 2005) or to promote transmission of progeny (viral) to new hosts (Krakauer and Payne, 1997, Stasiak et al., 2003).

The focus of this paper is to attempt to provide evidence and speculation on the possible linkage between the modes of action of parasitic wasp venoms and insect viruses. This review will present a comparison of what is currently known about how venoms from ectoparasitoids function at the cellular level with insect viruses that evoke similar host responses, and attempt to show the relationship of the action of wasp secretions and viral products to cell death.

Venom and Viral Suppression of Host Immune Responses

Background on Host Immunity

The host immune response represents perhaps one of the strongest selection pressures working on the host-parasitoid relationship (Godfray, 2000; Schmidt et al., 2001) because one member of this association will inevitably perish, either the host through parasitoid activity or the parasitoid as a result of the host immune system. Consequently, the initial stages of parasitism are critical to the survival of each organism. For example, the adult female parasitoid may inject factors into the host that overcome its defenses, rendering the host insect permissive to parasitoid larval feeding. Alternatively, the parasitoid may successfully complete oviposition but the eggs or larvae are eliminated by the innate immune responses well before the host insect suffers irreversible damage.

Invertebrate animals are known to posses an innate immune system capable of recognizing infectious eukaryotic animals like parasitic wasp eggs and larvae. Successful parasitism by parasitic wasps requires evasion or circumvention of this immune system. With parasitoids, the most intensively studied systems involve endoparasitic wasps and their insect hosts, which are frequently lepidopterans or dipterans (reviewed by Carton and Nappi, 2001; Schmidt et al., 2001; Strand and Pech, 1995). These endoparasitoids have evolved a variety of strategies, using both passive and active mechanisms, for overcoming host defense responses (Moreau, 2003; Nappi and Carton, 1996). Passive protection can be gained by the parasitoid developing in regions or tissues of the host that are inaccessible to cellular defenses (hemocyte attack), or by using a type of surface display equivalent to molecular mimicry in which the wasp's eggs or larvae are not recognized as non-self (Asgari and Schmidt, 1994; Asgari et al., 1998; Kinuthia et al., 1999). In contrast, active mechanisms generally involve suppression of host immunity through the activity of maternally derived factors that are injected by adult females at the time of oviposition. Venoms and ovarian proteins seem to be immunosuppressive in some systems (Luckhart and Webb, 1996; Asgari et al., 1998), whereas other endoparasitoids inject various polydnaviruses (ichnovirus, brachovirus) (Beckage, 1998; Webb, 1998; Dupas et al., 2003), non-polydnaviruses (ascovirus, entomopoxvirus) (Bigot et al., 1997; Lawrence, 2002), or other viruses or virus-like particles (VLP) still to be identified (Rizki and Rizki, 1990; Barratt et al., 1999; Labrosse et al., 2003), into hosts that function as immunosuppression factors, frequently targeting host hemocytes.

Schmidt et al (2001) have argued that ectoparasitism is conducive only to passive host defenses, because the wasp's offspring develop externally and therefore do not contend with the host fluid's directly, and because many ectoparasitoids are believed to kill their hosts almost immediately after envenomation (Beard, 1963; Quicke, 1997). This argument is likely true for most ectoparasitic species that rapidly paralyzes or kills their hosts. However, the reality is that few studies have been performed to examine the host immune responses in host-ectoparasitoid relationships. For those species that have been studied, maternally derived secretions, as well as those from feeding larvae, disable the host's capacity for clotting and wound healing (reviewed by Rivers, 2003). Such activities are consistent with active mechanisms of immunosuppression and are comparable to those used by endoparasitic species attempting to avoid encapsulation (Beckage, 1998). Thus, there is evidence that some ectoparasitoids display behaviors associated with evasion of host immunity that are thought to be typical of koinobiont endoparasitoids but not predicted to occur with ectoparasitism (Quicke, 1997).

Here, we will attempt to only make comparisons between endoparasitoids and ectoparasitoids with regard to maternal secretions that function in active suppression of host innate immunity. More specifically, comparisons will be restricted to viral and venom effects on cellular immune responses.

General Mechanisms of Viral Suppression of Host Immunity

Active suppression of host immunity by many species of endoparasitic wasps relies on the action of endosymbiotic viruses (entomopoxviruses, polydnaviruses) either alone or synergistically with venom. The primary function of these viruses is to disable the host's encapsulation response by infecting host hemocytes (Schmidt et al., 2001). In most hosts, which appear to be restricted to the orders Lepidoptera and Diptera (Beckage, 1998; Carton and Nappi, 2001; Lawrence and Akin, 1990; Rizki and Rizki, 1984), plasmatocytes and granular cells normally are the primary targets of virus infection, although capsule forming lamellocytes in *Drosophila melanogaster* appear to be the initial sites of cellular defenses attacked by virus-like particles from the figitid *Leptopilina boulardi* (Rizki and Rizki, 1990; Labrosse et al., 2003). Following oviposition by female wasps in which viruses are introduced into the hosts, encapsulation of parasitoid eggs, larvae, or any other foreign target located in the host hemocoel does not and apparently cannot occur. In some instances, suppression of encapsulation requires that the virus maintain transcriptional activity in the host (Davies et al., 1987; Strand and Noda, 1991), but at least with polydnaviruses, this apparently is not associated with increased virulence through viral replication in host tissue (Stoltz and Whitfield, 1992). Rather, physiological disruption of the host, including immunosuppression, follows after viral (wasp) genes have been transcribed in specific host tissues such as hemocytes and fat body (Cui et al., 1997; Fleming, 1992; Stoltz, 1993; Tanaka et al., 2003).

At the cellular level, the encapsulation response is abolished in virally infected hemocytes potentially due to rearrangment of the cytoskeleton, aggregation of hemocytes, loss of the ability to spread or adhere to non-self, a decline in the number of hemocytes in circulation, or due to cell death via apoptosis (Asgari et al., 1996; Strand and Noda, 1991; Strand and Pech, 1995). Relatively little is known about the mechanisms involved in the induction of these host immune alterations. It is entirely plausible that changes in the hemocyte cytoskeleton are linked to apoptotic pathways. For example, cytoplasmic bleb formation is frequently associated with plasma membranes of apoptotic cells (Trump and Berezesky, 1995). Formation of blebs usually involves disruption of the cytoskeletal-membrane interactions and is speculated to depend on the activation of Ca^{+2}-dependent proteases (Nicotera et al., 1986; Phelps et al., 1989). Protease activation, then, requires an elevation of intracellular calcium levels and perhaps also involvement of phospholipases (Troyer et al., 1986). Influx of Ca^{+2} through L-type calcium channels on the plasma membrane and mobilization of calcium from intracellular stores are primary effects of several animal viruses (Perez et al., 1998; Guo et al., 2003). Dib-Hajj et al (1993) found gene sequences from the polydnavirus of *Campoletis sonoriensis* showed a high degree of homology with α-conotoxins, potent calcium channel modulators produced by snails in the genus *Conus*. Collectively, these observations imply that virally induced morphological changes in hemocytes are attributed to calcium mobilization in the cytosol. As will be discussed later, apoptotic signals are not restricted to just immunosuppression as baculoviruses, ascoviruses and iridescent viruses exploit cell death pathways to prolong or reduce the life of a cell to maximize virion progeny production (Just and Essbauer, 2001; Karkauer and Payne, 1997; Miller, 1997; Ishikawa et al., 2003).

Some polydnavirus depend on the presence of endoparasitic wasp venom to abolish host immune responses (Guzo and Stoltz, 1985; Strand and Noda, 1991). A combination of venom and polydnavirus, but not either alone, from the braconid *Cotesia melanoscela* is necessary to allow successful development of *Hyposter fugitivus* (Ichneumonidae) without the threat of encapsulation in larvae of the tussock moth, *Orgyia leucostigma* (Guzo and Stolz, 1985). A similar synergism between polydnavirus and venom has been observed with *Microplitis demolitor*, *Cardiochiles nigriceps* and *C. kariyai* when parasitizing their respective lepidopteran hosts (Strand and Noda, 1991; Tanaka and Vinson, 1991). In the case of *Ca. nigriceps*, a venom protein with an approximate molecular mass of 66 kDa must interact with polydnavirus to evoke arrestment of host development and to presumably suppress host immune responses (Tanaka and Vinson, 1991). Coincidentally, a 66 kDa venom protein has been isolated from the ectoparasitoid *Euplectrus comstockii* (Eulophidae) that does not induce paralysis but does arrest host development (Coudron and Brandt, 1996). Similarly, a 66-69 kDa protein has been identified in crude venom from *N. vitripennis* that, in concert with 2-3 other venom proteins, appears to retard host development and suppress immune responses in susceptible fly species (Rivers et al., 2006). These are not isolated examples of the use of similar sized venom proteins, rather this may be the typical scenario: There are several examples of endo- and ectoparasitic wasps that use mid-range proteins derived either from venom glands or endosymbiotic viruses to disrupt host physiology (Doury et al., 1997; Leluk and Jones, 1989; Lawrence, 2002; Quistad, et al., 1994; Parkinson et al., 2002; Richards and Parkinson, 2000; Rizki and Rizki, 1991). The homology in size and action of these proteins suggests that venom and viral proteins in parasitic wasps manipulate host developmental and death pathways by similar mechanisms.

Ectoparasitoid Suppression of Host Immunity

Ectoparasitic wasps have been predicted to show only passive mechanisms for dealing with host defenses. Active suppression of innate immunity in host systems is generally believed to be absent in ectoparasitoids (Schmidt et al., 2001; Strand and Pech, 1995). Certainly the actual act or behaviors associated with oviposition by ectoparasitic wasps does not warrant disabling of the host's innate immune responses: Oviposition by ectoparasitoids results in egg laying on or near the host, separate from the internal environment of the host insect. Thus, if host immune responses functioned as a selection force driving the symbiogenesis between viruses and endoparasitoids (Federici and Bigot, 2003; Godfray, 2000; Schmidt et al., 2001), it would make sense that viral associations are not adaptive to host-ectoparasitoid relationships. However, the initial assumption concerning the absence of active suppression mechanisms among ectoparasitic species is not correct (Rivers, 2004). During the process of oviposition, typically during envenomation, many species of ectoparasitoids inject venom into the host that directly or indirectly suppresses the host immune system. Most of this is inferred from studies that describe paralysis in the host after venom injection with no inhibition of larval feeding in the immobilized hosts (Becchino and Sullivan, 1981; Doury et al., 1995; Morales-Ramos et al., 1995; Piek and Spanjier, 1986). With ectoparasitic species that inject nonparalyzing venoms, it is clear that suppression of clotting and wound healing in host insects are primary effects of venom (Richards and Edwards, 1999; Rivers et al., 1993, 1999a; 2002a; Shaw, 1981). Active suppression of host defenses is required by ectoparasitoids for at least two distinct reasons: to allow host feeding by synovigenic species (Bartlett, 1964) and to promote uninterrupted feeding by developing larvae. Host feeding is more characteristic of idiobiont parasitoids, which are more frequently ectoparasitoids than endoparasitic wasps (Quicke, 1997). Consequently, the former feature may be somewhat unique to ectoparasitism, and more specifically to gregarious species (Rivers et al., 1998).

Perhaps the most frequent association linked to ectoparasitism is the use of paralytic venoms. Numerous species of ectoparasitoids inject paralytic venoms into their insect hosts, regardless of the development stage of the host (Beard, 1963; Quicke, 1997). The effects of these paralyzing agents are the same: the host is immobilized almost immediately (Beard, 1964; Drenth, 1974), in many cases becoming flaccid (Piek and Spanjier, 1986), and in nearly all cases, host death ensues very rapidly (Quicke, 1997). For these species of wasps, developing larvae can acquire nutrients from a fixed resource, and perhaps most importantly, do not have to contend with host defenses. It is unlikely that paralytic venoms show specificity for immunosuppression since most are thought to target neuromuscular junctions (Piek and Spanjier, 1986). The resulting inhibition of cellular and humoral defenses can be attributed to a rapid deterioration and decay of all host tissues. Consequently, the feeding parasitoids must complete development quickly.

Nonparalytic venoms are also used by some ectoparasitic wasps for host immunosuppression, but these toxins are not used in a identical fashion by each species. For example, the eulophids *Euplectrus pennicornis* and *E. plathypenae* inject nonparalyzing venoms into their lepidopteran hosts, host development is delayed but not inhibited (Coudron et al., 1990; Marris and Edwards, 1995), wasp larval development is apparently unimpeded by host defenses, but in neither case has the venom been reported to suppress host defenses (Coudron et al., 1990; Richards and Edwards, 2000a). Richards and Edwards (2000a, b) have suggested that wasp larvae, not the adult females, are responsible for suppressing hemocyte recognition of non-self and reducing host phenoloxidase activity. Host immune activity has not been investigated in hosts parasitized by *E. playthpenae*.

Eupelmus orientalis larvae are also thought capable of paralyzing their weevil hosts (Doury et al., 1995, 1997), and hence, suppress host defenses. In contrast to *E. pennicornis*, adult females share this ability with larvae (Doury et al., 1997). This wasp is grouped with ectoparasitoids using nonparalyzing venoms because the venom appears to function more like a developmental arrestment factor as described for *N. vitripennis* (Rivers and Denlinger, 1994a) and *E. plathypenae* (Coudron et al., 1990) than as a true paralysant.

N. vitripennis has provided the clearest understanding of the impact of nonparalyzing venoms from ectoparasitoids on host defenses. Parasitism by *N. vitripennis* has an almost immediate impact on host hemocytes: the total numbers of circulating hemocytes decline abruptly by 60 min after parasite attack, plasmatocytes and granular cells lose the ability to be adhesive and spread, all plasmatocytes die within 3 h, and melanization and clotting in host hemolymph is retarded or completely inhibited when pharate adults of the flesh fly, *Sarcophaga bullata* (Sarcophagidae), serve as hosts (Rivers et al, 2002a). No obvious parasitism-mediated changes have been noted in two other hemocyte types found in these flies, oenocytoids and pro-hemocytes. All of the alterations in the host cellular immune response appear to be permanent as there was no evidence that any aspect of the hemocyte immune response recovered during the course of parasitism, a feature consistent with the effects that the ectoparasitic wasp *E. pennicornis* evokes on host hemocytes (Richards and Edwards, 1999).

In Diptera, wound healing and clotting typically require the active participation of at least two types of hemocytes, plasmatocytes and granular cells (Gotz, 1986; Lackie, 1988). Natural parasitism, envenomation, and isolated crude venom from *N. vitripennis* have been shown to have an immediate inhibitory effect on the ability of host hemocytes to adhere and spread (Rivers et al., 2002a). This inhibition of granular cells persists throughout the duration of parasitism while plasmatocytes are differentially targeted to die. This venom mode of action appears to be receptor-mediated as only hemocytes from insects within the wasps host range are susceptible to the venom (Zhang et al., 2005). As with the action of polydnaviruses, venom disruption of adhesion and spreading by host hemocytes is suggestive of altered cytoskeletal associations with plasma membranes. Rivers et al (1999b) have previously suggested that bleb formation in cultured insect cells exposed to lethal doses of *N. vitripennis* venom involved distortions of the cytoskeleton due to Ca^{+2}-dependent proteases. Additionally, mobilization of calcium from intracellular stores within mitochondria and endoplasmic reticulum appear to be linked to the cascade of events that lead to bleb formation, swelling, and oncotic death induced in susceptible cells by venom from *N. vitripennis* (Rivers et al., 2004; Rivers et al., 2005). Ectoparasitoid venom activity is thus consistent with polydnavirus-induced changes in hemocyte cytoskeleton and apoptotic pathways.

Venom and Viral Arrestment of Host Development

An arrestment or halt in host development is commonly associated with parasitism by parasitoids, regardless of where the wasp's eggs are deposited on or in the host (Beckage and Gelman, 2004; Marris et al., 2001). In many instances, it appears that host arrest results from an imbalance in host hormone levels (i.e., ecdysteroids, juvenile hormone), through a disruption or deterioration of host endocrine tissue, or possibly by inhibition of the ability of target tissues within the host insect to respond to hormonal signals. The exact mechanisms involved and host tissues targeted have not been fully revealed for most systems, but the usual suspects have been implicated as the causative agents: endosymbiotic viruses and/or venom with endoparasitic species and venom alone with ectoparasitoids. In recent years, research emphasis placed on examining host developmental changes has shifted more toward quantitative and qualitative analysis of viral gene expression in host tissues (Beliveau et al., 2003; Johner et al., 1999; Wyder et al., 2003) and viral mechanisms of immunosuppression (Webb, 1998). Consequently, the information presented in this section lacks many mechanistic details of how venom and viruses trigger developmental delay or arrest in susceptible hosts. At the cellular level, virtually nothing is known about the mechanisms of host arrest and this represents an area of host-parasite biology that is wide open to new investigations. Here, I have contained my comparisons to common developmental effects shared between viruses and venoms.

Viral-induced Host Arrest

Once delivered into the host, viral genes are transcribed by the host's cells, which can evoke major distortions of the host developmental program (Beckage, 1998; Webb, 1998). Polydnaviruses associated with *C. kariyai* trigger an elevation in host dopamine levels, which in turn is thought to promote an increase in the amount of growth-blocking peptide (GBP) present in the host, thereby delaying host pupation by GBP suppression of juvenile hormone esterase activity (Noguchi et al., 1995). Similarly, polydnaviruses from *C. congregata* prevents juvenile hormone levels from dropping in parasitized larvae of *Manduca sexta*, and thus host metamorphosis is blocked (Beckage and Gelman, 2001). These events may be linked to disintegration of the host's prothoracic gland (De Buron and Beckage, 1992), which is very likely the result of virally induced apoptosis (Le et al., 2003). Likewise, *P. includens* infected with polydnavirus from *M. demolitor* fail to complete metamorphosis apparently due to altered host juvenile hormone levels (Strand et al., 1988), but whether this occurs by a similar mechanism as induced by *C. congregata* or *C. kariyai* is not known. By contrast, the developmental arrest in *Anastrepha suspensa* (Tephritidae) induced by *Diachasmimorpha* (= *Biosteres*) *longicaudata* is associated with lowered ecdysteroid titers and/or alterations in epidermal cell activity (Lawrence, 1988), both presumably triggered by entomopoxvirus.

The precise cellular pathways leading to the virally mediated changes in the host endocrinological status have yet to be deciphered. Stage-specific expression of viral genes has been suggested to be responsible for induction of developmental delay or arrest of the host (Beliveau et al., 2000; Johner et al., 1999; Johner and Lanzrein, 2002). Though in many systems viral DNA is most abundant and expressed in the highest levels within host hemocytes (Hayakawa et al., 1994; Strand et al., 1992), Wyder et al. (2003) found abundant levels of *Chelonus inanitus* viral DNA in host fat body, nervous tissue, intestinal tract and hemocytes. Similarly, polydnavirus infection of tracheae, midgut, perineurium, epidermis and Malpighian tubules of host insects has been reported in other host-endoparasitoid associations (Stoltz and Vinson, 1979; De Buron and Beckage, 1992). Expression of viral genes in these tissues is not likely related to immunosuppression of host encapsulation responses. The fact that PDV genes from *C. congregata* and *Ch. inanitus* express proteins in host tissues "late" into parasitism, well after eggs have been deposited into the host, supports the proposal that these viral products are more likely associated with host arrestment than immunosuppression (Johner and Lanzrein, 2002; Le et al., 2003).

A cellular mechanism that may be involved in host arrestment is apoptosis. It is well known that many polydnaviruses have the capacity to trigger apoptotic death in host hemocytes to evade encapsulation (Le et al, 2003; Schmidt et al., 2001). However, this cellular death pathway has generally not been suggested to be involved in the induction or regulation of developmental delay or arrest in host tissues. There are several lines of evidence that argue a linkage between apoptosis and viral-mediated developmental arrest:

a. Viral proteins in calyx fluid of *Tranosema rostrale* induce apoptosis 24-72 h post-infection in non-hemocyte cultured cells (Cf-124T) from *Choristoneura fumiferana* (Beliveau et al., 2003);
b. Disintegration of prothoracic glands in virus-infected *M. sexta* and/or inhibition of neuromodulation in host brains impeding calcium-dependent prothoracicotropic hormone processing (Bikenbeil, 1998) argues that cell death is a feature of the parasitism pathway (Beckage and Du Jon, 1992; Beckage et al., 1998)
c. The requirement that some polydnaviruses maintain transcriptional activity in the host to elicit suppression of host immune responses and/or development (Davies et al., 1987; Strand and Noda, 1991) may be an indication of utilization of apoptotic signals as programmed cell death, but not oncosis, typically depends on induction of protein synthesis (Trump and Berezesky, 1995);
d. Polydnaviruses have been hypothesized to be derived from baculoviruses and ascoviruses (Federici and Bigot, 2003; Whitfield, 2002), and both of these viral groups exploit apoptotic pathways to prolong or reduce the life of a cell to maximize virion progeny production (Karkauer and Payne, 1997; Miller, 1997; Ishikawa et al., 2003).

Baculovirus infection has also been shown to lower ecdysteroid levels in developmentally arrested lepidopteran hosts (O'Reilly et al., 1998), a feature consistent with endocrine disruption in a number of host-parasitoid systems (Marris et al., 2001). In the case of viral infections by *Autographa californica*, a baculovirus ecdysteroid UDP-glucosyltransferase gene has been identified that is tentatively thought to function in delay or complete blockage of host development (O'Reilly et al., 1998). Enough circumstantial evidence to warrant further investigation into the possible role of apoptosis in virus-induced developmental arrest.

Venom-induced Host Arrest

The precise tissues targeted by ectoparasitic wasp venoms have not been clearly identified, nor have the pathways activated/inhibited by these venoms been determined. Neuromuscular junctions most likely are targeted by paralytic venoms. For example, venom from *H. hebetor* is thought to evoke paralysis by binding to a receptor associated with the excitatory glutamatergic system on pre-synaptic membranes (Walther and Rathmayer, 1974). Venom causes an accumulation of pre-synaptic neurotransmitter vesicles, thus blocking neurotransmitter release by inhibiting vesicle fusion with post-synaptic membranes (Walther and Rathmayer, 1983). Several proteins have been isolated from this wasp venom that possess paralytic/neurotoxic properties, including Brh-I, III and V (Quistad et al., 1994); A-MTX and B-MTX (Visser et al., 1983); and Brh TX-1 and Brh TX-2 (Windass et al., 1996). Amino acid sequence analysis of Brh TX-1 reveals low (31 and 29%, respectively) but significant homology to mouse membrane-associated mucin muc3 and mouse hepatocyte growth factor (Moreau and Guillot, 2005). In mouse, these proteins can elicit many effects including anti-apoptotic activities in a wide range of cells (Wang et al., 2002; Yoshida et al., 2003).

A venom protein (27 kDa) isolated from the endoparasitoid *Pimpla hypochondriaca* can induce paralysis in adult *Musca domestica* (Parkinson and Weaver, 1999). This fly is not a natural host and thus the venom action may not represent a typical host response. However, paralysis has been characterized for at least two other endoparasitoid venoms (*A. tabida* and *P. turionellae*)(Moreau et al., 2002; Uçkan et al., 2004). *P. hypochondriaca* venom has also been shown to be cytotoxic in vitro and to possess phenoloxidase and laccase activity (Parkinson et al., 2001, 2003), features that appear to be unique for parasitic wasp venom as only one other species (*N. vitripennis*) is known to possess PO activity (as well as cytolytic) in venom. Preliminary analysis of the mode of action of venom from *P. hypochondriaca* suggests that in vitro, cell death is induced rapidly (less than 1 h) and that separate populations of dying cells displayed apoptotic and oncotic morphologies (Parkinson et al., 2004).

Non-paralytic venoms from ectoparasitic wasps seem to target a variety of tissues, most likely in a receptor-mediated fashion. Inhibition of larval and pupal ecdysis by *E. comstockii*, *E. plathypenae*, and *N. vitripennis*, respectively, suggests that these venoms disrupt tissue responsiveness to ecdysteroids, a condition that is not corrected by application of exogenous 20-hydroxyecdysone (Coudron et al., 1990, 1994; Rivers and Denlinger, 1994a; Rivers et al., 2004). This effect implies that some aspect of the receptor-hormone complex is disrupted or that the complex is not able to bind to and/or activate response elements within the cell nucleus. Venom from *N. vitripennis* induces fly hosts to enter a developmental arrest that is characterized by a suppression of respiratory metabolism (Rivers and Denlinger, 1994b) and altered protein expression (Rivers, unpublished), both of which show tremendous similarity to the diapause physiology of the host. The latter observation has lead to speculation that the venom targets the host brain to redirect the cell-cycle control system.

Consistent with this prediction is the observation that heat shock protein synthesis (hsp 23 and hsp 70) in brains of the flesh fly, *S. crassipalpis*, is highly upregulated following injection of crude venom from *N. vitripennis* (Rinehart et al., 2002). The host response differs from a typical stress response (Feder, 1996) in that all hsps are not uniformly upregulated due to envenomation by *N. vitripennis*, and the response persists for an extended duration (Rinehart et al., 2002). The actual role of heat shock proteins during parasitism is not clear but they may be necessary to arrest host development. For example, regulation of hsp expression has been shown to directly influence induction of apoptosis in specific mammalian immune cells (Hisaeda and Himeno, 1997), arguing that like with polydnaviruses, apoptotic signals may be linked to developmental arrest. Evidence from *D. melanogaster* indicates that expression of hsp70 during non-stress conditions leads to slowed development (Krebs and Feder, 1997). As well, the cell cycle in cultured cells from *D. melanogaster* can be arrested in the presence of small hsps (Berger and Woodward, 1983). The developmental status of envenomated flies is also very similar to pupal diapause, another form of developmental arrest, in that hsps are upregulated. In both cases, hsp23 and hsp70 are upregulated, hsp90 is downregulated, and hsc70 is unaltered (Denlinger et al., 2001).

Tamerriello (2001) has demonstrated that the cell cycle in *S. crassipalpis* is arrested during pupal diapause, and agents that trigger apoptosis can induce the arrestment. Thus, if the host arrest elicited by venom from *N. vitripennis* is in fact similar to that of the short day response involved in programming pupal diapause, then venom may be a trigger of apoptotic signals.

Venom and Viral Cellular Death Signals

Cell death is a feature of host-parasitoid relationships. Though there are numerous examples of PDV and/or venom mediated apoptosis in host hemocytes (Cui et al., 1997; Le et al., 2003; Richards and Parkinson, 2000; Schmidt et al., 2001; Strand et al., 1992), there is very little evidence for this form of cell death being induced in other host tissues. Nakamatsu and Tanaka (2003) have argued that apoptosis follows venom-induced elevations of host fat body lipid levels, thus releasing fat into the host hemolymph to promote parasitoid larval feeding. Likewise, venom from *N. vitripennis* is believed to evoke de novo lipid synthesis in *S. bullata* at the expense of host respiratory metabolism (Rivers and Denlinger, 1994b, 1995), and conceivably the cytolytic action of venom proteins triggers lipid release for parasitoid larval consumption (Rivers et al., 1999). Many parasitoids use venom to induce host paralysis (Deyrup et al., 2006; Doury et al., 1997; Parkinson and Weaver, 1999; Quistad et al., 1994; Uçkan et al., 2004), and ultimately the host dies. What is not clear is the role of cell death in the host-parasitoid relationship. Somewhat surprisingly, in most instances when apoptosis or oncosis has been identified in host tissue, investigations of the pathways involved or specific intracellular changes have apparently not been conducted. Consequently, apoptosis is known to be induced by polydnaviruses, but what relationships exist between this form of cell death and those triggered by other insect viruses are unclear, nor have any attempts been made to compare mechanisms of oncosis evoked by other viral groups and some ectoparasitoid venoms (Rivers et al., 2002b, 2005). What follows is an attempt to show homology in the modes of death evoked by pathogenic viruses, symbiotic viruses, and wasp venoms. The two forms of death induced, oncosis and apoptosis, are intimately linked by common signal transduction pathways and by mitochondria (Petit et al., 1996).

Apoptotic Death Signals of Insect Viruses

Apoptosis is a form of programmed cell death that is frequently characterized by plasma membrane blebbing, cytoplasmic vacuolization, formation of apoptotic bodies that usually contain organelles, a loss of cell volume giving a shrunken appearance to the cell, increased caspase activity, and degradation of chromosomal DNA into oligonucleosome-sized DNA fragments (Hale et al., 1996). Bleb formation typically results from distortions of cytoskeletal-membrane interactions and appears to require the activation of Ca^{+2}-dependent proteases (Nicotera et al., 1986; Phelps et al., 1989). Protease activation, in turn, is dependent on an elevation of intracellular calcium and perhaps also involvement of phospholipases (Troyer et al., 1986). Increases in cytosolic $[Ca^{+2}]_i$ occurs as a result of the release of calcium from intracellular stores located in endoplasmic reticulum (Gomperts et al., 2002), or through influx of Ca^{+2} through L-type calcium channels on the plasma membrane (Perez et al., 1998). Many polydnaviruses and entomopoxviruses evoke apoptosis in host hemocytes as a means to suppress cellular immune responses. The details of apoptotic cellular pathways triggered by PDV infections are lacking, as are the pathological changes associated with symbiotic viral apoptosis. In contrast, anti-apoptotic genes are associated with baculoviruses and ascoviruses for suppressing host apoptosis (Du and Thiem, 1997; Miller, 1997). As well, many of the events that occur within cells infected by bacuolviruses have been well characterized. Thus, an examination of these cellular responses will be presented to potentially shed light on the pathways targeted by polydnaviruses.

Baculoviruses posses at least two distinct classes of anti-apoptotic genes which block apoptosis in host cells, thereby arresting host development and increasing the time of viral infectivity, as well as increasing the range of host cells that can be infected (Miller, 1997). One of these genes, p35, encodes a broadly acting inhibitor of the caspase family of cysteine proteases involved in the signaling pathways leading to apoptosis (Bertin et al., 1996; Liu et al., 2003; Miller, 1997). Activation of caspases is a central, highly conserved step in apoptotic pathways, and p35 functions in a wide range of organisms (Hay et al., 1994; Sugimoto et al., 1994; Beidler et al., 1995). A second class of genes, IAPs, appears to function earlier (upstream) in apoptotic pathways than p35 (Du and Thiem, 1997). IAPs seem to have a more limited role in blocking apoptotic events as the action of some of these proteins is limited to specific signal transduction pathways. Surprisingly, both IAPs and p35 bind to and inhibit caspase proteases (Luque et al., 2002; Miller, 1997), yet these gene products do not have any homology in either nucleotide or amino acid sequence (Crook et al., 1993; Freisen and Miller, 1987).

At the cellular level, these pathogenic insect viruses frequently shut off host gene expression. For example, *A. californica* virus promotes a steady decline in mRNA levels in host cells during the first 12-h after infection (Ooi and Miller, 1988). Host protein synthesis is also altered, at first becoming reduced, and within 24 h of infection, it is completely inhibited (Dobos and Cochran, 1980). The exact mechanisms eliciting these cellular responses have not been identified. However, the inhibition of protein synthesis does not appear to be linked to apoptosis (Du and Thiem, 1997).

Abolishing host apoptosis is believed to be an adaptive "reproductive" strategy used by several different viral families as means to maximize the number of virions produced in infected host cells, which ultimately increases viral fitness (Krakauer and Payne, 1997). The host, on the other hand, is expected to trigger apoptosis in infected cells to slow or stop viral replication and block further spread of the virus (Krakauer and Payne, 1997). The fact that polydnaviruses stimulate rather than suppress apoptosis in lepidopteran hosts of endoparasitic wasps is consistent with this adaptive reproductive strategy: Virions are produced in the wasp not in the lepidopteran hosts (Fleming, 1992). Thus host cell death is not limiting to viral fitness. Complicating this seemingly simple explanation is the view that polydnaviruses evolved from other insect viruses but they are now no longer viruses (Federici and Bigot, 2003). Federici and Bigot (2003) have proposed that polydnaviruses should be viewed more correctly as wasp organelle systems. In essence, the virions are delivery systems containing wasp DNA that can be expressed in host cells that take up or are invaded by the "polydnaviruses." If this theory is correct, an expansion of this theory may be that apoptosis inducer genes from parasitic wasps have been packaged into the virions and are expressed in host hemocytes and likely other tissues. This extension would also be in agreement with the adaptive reproductive strategy outlined by Kraukauer and Payne (1997), but it is wasp rather viral fitness that has been maximized through the action of polydnavirus.

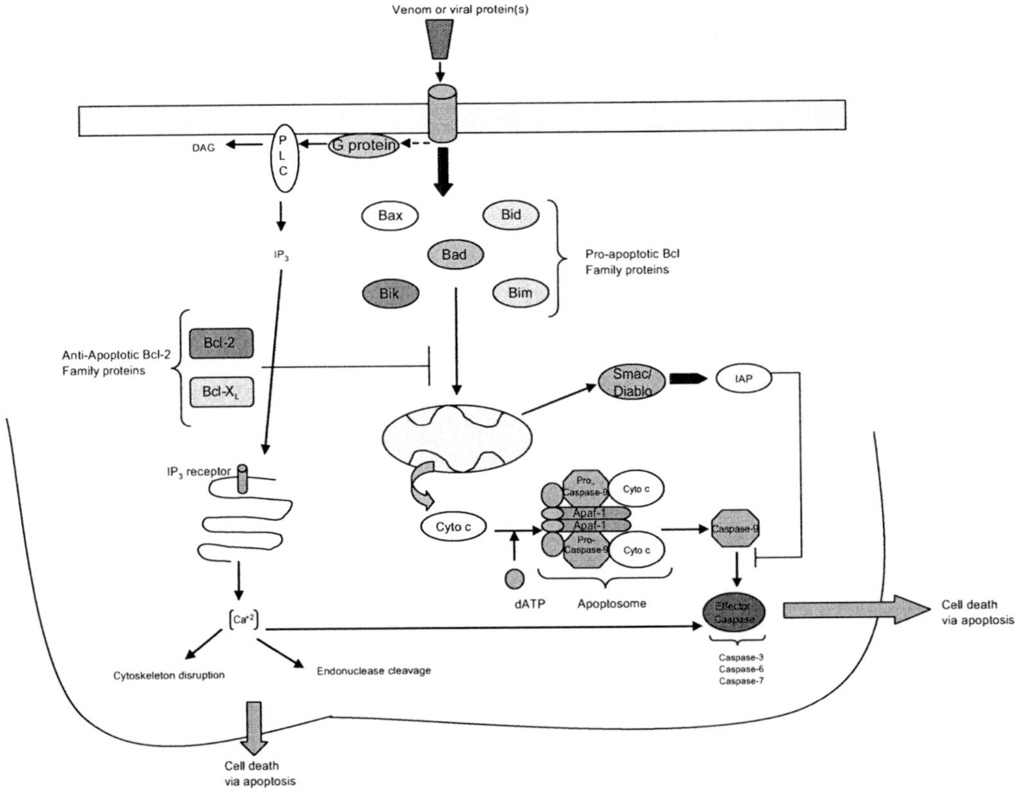

Figure 1. Apoptotic death pathways exploited by parasitic wasp venoms and/or insect viruses. Details of the pathways used to induce cell death in host-parasitoid systems have not been deciphered, but induced programmed cell death does occur with some endoparasitoid venoms and viral proteins from polydnaviruses. Presumably both operate on cell surface receptors to initiate proapoptotic protein formation leading to caspase activation. Proapoptotic proteins interact with mitochondria to promote a decrease in mitochondrial membrane potential and unregulated release of cytochrome c (Cyto C). With the appropriate nucleotide (dATP), Cyto C forms an apoptotic complex (apoptosome) that yields active Apaf-1. Apaf-1 binds downstream to caspases, which in turn become proteolytically active and results in a caspase cascade that culminates in apoptosis. The above scheme is base on the mechanism of apoptosis deduced from *C. elegans*.

Perhaps adding to the already intriguing relationship between apoptosis and viruses are baculoviruses that induce programmed cell death in insect cells. At least 7 different baculovirus species are known to induce apoptosis in lepidopteran cells (Ishikawa et al., 2003), and they appear to do so by a direct interaction with the same caspases that can be inhibited by other insect viruses. In fact, caspase activation can account for some of the cellular effects attributed to pathogenic ascoviruses (Federici et al., 2000), entomopoxvirus infections in *A. suspensa* (Lawrence and Akin, 1990; Shi et al., 1999), and distortion of the cytoskeleton in lamellocytes from *D. melanogaster* by virus-like particles from *L. heterotoma* (Rizki and Rizki, 1984, 1990). Signal transduction pathways leading to caspase activation can even link viral induced oncosis (Perez et al., 1998) and the oncotic death elicited by ectoparasitic wasp venom (Rivers et al., 2004a).

Oncotic Death Signals of Wasp Venom

All ectoparasitic wasp venoms have the ability to kill the host. Despite the lethal action of these venoms, the molecular mechanisms exploited by venom proteins to induce death remain unknown (Moreau and Guillot, 2005). Venom from two species of *Pimpla* (*P. hypochondriaca* and *P. turionellae*) trigger host paralysis and appear to operate by cytotoxic and/or cytolytic mechanisms (Parkinson and Weaver, 1999; Ergin et al., 2006). Parkinson et al (2002) identified a heterodimeric protein named pimplin (approximate molecular weight of 27 kDa) that elicits paralysis when injected into the hemocoel of adult *M. domestica*. A smaller protein (13 kDa) has been isolated from this wasp venom, which appears to be cytotoxic toward *Spodoptera frugiperda* cells (Sf21) (Parkinson and Weaver, 1999). The cytolytic action of *P. hypochondriaca* venom also extends to hemocytes collected from *Lancanobia oleracea* (Richards and Parkinson, 2000). Prior to cell death, Sf21 cells and hemocytes display degranulation and extensive cytoplasmic vacuolization (Parkinson and Weaver, 1999; Richards and Parkinson, 2000), features that are consistent with apoptotic mechanisms. Similar morphological changes have been observed in cultured cells of *Trichoplusia ni* (BTI-TN-5B1-5) when incubated with cytotoxic doses of *P. turionellae* venom (Ergin et al, 2006; Keenan et al., unpublished). Additionally, venom-induced increases in intracellular (cytosol) calcium precede plasma membrane blebbing and cytoplasmic vacuolization. The source of the calcium elevations has yet to be discovered. In vitro assays using BTI-TN-5B1-4 cells incubated with venom from *P. turionellae* suggest that a loss of membrane integrity occurs prior to the increase in $[Ca^{+2}]_i$ (Keenan et al, unpublished). This observation would argue that an influx of extracellular calcium is a key signal leading to cell death. However, though a variety of calcium channel blockers and osmotic protectants could slow the eventual demise of venom-treated cells, the rise in intracellular calcium was not attenuated (Keenan et al., unpublished). The exact mechanisms leading to cell death triggered by these venoms are the subject of further investigations.

Biochemical characterization of venom from *P. turionellae* indicates the presence of histamine, serotonin, noradrenaline, phospholipase A_2 and apamine, components common in social Hymenoptera that produce strongly paralytic venoms (Schmidt, 1982). Similarly, both adult females and larvae of the ectoparasitoid *Eupelmus orientalis* produce paralytic secretions (venom and saliva, respectively) that are composed of phospholipases and hyaluronidase, constituents also common in most species of Hymenoptera that use their venom for defense (Schmidt, 1982). Wasp venoms containing phospholipases are expected to evoke cell death via an oncotic lytic mechanism since this class of enzymes interact with plasma membrane phospholipids to produce trans-membrane pores that promote ion and water movement (Doury et al., 1997), which culminate in cell swelling and eventual lysis (Knowles and Ellar, 1987). In turn, hyaluronidase generally functions to promote distribution or 'spreading' of active venom components (Banks and Shipolini, 1986), and is common in venoms that evoke indiscriminate cell lysis (Piek and Spanjier, 1986).

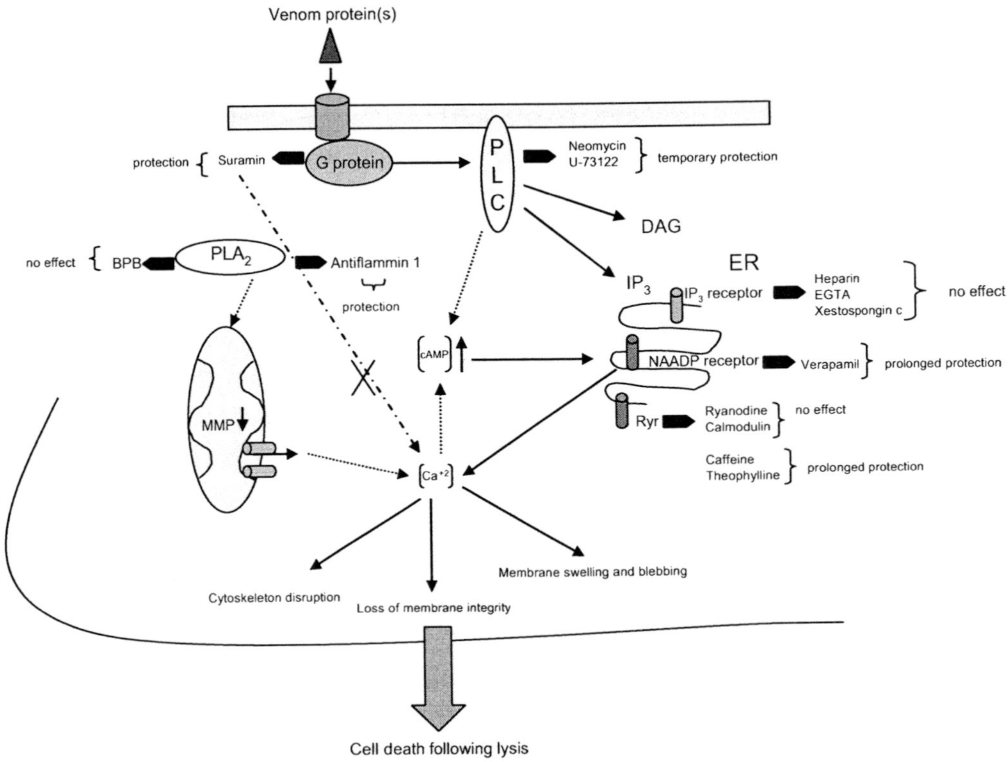

Figure 2. Oncotic death pathways exploited by parasitic wasp venoms. The details of venom-induced cell death are derived from work with the ectoparasitoid N. vitripennis. Venom proteins are thought to bind to cell surface receptors that then activate a G-protein-dependent transduction pathway. Mitochondrial membrane potential decreases, presumably allowing unregulated efflux of Ca^{+2} into the cytosol. Elevated free cytosolic Ca^{+2} is known to activate an array of proteolytic enzymes, which in turn trigger membrane blebbing and cytoskeletal movement/disruption. The result is a loss of membrane integrity, allowing a net movement of ions into the cell, followed by water, and yielding swollen cells that die by an oncotic lytic mechanism.

Venom from the eulophid ectoparasitoid *Melittobia digitata* also is paralytic and displays a broad range of activity toward factitious and natural hosts (Deyrup et al., 2006). The paralytic action of the venom seems to be associated with induction of oncotic cell death in somatic musculature (Deyrup et al., 2006), and not an inhibition of neuromuscular junctions as typical for the well-characterized paralysants found in venom of *H. hebetor* (Quistad et al., 1994; Visser et al., 1983). Initial analysis from in vitro assays indicates that the venom triggers cellular swelling and is cytolytic toward cultured insect cells (BTI-TN-5B1-4) (Rivers and Meinhold, unpublished). Other details of the venom's mode of action are unknown.

The host death pathways exploited by venom from *N. vitripennis* are the only ones that have been investigated thoroughly. However, Rivers et al (1998, 1999) have speculated that other closely related pteromalids likely use similar mechanisms to elicit host death, and indeed, venom for endo- and ectoparasitic members of this family appear to evoke similar morphological changes in vitro (Zhang et al., 2005). When incubated with natural and factitious hosts, crude venom from *N. vitripennis* induces morphological changes in susceptible cells that include retraction of cytoplasmic extensions, rounding, blebbing of plasma membranes, and swelling of both nuclear and plasma membranes (Rivers et al., 1999). Formation of cytoplasmic blebs at the plasma membrane has already been addressed as a feature of apoptosis that suggests involvement of intracellular calcium through disruption of cytoskeletal proteins (Trump and Berezesky, 1995). Phalladicin staining of cultured cells (NIH-SaPe4) from the flesh fly, *S. peregrina*, indicates that the cytoskeleton is altered within a few minutes after incubation with venom (Rivers and Evangelista, unpublished). Bleb formation can occur in media lacking extracellular calcium (Rivers et al., 1999), implying calcium is mobilized from intracellular stores. Indeed, efflux of calcium from mitochondrial stores appears to lead to cytosolic increases in intracellular calcium, which is augmented by an elevation of cAMP (Rivers et al., 2005). Ultimately, susceptible cells die following swelling and lysis. This form of cell death is oncosis (Manjo and Joris, 1995, 1996), and the swelling and lysis typify a colloid osmotic lytic mechanism (Knowles and Ellar, 1987) in which venom irreversibly injures the cells by triggering a net movement of ions followed by water into the cytosol until lysis occurs.

Oncotic death is hypothesized to occur through a pathway that involves G-protein activation of phospholipase C, eventually leading to a loss of mitochondrial membrane potential attributed possibly to a altered expression of uncoupling protein 2 (Rivers et al., 2005). Eventually mitochondria are believed to loose membrane integrity and thus allow unregulated efflux of intracellular calcium stores into the cytosol (Rivers et al., 2005). Regardless of the mechanism, cytosolic cAMP levels rise and are presumed to trigger mobilization of calcium from endoplasmic reticulum via NAADP-sensitive pools (Rivers et al., 2005). An elevation of intracellular cAMP can activate a cytosolic form of ADP-ribosyl cylase, which in turn acts upon $NADP^+$ to produce nicotinic acid adenine dinucleotide phosphate (NAADP) (Gomperts et al., 2002). This nucleotide can stimulate release of intracellular calcium from endoplasmic reticulum stores independent of IP_3 and cyclic ADP-ribose (Gomperts et al., 2002).

The rise in intracellular Ca^{+2} can strongly promote formation of cAMP, which would satisfy the initiation of a NAAD-dependent pathway. What, then, is the source of the initial rise in intracellular calcium? It appears that the mitochondria connect the cellular stress signals evoked by the action of wasp venom. In oncosis, cellular energy depletion following metabolic insults such as ischemia produce dramatic reductions in mitochondrial respiration and ATP synthesis, resulting in a loss of ionic homeostasis and triggering cellular swelling (Majno and Joris, 1995, 1996). Mitochondrial generation of ATP requires the presence of an electrochemical gradient. Consequently, a reduction in mitochondrial membrane potential ($\Delta\Psi_m$) has been suggested to precede the morphological changes associated with apoptosis and oncosis (Petit et al., 1996). Consistent with this view is the observation that venom from *N. vitripennis* triggers a reduction in $\Delta\Psi_m$ and the timing of mitochondrial membrane potential decay correlates with increases in free $[Ca^{+2}]_i$ (Rivers et al., 2005). Importantly, both events precede the elevations of cAMP (Rivers et al., 2002b, 2005). The final blow that evokes cell death is not clear, but a multitude of pathways can be altered by such dramatic shifts in calcium homeostasis (Gomperts et al., 2002).

Conclusion

What is the connection between viral and venom mediated cell death? The morphological changes induced by venom from *N. vitripennis*, including the dependence on calcium involvement, are nearly identical to those described for viral-potentiated apoptotic pathways and quite likely account for cell death induced by venom from the ectoparasitoid *Eupletrus* sp. near *plathypenae* (Nakamatsu and Tanaka, 2003). Rotaviruses, though they target vertebrate cells, elicit parallel effects on host morphology, calcium mobilization, and oncotic death as this wasp venom (Perez et al., 1998). Interestingly, rotaviruses also shutdown host cell protein synthesis over a time course that resembles baculovirus infection (Dobos and Cochran, 1980; Perez et al., 1998). Baculoviruses, already shown to have the ability to block and induce apoptosis, may also produce oncotic signals that turn on cellular death pathways. Infection of *S. frugiperda* (SF9) cells by *A. californica* multiple nucleoployhedrovirus can stimulate increases in lipid hydroperoxides, protein carbonyls, and increases in oxidized glutathione, features that are consistent with oxidative stress (Wang et al., 2001). Glutathione-dependent oxidative stress is associated with increased expression of uncoupling protein 2 (UCP-2) in mitochondria (Mills et al., 2002), which leads to a drop in mitochondrial membrane potential, suppressed ATP synthesis, and unregulated calcium efflux into the cell cytosol. This, of course, is the same pathway predicted to be targeted by venom from *N. vitripennis* (Rivers et al., 2005). Expression of UCP-2 is thought to be associated with a genetic program that modulates cellular responses to oncotic and apoptotic stimuli (Mills et al., 2002). Further investigation is underway to determine if this pathway is key to linking Ca^{+2} mobilization with oncotic death induced by venom from *N. vitripennis* and virally mediated apoptosis.

Acknowledgements

This research was supported in part by grants from the USDA-NRICGP Seed Grants Program (2001-1005) and Holcomb Research Foundation.

References

Asgari, S., 2006. Venom proteins from the polydnavirus-producing endoparasitoids: their role in host-parasite interactions. *Archives of Insect Biochemistry and Physiology* 61, 146-156.

Asgari, S., Hellers, M., Schmidt, O., 1996. Host haemocyte inactivation by an insect parasitoid: transient expression of a polydnavirus gene. *Journal of General Virology* 77, 2653-2662.

Asgari, S., Schmidt, O., 1994. Passive protection of eggs from the parasitoid, *Cotesia rubecula*, in the host, *Pieris rapae*. *Journal of Insect Physiology* 40, 789-795.

Asgari, S., Theopold, U., Wellby, C., Schmidt, O., 1998. A protein with protective properties against the cellular defense reactions in insects. *Proceedings of the National Academy of Sciences USA* 95, 3690-3695.

Askew, R.R., Shaw, M.R., 1986. Parasitoid communities: their size, structure and development. In: Waage, J., Greathead, D. (Eds.), *Insect parasitoids*. Academic Press, London, pp. 225-264.

Barratt, B.I.P., Evans, A.A., Stoltz, D.B., Vinson, S.B., 1999. Virus-like particles in the ovaries of *Microctonus aethiopoides* Loan (Hymenoptera: Braconidae), a parasitoid of adult Weevils (Coleoptera: Curclionidae). *Journal of Invertebrate Pathology* 73, 182-188.

Bartlett, B.R., 1964. Patterns in the host feeding habit of adult parasitic Hymenoptera. *Annals of the Entomological Society of America* 57, 344-350.

Beard, R.L., 1963. Insect toxins and venoms. *Annual Review of Entomology* 8, 1-18.

Beard, R.L., 1964. Pathogenic stinging of housefly pupae by *Nasonia vitripennis* (Walker). *Journal of Insect Pathology* 6, 1-7.

Beckage, N.E., 1998. Modulation of immune responses to parasitoids by polydnaviruses. *Parasitology* 116, S57-S64.

Beckage, N.E., Gelman, D.B., 2001. Parasitism of *Manduca sexta* by *Cotesia congregata*: A multitude of disruptive endocrine effects. In: Edwards, J., Weaver, R. (Eds.), *Endocrine interactions of insect parasites and pathogens*. BIOS Scientific Publishers, Oxford, pp. 59-81.

Beckage, N.E., Gelman, D.B., 2004. Wasp parasitoid disruption of host development: Implications for new biologically based strategies for insect control. *Annual Review of Entomology* 49, 299-330.

Beckage, N.E., Reed, D.A., Gelman, D.B., Kelly, T.J., 1998. Effects of parasitization by *Cotesia congregata* on the brain-prothoracic gland axis of its host, *Manduca sexta*. *Journal of Insect Physiology* 44, 323-332.

Beidler, D.R., Tewari, M., Friesen, P.D., Poirier, G., Dixit, V.M., 1995. The baculovirus P35 protein inhibits Fas- and tumor necrosis factor-induced apoptosis. *Journal of Biological Chemistry* 270, 16526-16528.

Beliveau, C., Laforge, M., Cusson, M., Bellemare, G., 2000. Expression of a *Tranosema rostrale* polydnavirus gene in the spruce budworm, *Choristoneura fumiferana*. *Journal of General Virology* 81, 1871-1880.

Beliveau, C., Levasseur, A., Stoltz, D.B., Cusson, M., 2003. Three related TrIV genes: comparative sequence analysis and expression of host larvae and Cf-124T cells. *Journal of Insect Physiology* 49, 501-511.

Berger, E.M., Woodward, M.P., 1983. Small heat shock proteins in *Drosophila* confer thermal tolerance. *Experimental Cell* Research 147, 437-442.

Bertin, J., Mendrysa, S.M., LaCount, D.J., Guar, S., Krebs, J.F., Armstrong, R.C., Tomaselli, K.J., Friesen, P.D., 1996. Apoptotic suppression by baculovirus P35 involves cleavage by and inhibition of a virus-induced CED-3/ICE-like protease. *Journal of Virology* 70, 6251-6259.

Bigot, Y., Rabouille, A., Doury, G., Sizaret, P.-Y., Delbost, F., Hamelim, M.-H., Periquet., G., 1997. Biological and molecular features of the relationships between *Diadromus pulchellus* ascovirus, a parasitoid hymenopteran wasp (*Diadromus pulchullus*) and its lepidopteran host, *Acrolepiosis assectella*. *Journal of General Virology* 78, 1149-1163.

Birkenbeil, H., 1998. Intracellular calcium in prothoracic glands of *Manduca sexta*. *Journal of Insect Physiology* 44, 279-286.

Bocchino, F.J., Sullivan, D.J., 1981. Effects of venoms from two aphid hyperparasitoids, *Asaphes lucens* and *Dendrocerus carpenteri* (Hymenoptera: Pteromalidae and Megaspilidae),on larvae of *Aphidus smithi* (Hymenoptera: Aphididae). *Canadian Entomologist* 113, 887-889.

Carton, Y., Nappi, A.J., 2001. Immunogenetic aspects of the cellular immune response of *Drosophila* against parasitoids. *Immunogenetics* 52, 157-164.

Coudron, T.A., Brandt, S.L., 1996. Characteristics of a developmental arrestant in the venom of the ectoparasitoid wasp *Euplectrus comstockii*. *Toxicon* 34, 1431-1441.

Coudron, T.A., Kelly, T.J., and Puttler, B., 1990. Developmental responses of *Trichoplusia ni* (Lepidoptera: Noctuidae) to parasitism by the ectoparasite *Euplectrus plathypenae* (Hymenoptera: Eulophidae). *Archives of Insect Biochemistry and Physiology* 13, 83-94.

Crook, N.E., Clem, R.J., Miller, L.K., 1993. An apoptosis-inhibiting baculovirus gene with a zinc finger-like motif. *Journal of Virology* 67, 2168-2174.

Cui, L., Soldelvila, A., Webb, B.A., 1997. Expression and hemocyte-targeting of a *Campolestis sonorensis* polydnvirus cysteine-rich gene in *Heliothis virescens* larvae. *Archives of Insect Biochemistry and Physiology* 36, 251-271.

Davies, D.H., Strand, M.R., Vinson, S.B., 1987. Changes in differential haemocyte count and *in vitro* behaviour of plasmatocytes from host *Heliothis virescens* caused by *Campoletis sonorensis* polydnavirus. *Journal of Insect Physiology* 33, 143-153.

De Buron, I., Beckage, N.E., 1992. Characterization of a polydnavirus and virus-like filamentous particle (VLFP) in the braconid wasp *Cotesia congregata* (Hymenoptera: Braconidae). *Journal of Invertebrate Pathology* 59, 315-327.

Denlinger, D.L., Rinehart, J.P., Yocum, G.D., 2001. Stress proteins: a role in insect diapause? In: Denlinger, D.L., Giebutowitz, J.M., Saunders, D.S. (Eds.), Insect timing: circadian rhythmicity to seasonality. *Elsevier, Amsterdam,* pp. 155-171.

Deyrup, L.D., Rivers, D.B., Matthews, R.W. 2006. Venom from the ectoparasitic wasp *Melittobia digitata* Dahms (Hymenoptera: Eulophidae) induces paralysis and developmental delay in natural and factitious hosts. *Annals of the Entomological Society of America,* In press.

Dib-Hajj, S.D., Webb, B.A., Summers, M.D., 1993. Structure and evolutionary implication of a "cysteine-rich" Campoletis sonorensis polydnavirus gene family. *Proceeding of the National Academy of Sciences* USA 90, 3765-3769.

Doury, G., Rojas-Rousse, D., Periquet, G., 1995. Ability of *Eupelmus orientalis* ectoparasitoid larvae to develop on an unparalysed host in the absence of female stinging behaviour. *Journal of Insect Physiology* 41, 287-296.

Doury, G., Bigot, Y., Periquet, G., 1997. Physiological and biochemical analysis of factors in the female venom gland and larval salivary secretions of the ectoparasitoid wasp *Eupelmus orientalis*. *Journal of Insect Physiology* 43, 69-81.

Drenth, D., 1974. Susceptibility of different species of insects to an extract of the venom gland of the wasp *Microbracon hebetor* (Say). *Toxicon* 12, 189-192.

Du, X., Thiem, S.M., 1997. Responses of insect cells to baculovirus infection: protein synthesis shutdown and apoptosis. *Journal of Virology* 71, 7866-7872.

Dubos, P., Cochran, M.A., 1980. Protein synthesis in cells infected by *Autographa californica* nuclear polyhedrosis gene (AcNPV): the effect of cytosine arabinoside. *Virology* 103, 446-464.

Dupas, S., Turnbull., M.W., Webb, B.A., 2003. Diversifying selection in a parasitoid's symbiotic virus among genes involved in inhibiting host immunity. *Immunogenetics* 55, 351-361.

Ergin, E., Uçkan, F., Rivers, D.B., Sak, O., 2006. In vivo and in vitro activity of venom from the endoparasitoid *Pimpla turionellae* (L.)(Hymenoptera: Ichneumonidae). *Archives of Insect Biochemistry and Physiology* 61, 87-97.

Feder, M.E., 1996. Ecological and evolutionary physiology of stress proteins and the stress response: the *Drosophila melanogaster* model. In: Johnston, I.A., Bennett, A.F. (Eds.), *Animals and temperature: phenotypic and evolutionary adaptations*. Cambridge University Press, Cambridge, pp. 79-102.

Federici, B.A., Bigot, Y., 2003. Origin and evolution of polydnaviruses by symbiogenesis on insect DNA viruses in endoparasitic wasps. *Journal of Insect Physiology* 49, 419-432.

Federici, B.A., Bigot, Y., Hamm, J.J., Granafos, R.R., Vlak, J.M., Miller, L.K., 2000. Family Ascoviridae. In: van Regenmortel, M.H.V., Fauquet, C.M., Bishop, D.H.L., Carstens, E.B.,Estes, M.K., Lemon, S.M., Manilof, J., Mayo, M.A., McGeoch, D.J., Pringle, C.R., Wickner, R.B. (Eds.), *Virus taxonomy*. Seventh report of the International Committee on Taxonomy of Viruses. Academic Press, San Diego, pp. 261-265.

Fleming, J.A.G., 1992. Polydnavirus-mutualists and pathogens. *Annual Review of Entomology* 37, 401-425.

Friesen, P.D., Miller, L.K., 1987. Divergent transcription of early 35- and 94-kilodalton protein genes encoded by the HindIII K genome fragment of the baculovirus *Autographa californica* nuclear polyhedrosis virus. *Journal of Virology* 61, 2264-2272.

Gauld, I.D., Hanson, P.E., 1995. Carnivory in the larval Hymenoptera. In: Hanson, P., Gauld, I.D. (Eds.), The Hymenoptera of Costa Rica, Oxford University Press, Oxford, pp. 40-45.

Godfray, H.C.J., 1994. *Parasitoids: Behavioral and evolutionary ecology*. Princeton University Press, Princeton.

Godfray, H.C.J., 2000. Host resistance, parasitoid virulence, and population dynamics. In: Hochberg, M., Ives, A.R. (Eds.), *Parasitoid population biology*. Princeton University Press, Princeton, pp. 121-138.

Gomperts, B.D., Kramer, I.M., Tatham, P.E.R., 2002. *Signal transduction*. Academic Press, San Diego.

Gotz, P., 1986. Mechanisms of encapsulation in dipteran hosts. In: Lackie, A.M. (Ed.), *Immune mechanisms in invertebrate vectors*. Clarendon, Oxford, pp. 1-19.

Guo, Y.X., Wei, T., Dallmann, K., Kwang, J., 2003. Induction of caspase-dependent apoptosis by betanodaviruses GGNNV and demonstration of protein alpha as an apoptosis inducer. *Virology* 30, 74-82.

Guzo, D., Stoltz, D.B., 1985. Obligatory multiparasitism in the tussock moth, *Orgyia leucostigma*. *Parasitology* 90, 1-10.

Hale, A.J., Smith, C.A., Sutherland, L.C., Stoneman, V.E.A., Longthorne, V.L., Culhane, A.C., Williams, G.T. 1996. Apoptosis: molecular regulation of cell death. *European Journal of Biochemistry 236*, 1-26.

Hay, B.A., Wolff, T., Rubin, G.M., 1994. Expression of baculovirus p35 prevents cell death in Drosophila. *Development* 120, 2121-2129.

Hayakawa, Y., Yazaki, K., Yamanaka, A., Tanaka, T., 1994. Expression of polydnavirus genes from the parasitoid *Cotesia kariyai* in two noctuid hosts. *Insect Molecular Biology* 3, 97-103.

Hisaeda, H., Himeno, K., 1997. The role of host-derived heat shock protein in immunity against *Toxoplasma gondii* infection. *Parasitology Today* 13, 465-468.

Ishikawa, H., Ikeda, M., Yanagimoto, K., Alves, C.A., Katou, Y., Lavina-Caoili, B.A., Kobayashi, M., 2003. Induction of apoptosis in an insect cell line, IPLB-Ld652Y, infected with nucleopolyhedroviruses. Journal of General Virology 84, 705-714. Jervis, M.A., Copeland, J.W. 1996. The life cycle. In: Jervis, M.A., Kidd, N.A.C. (Eds.), *Insect natural enemies: practical approaches to their study and evaluation.* Chapman and Hall, London, pp. 63-161.

Johner, A., Lanzrein, B., 2002. Characterization of two genes of the polydnavirus of *Chelonus inanitus* and their stage-specific expression in the host *Spodoptera littoralis. Journal of General Virology 83*, 1075-1085.

Johner, A., Stettler, P., Gruber, A., Lanzrein, B., 1999. Presence of polydnavirus transcripts in an egg-larval parasitoid and its lepidopteran host. *Journal of General Virology* 80, 1847-1857.

Just, F.T., Essbauer, S.S., 2001. Characterization of an iridescent virus isolated from *Gryllus bimaculatus* (Orthoptera: Gryllidae). *Journal of Invertebrate Pathology* 77, 51- 61.

Kinuthia, W., Li, D., Schmidt, O., Theopold, U., 1999. Is the surface of endoparasitic wasp egg and larvae covered by a limited coagulation reaction? *Journal of Insect Physiology* 45, 501-506.

Knowles, B.H., Ellar, D.J., 1987. Colloid-osmotic lysis is a general feature of the mechanism of action of *Bacillus thuringiensis* delta-endotoxins with different insect specificity. *Biochimica Biophysica Acta* 924, 509-518.

Krakauer, D.C.,Payne, R.J., 1997. The evolution of virus-induced apoptosis. *Proceeding of the Royal Society of London B: Biological Sciences* 264, 1757-1762.

Krebs, R.A., Feder, M.E., 1997. Deleterious consequences of Hsp70 overexpression in *Drosophila melanogaster* larvae. *Cell Stress Chap.* 2, 60-71.

Labrosse, C., Carton, Y., Dubuffet, A., Drezen, J.M., Poirie, M., 2003. Active suppression of *D. melanogaster* immune response by long gland products of the parasitic wasp *Leptopilina boulardi. Journal of Insect Physiology* 49, 513-522.

Labrosse, C., Eslin, P., Doury, G., Drezen, J.M., Poirie, M., 2005. Haemocyte changes in *D. melanogaster* in response to long gland components of the parasitoid wasp *Leptopilina boulardi*: a Rho-GAP protein as an important factor. *Journal of Insect Physiology* 49, 513-522.

Lackie, A.M., 1988. Haemocyte behavior. In: Beeridge, M.J., Treherne, J.E., Wigglesworth, V.B. (Eds.), *Advances in Insect Physiology,* Academic Press, New York, pp. 85-178.

Lawrence, P.O., 1988. Ecdysteroid titers and integument changes in superparasitized puparia of *Anastrepha suspensa* (Diptera: Tephritidae). *Journal of Insect Physiology* 34, 603-608.

Lawrence, P.O., 2002. Purification and partial characterization of an entomopoxvirus (DIEPV) from a parasitic wasp of tephritid fruit flies. *Journal of Insect Science,* pp. 1-12. Insectsuerle.org/2.10.

Lawrence, P.O., Akin, D., 1990. Virus-like particles from the poison gland of the parasitic wasp *Biosteres lonicaudatus* (Hymenoptera: Braconidae). *Canadian Journal of Zoology* 68, 539-546.

Le, N.T., Asgari, S., Amaya, K., Tan, F.F., Beckage, N.E., 2003. Persistence and expression of *Cotesia congregata* polydnavirus in host larvae of the tobacco hornworm, *Manduca sexta. Journal of Insect Physiology* 49, 533-543.

Leluk, J., Jones, D., 1989. *Chelonus* sp. near *curvimaculatus* venom proteins: analysis of their potential role and processing during development of host *Trichoplusia ni. Archives of Insect Biochemistry and Physiology* 10, 1-11.

Liu, Q., Qi., Y., Chejanovsky, N., 2003. Identificaiton and classification of the *Spodoptera littoralis* nucleopolyhedrovirus inhibitor of apoptosis gene. *Virus Genes* 26, 143-149.

Luckhart, S., Webb, B.A., 1996. Interaction of a wasp ovarian protein and polydnavirus in host immune suppression. *Developmental and Comparative Immunology* 20, 1-21.

Luque, L.E., Grape, K.P., Junker, M., 2002. A highly conserved arginine is critical for the functional folding of inhibitor of apoptosis (IAP) BIR domains. Biochemistry 19, 13663-13667. Majno, G., Joris, I., 1995. Apoptosis, oncosis, and necrosis. *An overview of cell death. American Journal of Pathology* 146, 3-15.

Majno, G., Joris, I., 1996. Cells, *Tissues, and Disease.* Blackwell Science, Oxford.

Marris, G.C., Edwards, J.P., 1995. The biology of the ectoparasitoid wasp *Eulophus pennicornis* (Hymenoptera: Eulophidae) on host larvae of the tomato moth, *Lacanobia oleracea* (Lepidoptera: Noctuidae). *Bulletin of Entomological Research* 85, 507-513.

Marris, G.C., Weaver, R.J., Edwards, J.P., 2001. Endocrine interactions of ectoparasitoid Wasps with their hosts: an overview. In: Edwards, J.P., Weaver, R.J. (Eds.), *Endocrine interactions of insect parasites and pathogens.* Bios, Oxford, pp. 133-151.

Miller, L.K., 1997. Baculovirus interaction with host apoptotic pathways. *Journal of Cell Physiology* 173, 178-182.

Mills, E.M., Xu, D., Fergusson, M.M., Combs, C.A., Xu, Y., Finkel, T., 2002. Regulation of cellular oncosis by uncoupling protein 2. *Journal of Biological Chemistry* 277, 27385-27392.

Morales-Ramos, J.A., Rojas, M.G., King, E.G., 1995. Venom of *Catolaccus grandis* (Hymenoptera: Pteromalidae) and its role in parasitoid development and host regulation. *Annals of the Entomological Society of America* 88, 800-808.

Moreau, S.J.M., 2003. Relationships between parasitoid Hymenoptera and the immune system of their hosts: the "active" and "passive" mechanisms redefined. *Annals of the Entomological Society of America* 88, 800-808.

Moreau, S.J.M., Guillot, S., 2005. Advances and prospects on biosynthesis, structures and functions of venom proteins from parasitic wasps. *Insect Biochemistry and Molecular Biology* 35, 1209-1223.

Nakamatsu, Y., Tanaka, T., 2003. Venom of ectoparasitoid, *Euplectrus*. sp. near *plathypenae* (Hymenoptera: Eulophidae) regulates the physiological state of *Pseudaletia separata* (Lepidoptera: Noctuidae) host as a food resource. *Journal of Insect Physiology* 49,149-159.

Nappi, A.J., Carton, Y., 1996. Cellular immune responses and their genetic aspects in *Drosophila*. In: Brehelin, M. (Ed.), *Immunity in invertebrates,* Springer-Verlag, Berlin, pp. 171-187.

Nicotera, P., Hartzell, P., Davis, G.,Orrenius, S., 1986. The formation of plasma membrane blebs in hepatocytes exposed to agents that increase cytosolic Ca^{2+} is mediated by activation of a non-lysosomal proteolytic system. *FEBS Letters* 209,139.

Noguchi, H., Hayakawa, Y., Downer, R.G.H., 1995. Elevation of dopamine levels in parasitized insect larvae. *Insect Biochemistry and Molecular Biology* 25, 197-201.

Ooi, B.G., Miller, L.K., 1988. Regulation of host RNA levels during baculovirus infection. *Virology* 166, 515-523.

O'Reilly, D.R., Hails, R.S., Kelly, T.J., 1998. The impact of host developmental status on baculovirus replication. *Journal of Invertebrate Pathology* 72, 269-275.

Parkinson, N.M., Conyers, C.M., Keen, J.N., MacNicoll, A.D., Smith, I., Weaver, R.J., 2003. cDNAs encoding large venom proteins from the parasitoid wasp Pimpla hypochondriaca identified by random sequence analysis. *Comparative Biochemistry and Physiology C* 134, 513-520.

Parkinson, N.M., Weaver, R.J., 1999. Noxious components of venom from the pupa- Specific parasitoid *Pimpla hypochondriaca. Journal of Invertebrate Pathology* 73, 74- 83.

Parkinson, N., Smith, I., Audsley, N., Edwards, J.P., 2002. Purification of pimplin, a paralytic heterodimeric polypeptide from venom of the parasitoid wasp *Pimpla hypochondriaca*, and cloning of the cDNA encoding one of the subunits. *Insect Biochemistry and Molecular Biology* 32, 1769-1773.

Perez, J.F., Chemello, M.E., Liprandi, F., Ruiz, M.C., Michelangeli, F., 1998. Oncosis in MA104 cells is induced by rotovirus infection through an increases in intracellular Ca^{+2} concentration. *Virology* 252, 17-27.

Petit, P.X., Susin, S.-A., Zamzami, N., Mignotte, B., Kroemer, G., 1996. Mitochondria and programmed cell death: back to the future. *FEBS Letters* 396, 7-13.

Phelps, P.C., Smith, M.W., Trump, B.F., 1989. Cytosolic ionized calcium and bleb formation following acute cell injury of cultured rabbit renal tubular cells. *Laboratory Investigations* 60, 630-642.

Piek, T., Spanjier, W., 1986. Chemistry and pharmacology of solitary wasp venoms. In: Piek, T. (Ed.), *Venoms of the Hymenoptera.* Academic Press, NY, pp. 161-308.

Quicke, D.L.J., 1997. *Parasitic wasps,* Chapman and Hall, London.

Quistad, G.B., Ngyuyen, Q., Bernasconi, P., Leisy, D.J., 1994. Purification and characterization of insecticidal toxins from venom glands of the parasitic wasp, *Bracon hebetor. Insect Biochemistry and Molecular Biology* 24, 955-961.

Richards, E.H., Edwards, J.P., 1999. Parasitization of Lacanobia oleracea (Lepidoptera: Noctuidae) by the ectoparasitic wasp, Eulophus pennicornis: Effects of parasitization, venom and starvation on host haemocytes. *Journal of Insect Physiology* 45, 1073- 1083.

Richards, E.H., Edwards, J.P., 2000a. Parasitization of *Lacanobia oleracea* (Lepidoptera) by the ectoparasitic wasp, *Eulophus pennicornis*, suppresses haemocyte- mediated recognition of non-self and phagocytosis. *Journal of Insect Physiology* 46, 1- 11.

Richards, E.H., Edwards, J.P., 2000b. Parasitism of *Lacanobia oleracea* (Lepidoptera) by the ectoparasitoid, *Eulophus pennicornis*, is associated with a reduction in host haemolymph phenoloxidase activity. *Comparative Biochemistry and Physiology Part B* 127, 289-298.

Richards, E.H., Parkinson, N.M., 2000. Venom from the endoparasitic wasp *Pimpla hypochondriaca* adversely affects the morphology, viability, and immune function of hemocytes from larvae of the tomato moth, *Lacanobia oleracea*. *Journal of Invertebrate Pathology* 76, 33-42.

Rinehart, J.P., Rivers, D.B., Denlinger, D.L., 2002. Upregulation of transcripts encoding select heat shock proteins in the flesh fly *Sarcophaga crassipalpis* in response to venom from the ectoparasitoid wasp *Nasonia vitripennis*. *Journal of Invertebrate Pathology* 79, 62-63.

Rivers, D.B., 2003. Evasion of host defenses by ectoparasitoids. *Trends in Entomology*, 3, 131-141.

Rivers, D.B., 2004. Venoms of ectoparasitic wasps. In: Capinera, J. (Ed.), *Encyclopedia of Entomology*. Kulwer Publishers, NY.

Rivers, D.B., Crawley, T., Bauser, H., 2005. Localization of intracellular calcium release in cells injured by venom from the ectoparasitoid *Nasonia vitripennis* (Walker) (Hymenoptera: Pteromalidae) and dependence of calcium mobilization on G-protein activation. *Journal of Insect Physiology* 51, 149-160.

Rivers, D.B., Denlinger, D.L., 1994a. Developmental fate of the flesh fly, *Sarcophaga bullata* (Diptera: Sarcophagidae), envenomated by the pupal ectoparasitoid, *Nasonia vitripennis* (Hymenoptera: Pteromalidae). *Journal of Insect Physiology* 40, 121-127.

Rivers, D.B., Denlinger, D.L., 1994b. Redirection of metabolism in the flesh fly, *Sarcophaga bullata*, following envenomation by the ectoparasitoid *Nasonia vitripennis* and correlation of metabolic effects with the diapause status of the host. *Journal of Insect Physiology* 40, 207-215.

Rivers, D.B., Denlinger, D.L., 1995. Venom-induced alterations in fly lipid metabolism and its impact on larval development of the ectoparasitoid *Nasonia vitripennis* (Walker) (Hymenoptera: Pteromalidae). *Journal of Invertebrate Pathology*. 66, 104-110.

Rivers, D.B., Hink; W.F., Denlinger, D.L., 1993. Toxicity of the venom from *Nasonia vitripennis* (Hymenoptera: Pteromalidae) toward fly hosts, nontarget insects, different developmental stages, and cultured insect cells. *Toxicon* 31, 755-765.

Rivers, D.B., Genco, M., Sanchez, R.A., 1999. In vitro analysis of venom from the wasp *Nasonia vitripennis*: susceptibility of different cell lines and venom-induced changes in plasma membrane permeability. In *Vitro Cellular and Developmental Biology-Animal* 35, 102-110.

Rivers, D.B., Pagnotta, M.A. Huntington, E.R., 1998. Reproductive strategies of 3 species of ectoparasitic wasps are modulated by the response of the fly host, *Sarcophaga bullata* (Diptera: Sarcophagidae) to parasitism. *Annals of the Entomological Society of America* 91, 458-465.

Rivers, D.B., Rocco, M.A., Frayha, A.R., 2002a. Venom from the ectoparasitic wasp *Nasonia vitripennis* increases Na^+ influx and activates phospholipase C and phospholipase A2 dependent signal transduction pathways in cultured insect cells. *Toxicon* 40, 9-21.

Rivers, D.B., Ruggiero, L., Hayes, M., 2002b. The ectoparasitic wasp *Nasonia vitripennis* (Walker) (Hymenoptera: Pteromalidae) differentially affects cells mediating the immune response of its flesh fly host, Sarcophaga bullata Parker (Diptera: Sarcophagidae), *Journal of Insect Physiology* 48, 1053-1064.

Rivers, D.B., Uçkan, F., Ergin, E. 2006. Characterization and biochemical analyses of venom from the ectoparasitic wasp *Nasonia vitripennis* (Walker)(Hymenoptera: Pteromalidae). *Archives of Insect Biochemistry and Physiology* 61, 24-41.

Rivers, D.B., Zdarek, J., Denlinger, D.L., 2004. Disruption of pupariation and eclosion behavior in the flesh fly, *Sarcophaga bullata* Parker (Diptera: Sarcophagidae), by venom from the ectoparasitic wasp *Nasonia vitripennis* (Walker) (Hymenoptera: Pteromalidae). *Archives of Insect Biochemistry and Physiology*, 57, 78-91.

Rizki, R.M., Rizki, T.M., 1984. Selective destruction of a host blood cell type by a parasitoid wasp. *Proceedings of the National Academy of Science* USA 81, 6154- 6158.

Rizki, R.M., Rizki, T.M., 1990. Parasitoid virus-like particles destroy *Drosophila* cellular immunity. *Proceeding of the National Academy of Sciences* USA 87, 8388- 8392.

Rizki, R.M., Rizki, T.M., 1991. Effects of lamellolysin from a parasitoid on *Drosophila* blood cells in vitro. *Journal of Experimental Zoology* 257, 236-244.

Schmidt, J.O. 1982. Biochemistry of insect venoms. *Annual Review of Entomology* 27, 339-368.

Schmidt, O., Theopold, U., Strand, M., 2001. Innate immunity and its evasion and suppression by hymenopteran endoparasitoids. *BioEssays* 23, 344-351.

Shaw, M.R., 1981. Delayed inhibition of host development by the non-paralyzing venoms of parasitic wasps. *Journal of Invertebrate Pathology* 37, 215-221.

Shi, X., Gomez, S.P., Lawrence, P.O., 1999. A 24-kDa parasitism-specific protein from the Caribbean fruit fly *Anastrepha suspensa*: cDNA and deduced amino acid sequence. *Insect Biochemistry and Molecular Biology* 29, 749-755.

Stasiak, K., Renault, S., Demattei, M.V., Bigot, Y., Federici, B.A., 2003. Evidence for the evolution of ascoviruses from iridoviruses. *Journal of General Virology* 84, 2999- 3009.

Stoltz, D.B., 1993. Polydnavirus life cycle. In: Beckage, N.E., Thompson, S.N., Federici, B.A. (Eds.), *Parasites and pathogens of insects,* vol. 1 parasites. Academic Press, NY, pp.167-187.

Stoltz, D.B. and Vinson, S.B., 1979. Viruses and parasitism in insects. *Advances in virus research 24*. Academic Press, Inc., pp. 125-171.

Stoltz, D.B., Whitfield, J., 1992. Viruses and virus-like entities in the parasitic Hymenoptera. *Journal of Hymenoptera Research* 1, 125-139.

Strand, M.R., 2000. Developmental traits and life-history evolution in parasitoids. In: Hochberg, M., Ives, A.R. (Eds.), *Parasitoid population biology.* Princeton University Press, Princeton, pp. 139-162.

Strand, M.R., Johnson, J.A., Culin, J.D., 1988. Developmental interactions between the parasitoid *Microplitis demolitor* (Hymenoptera: Braconidae) and its host *Heliothis virescens* (Lepidoptera: Nocutidae). *Annals of the Entomological Society of America* 81, 822-830.

Strand, M.R., Mckenzie, D.I., Grassl, V., Dover, B.A., Aiken, J.M., 1992. Persistence and expression of *Microplitis demolitor* polydnavirus in *Pseudoplusia includens*. *Journal of General Virology* 73, 1627-1635.

Strand, M.R., Noda, T., 1991. Alterations in the hemocytes of *Pseudoplusia includens* after parasitism by *Microplitis demolitor*. *Journal of Insect Physiology* 37, 839-850.

Strand, M.R., Pech, L.L., 1995. Immunological basis for compatibility in parasitoid-host relationships. *Annual Review of Entomology* 40, 31-56.

Sugimoto, A., Friesen, P.D., Rothman, J.H., 1994. Baculovius p35 prevents developmentally programmed cell death and rescues a ced-9 mutant in the nematode *Caenorhabditis elegans*. *EMBO Journal* 13, 2023-2028.

Tammariello, S.P., Landreth, G.E., Estus, S., 2001.. The role of Jun-Kinases in Apoptosis. In: *Programmed Cell Death*. Volume I. Elsevier. Amsterdam, pp.197-214.

Tanaka, K., Tsuzuki, S., Matsumoto, H., Hayakawa, Y., 2003. Expression of *Cotesia kariyai* polydnavirus genes in lepidopteran hemocytes and Sf9 cells. Journal of Insect Physiology 49, 433-440.

Tanaka, T., Vinson, S.B., 1991. Depression of thoracic gland activity of *Heliothis virescens* by venom and calyx fluids from the parasitoid, *Cardiochiles nigriceps*. *Journal of Insect Physiology* 37, 139-144.

Thompson, S.N., 1993. Redirection of host metabolism and effects on parasite nutrition. In: Beckage, N.E., Thompson, S.N., Federici, B.A. (Eds.), *Parasites and pathogens of insects, vol 1 parasites*. Academic Press, NY, pp.125-144.

Troyer, D.A.; Kreisberg, J.I.; Venkatachalam, M.A., 1986. Lipid alterations in LLC-PK1 cells exposed to mercuric chloride. *Kidney International*. 29, 530-538.

Trump, B.F., Berezesky, I.K., 1995. Calcium-mediated cell injury and cell death. *FASEB Journal 9*, 219-228.

Uçkan, F., Sinan, S., Savasci, S., Ergin, E., 2004. Determination of venom components from the endoparasitoid wasp *Pimpla turionellae* L. (Hymenoptera: Ichneumonidae). *Annals of the Entomological Society of America* 97, 775-780.

Vinson, S.B., Iwantsch, G.F., 1980. Host regulation by insect parasitoids. *Quarterly Review of Biology* 55, 143-165.

Visser, B.J., Labruyère, W.T., Spanjier, W., Piek, T., 1983. Characterization of two paralyzing protein toxins (A-MTX and B-MTX), isolated from a homogenate of the wasp *Microbracon hebetor* (Say). *Comparative Biochemistry and Physiology* B 75, 523-530.

Walther, C., Rathmayer, W., 1974. The effect of *Habrobracon* venom on excitatory neuromuscular transmission in insects. *Journal of Comparative Physiology* 89, 23-38.

Walther, C., Rathmayer, W., 1983. Block of synaptic vesicle exocytosis without block of Ca^{+2}-influx. An ultrastructural analysis of the paralysing action of *Habrobracon* venom on locust motor nerve terminals. *Neuroscience* 9, 213-224.

Wang, Y., Oberley, L.W., Murhammer, D.W., 2001. Evidence of oxidative stress following the viral infection of two lepidopteran insect cell lines. *Free Radical Biology and Medicine* 31, 1448-1455.

Wang, X.-H., Khatri, I.A., Forstner, J.F., 2002. C-terminal domain of rodent intestinal mucin Muc3 is proteolytically cleaved in the endoplasmic reticulum to generate extracellular and membrane components. *Biochemistry Journal* 366, 623-631.

Webb, B.A., 1998. Polydnavirus biology, genome structure and evolution. In: Miller, L.K., Ball., L.A. (Eds.), *The insect viruses.* Plenum Press, NY, pp. 105-139.

Whitfield, J.B., 1998. Phylogeny and evolution of host-parasitoid interactions in Hymenoptera. *Annual Review of Entomology* 43, 129-151.

Whitfield, J.B., 2002. Estimating the age of polydnavirus/braconid wasp symbiosis. *Proceeding of the National Academy of Sciences* USA 99, 7508-7513.

Windass, J.D., Duncan, R.E., Christian, P.D., Baule, V.J., 1996. International patent-WO 96/16171.

Wyder, S., Blank, F., Lanzrein, B., 2003. Fate of polydnavirus DNA of the egg-larval parasitoid *Chelonus inanitus* in the host *Spodoptera littoralis*. *Journal of Insect Physiology* 49, 491-500.

Yoshida, S., Yamaguchi, Y., Itami, S., Yoshikawa, K., Tabata, Y., Matsumoto, K., Nakamura, T., 2003. Neutralization of hepatocyte growth factor leads to retarded cutaneous wound healing associated with decreased neovascularization and granulation tissue formation. *Journal of Investigative Dermatology* 120, 335-343.

Zhang, Z., Ye, G., Cai, J., Hu, C., 2005. Comparative venom toxicity between *Pteromalus puparum* and *Nasonia vitripennis* (Hymenoptera: Pteromalidae) toward the hemocytes of their natural hosts, non-target insects and cultured insect cells. *Toxicon* 46, 337-349.

In: New Developments in Cell Apoptosis Research
Editor: Alan J. Corvin, pp. 97-120
ISBN: 1-60021-453-3

Chapter IV

Protein Kinase C δ: The Determinant of Cell Fate in Response to DNA Damage

Kiyotsugu Yoshida*
Medical Research Institute, Tokyo Medical and Dental University, Tokyo, Japan

Abstract

The cellular response to genotoxic stress that damages DNA includes cell cycle arrest, activation of DNA repair, and in the event of irreparable damage, induction of apoptosis. However, the signals that determine cell fate, that is, survival or apoptosis, are largely unknown. The δ isoform of protein kinase C (PKCδ) has been implicated in many important cellular processes, including regulation of apoptotic cell death. The available information supports a model in which certain sensors of DNA lesions activate PKCδ. This activation is triggered in part by tyrosine phosphorylation of PKCδ by c-Abl tyrosine kinase. PKCδ is further proteolytically activated by caspase-3. The cleaved catalytic fragment of PKCδ translocates to the nucleus and induces apoptosis. Importantly, accumulating data have revealed the nuclear targets for PKCδ in the induction of apoptosis. A pro-apoptotic function of activated PKCδ is mediated by at least several downstream effectors known to be associated with the elicitation of apoptosis. Recent findings also demonstrated that PKCδ is involved in cell cycle-specific activation and induction of apoptotic cell death. Moreover, previous studies have shown that PKCδ regulates transcription by phosphorylating various transcription factors, including the p53 tumor suppressor that is critical for cell cycle arrest and apoptosis in response to DNA damage. These findings collectively support a pivotal role for PKCδ in

* Correspondence: Kiyotsugu Yoshida, Medical Research Institute, Tokyo Medical and Dental University, 1-5-45 Yushima, Bunkyo-ku, Tokyo 113-8510, Japan. Phone: 81-3-5803-5826; Fax: 81-3-5803-0242; E-mail: yos.mgen@mri.tmd.ac.jp

the induction of apoptosis with significant impact. This review is focused on the current views regarding the regulation of cell fate by PKCδ in response to DNA damage.

Introduction

Genotoxic stress that damages DNA induces cell cycle arrest, activation of DNA repair, and in the event of irreparable damage, induction of apoptosis. The signaling mechanisms responsible for regulation of the DNA damage response are largely unknown. Certain insights have been derived from the finding that isoforms of the protein kinase C (PKC) family are activated in response to DNA damage. The calcium-independent PKCδ isoform is cleaved to a 40 kDa catalytically active fragment by caspase-3 in cells treated with DNA-damaging agents [1, 2]. The finding that overexpression of the PKCδ catalytic fragment induces chromatin condensation and DNA fragmentation has supported a role for PKCδ cleavage in the induction of apoptosis [3]. PKCδ-deficient cells are impaired in their response to DNA damage-induced apoptosis, suggesting that PKCδ is required to mount an appropriate apoptotic response under conditions of stress [4]. In the past 10 years, understanding of the cellular mechanisms of apoptosis mediated by PKCδ has advanced considerably, and the primary focus of this review is to provide an overview of PKCδ, its mode of action and its physiological role in genotoxic stress-induced apoptosis.

Protein Kinase C

Protein kinase C (PKC) was first described as a calcium-activated, phospholipid-dependent serine/threonine protein kinase [5]. PKC is activated diacylglycerol (DAG) hydrolyzed from phosphatidylinositol (PI) by phospholipase C (PLC) under a different cell-signaling system [6-9]. It has attracted attention as an intracellular receptor for tumor-promotor phorbol esters, such as 12-O-tetradecanoyl-13- phorbol acetate (TPA) [10]. Although PKC had been recognized as a protein kinase, subsequent studies have revealed that it belongs to a family of serine/threonine-specific protein kinases and is involved in diverse cellular functions, such as cell proliferation, tumor promotion, differentiation and apoptotic cell death [6-9, 11-15].

PKC consists of at least 11 isozymes (α, βI, βII, γ, δ, ε, ζ, η, θ, ι/λ and μ) with selective tissue distribution, activators, and substrates (ref). PKC isozymes have been categorized into three groups: i) the classical/conventional PKCs (cPKCs: α, βI, βII γ), which are calcium dependent and activated by DAG; ii) the novel PKCs (nPKCs: δ, ε, θ, μ), which are calcium-independent and activated by DAG; and iii) the atypical PKCs (aPKCs: ζ, λ), which are calcium-independent and not activated by DAG [7-9, 11, 13, 15].

The cell-specific expression and subcellular localization of individual PKC isozymes indicate important isozyme-specific functions. To elucidate these functions, it will be necessary to study in vitro and/or in vivo the individual features of each isozyme, such as expression, post-translational modification, substrate specificity, subcellular localization and signaling cross-talk with other proteins. Moreover, the involvement of a PKC isozyme in a signaling pathway resulting in a specific cellular response can be investigated by several distinct methods such as overexpression of the enzyme or inhibition of enzyme expression or activity.

Molecular Cloning and Genomic Structure of PKCδ

PKCδ was originally discovered in mouse epidermis [16] and cloned from a rat brain cDNA library using fragments encoding cPKC isoforms as probes [17]. Subsequently, PKCδ was detected in different mammalian species such as mouse [18, 19] and human [20] and classified as a member of the nPKC subgroup [9, 21, 22]. The open reading frames of rat, mouse and human clones encode proteins for 673, 674 and 676 amino acid residues, respectively, all of which are highly homologous and have an almost identical calculated molecular mass of 77.5 kDa. The phylogenic tree of the PKC isoforms shows that the primary structure of PKCδ is most closely related to another nPKC isoform PKCθ [23]. PKCθ is expressed predominantly in muscle and hematopoietic cells, whereas PKCδ is widely distributed among cells and tissues, suggesting that PKCδ plays universal rather than cell-type-specific roles in mammals.

The PKCδ gene is localized on chromosomes 3, 14 and 19 of human, mouse, and rat, respectively [24, 25]. Functional studies of the promoter regions will clarify the regulatory mechanisms for the expression of PKCδ. In addition, two cDNA clones encoding possible splicing variants of PKCδ have been reported; one contains an insertion in the middle of the gene [26], while the other is a truncated enzyme lacking its carboxy-terminal half [27]. Their physiological roles are yet to be elucidated.

Structural Properties of PKCδ

PKCδ contains a C-terminal catalytic domain with two conserved regions, and an ATP-binding region (C3), a catalytically active/substrate binding region (C4), an N-terminal regulatory domain with an inhibitory pseudosubstrate sequence, and two cysteine-rich zinc-finger-like sequences (C1A and C1B) in the C1 region but not in the C2 region (Figure 1). The C1B motif binds the phorbol ester PDBu with high affinity, whereas no significant binding is seen for the C1A motif [21]. Functional studies have revealed that the C1B region may play a critical role in the translocation of cytosolic PKCδ into cellular membranes following activation by phorbol ester. Five of six cysteine residues and two histidine residues interact with Zn^{2+} to form a specific coordination and attract phorbol ester binding [21]. The crystal structure of the C1B region of PKCδ has been determined in a complex with phorbol ester, which binds to a pocket between two pulled-apart β sheets at the rip of the region [28]. PKCδ does not require Ca^{2+} for its activity, and crystallographic analysis has consistently indicated that the C2-like region of PKCδ (amino acid residues 1-123) is a variant of the C2 fold that lacks sequences necessary for Ca^{2+} coordination [29]. Furthermore, two dimensional crystal structures of intact PKCδ and its regulatory domain on the lipid monolayer show a donut-shaped structure, to which the crystal structure of the C1B region is accommodated by overlaying the images. Precise structural information on intact PKCδ is essential to elucidate the regulatory mechanism of this enzyme.

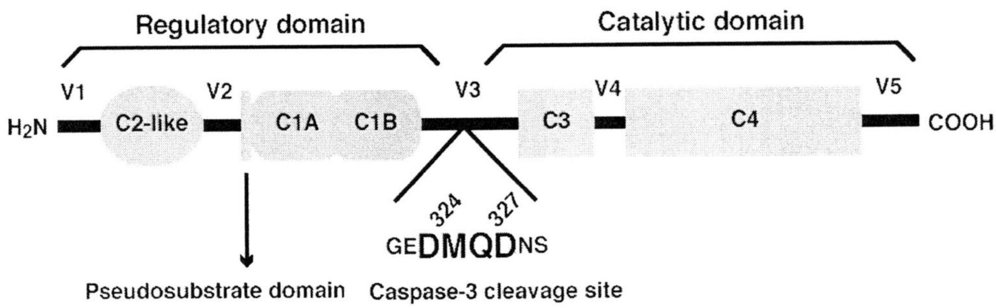

Figure 1. The structure of PKCδ. The domain structure of PKCδ is schematically shown, along with the caspase-3 cleavage sites.

Activation Mechanisms of PKCδ

PKCδ is activated by a variety of stimuli including ionizing radiation, anti-cancer agents, reactive oxygen species (ROS), ultraviolet radiation, growth factors and cytokines [30-35]. Molecular mechanisms such as tyrosine phosphorylation and proteolytic cleavage by caspase-3 are of importance for understanding the pro-apoptotic role for PKCδ activation.

(1) Tyrosine Phosphorylation

Tyrosine phosphorylation of PKCδ is important in modulating PKCδ activity (Table 1). Tyrosine 52, 64, 155, 187, 311, 332, 512, 523 and 565 of PKCδ are phosphorylated and regulate kinase activity [36-38]. The catalytic activity of PKCδ was shown to be reduced by tyrosine phosphorylation in *v-ras*-transformed keratinocytes [39] and in epidermal cells treated with epidermal growth factor [33] or phosphorylated by Src family kinases [40]. On the other hand, tyrosine phosphorylation enhances the enzymatic activity in various cells stimulated with substances such as phorbol ester, growth factors and hormones [36, 41-48]. In some cases, the enzymatic specificity is altered by this modification [49-51]. Importantly, tyrosine phosphorylation at positions 311, 332 and 512 induces activation of PKCδ in HaCaT and COS-7 cells following exposure to hydrogen peroxide and UV-B, respectively [30, 52]. Tyrosine phosphorylation at these positions may cause conformational changes and open the catalytic domain [30, 52], because two major tyrosine phosphorylation sites (Y311 and Y332) are strategically located in the hinge region of PKCδ. Upon phosphorylation of these residues, PKCδ may undergo conformational change and expose the catalytic domain, yet further studies are necessary to confirm this hypothesis. Etoposide also induces tyrosine phosphorylation of PKCδ, and sites 64 and 187 are critical for its increased activity [53]. Several tyrosine kinases, including Src [37, 54-57], Fyn [33, 36, 55], Lyn [35, 37, 41], c-Abl

[58, 59], PYK2 [60], Lck [30] and growth factor receptors [33, 36], are involved in the phosphorylation of PKCδ and some of the phosphorylation sites seem to be isoform-specific [61]. It is noteworthy that the targets of the tyrosine-phosphorylated PKCδ have been analyzed [53, 62, 63]. In addition, a receptor-type tyrosine phosphatase, CD45, is shown to be involved in dephosphorylation of PKCδ [64], although the treatment of cells with protein-tyrosine phosphatase inhibitors does not induce tyrosine phosphorylation or generation of the active PKCδ [65].

Table 1. Tyrosine Phosphorylation of PKC δ

Tyr	Stimulus	Kinases	Activity	Function
52	IgE antigen	Lyn	Reduced	association with Lyn
64	Etoposide			required for apoptosis
155	TPA	Lyn		required for differentiation
187	PDGF Etoposide IgE antigen		Increased	Proliferation nuclear translocation
311	Hydrgen peroxide	Src	Increased	degradation of PKCδ
332	Hydrgen peroxide		Increased	mitochondria translocation
512	Hydrgen peroxide	c-Abl	Increased	
	Ultraviolet radiation			
523	Hydrgen peroxide			
565	Hydrgen peroxide	Lyn		prevents differentiation

Antisense PKCδ inhibits differentiation of mouse keratinocytes whereas overexpression of PKCδ in the presence of a Src kinase inhibitor enhances differentiation [40]. Plasmids encoding phenylalanine mutants of PKCδ tyrosine residues 64 and 565 induce differentiation in the absence of a Src inhibitor, suggesting that phosphorylation of these sites by Src suppresses differentiation. In contrast, phenylalanine mutants of tyrosine residues 52, 155 and 187 are inactive, indicating that these sites are required for differentiation. Thus, Src-induced post-translational modification of PKCδ on specific tyrosine residues in mouse keratinocytes apparently suppresses PKCδ-mediated differentiation.

Many of the phosphorylation site studies have used Tyr→Phe mutations to block tyrosine phosphorylation and thus implicate various sites as critical residues for tyrosine phosphorylation. It is possible that a mutation at a particular site blocks phosphorylation at other sites. Therefore, it is not always clear which and how many sites are actually phosphorylated in response to a given stimulus. The ability to alter substrate specificity and phenotype with these mutations indicates that many permutations of differential tyrosine phosphorylation by different tyrosine kinases in response to different stimulation provide PKCδ the ability to be used in many different ways under different circumstances. The reported differential effects of tyrosine phosphorylation on the activity of PKCδ have also suggested that PKCδ can be manipulated in different ways through tyrosine phosphorylation. However, PKCδ activity assays are performed in vitro and may not reflect the actual in vivo activity. There remains much to be learned about the apparent multiple effects of tyrosine phosphorylation on PKCδ activity, substrate specificity and subcellular localization.

(2) Caspase-3 Cleavage

Recent studies have demonstrated a novel form of PKCδ activation in which the kinase is proteolytically cleaved in response to a variety of apoptotic stimuli in diverse cell types [31, 53, 66-68]. The first studies demonstrating a cleavage of PKC during apoptosis were by *Emoto et al.* who reported that PKCδ can be proteolytically cleaved by an ICE-like protease in response to ionizing radiation [1] or 1-β-D-arabinofuranosylcytosine (ara-C) [2]. Using the caspase-3 specific inhibitor Z-Asp-Glu-Val-Asp-fluoromethyl ketone (Z-DEVD-fmk), proteolytic cleavage of PKCδ was mediated by caspase-3 [69]. The caspase-3 cleavage site in PKCδ was identified as the V3-domain adjacent to the aspartic acid at the DMQD327Nsite, indicating that PKCδ is an endogenous substrate for caspase-3 (Figure 1). Caspase-3 cleaves the kinase to yield a 40 kDa catalytically active fragment and a 38 kDa regulatory fragments to permanently dissociate the regulatory domain from the catalytic fragment and persistently activate the kinase. The proteolytic activation of PKCδ has been implicated primarily in the apoptotic cell death of many cellular systems. The importance of the PKCδ catalytic fragment in the apoptotic effect of PKCδ was demonstrated by the finding that overexpression of the PKCδ catalytic domain induces nuclear fragmentation and apoptotic cell death whereas no apoptosis is observed in cells overexpressing the kinase-dead catalytic domain [3]. Similar results were obtained in human keratinocytes [70], in a dopaminergic neuronal cell lines [71], and in HL-60 cells overexpressing the PKCδ catalytic domain [72]. In agreement with these studies, it has been reported that the catalytic fragment of PKCδ localizes to the mitochondria in keratinocytes [70] and to the nucleus in HL-60 cells [72], similar to the localization of PKCδ in these cells in response to apoptotic stimuli.

In most systems, apoptotic cell death is associated with the caspase-3-dependent cleavage of PKCδ. Indeed, DNA damaging agents, such as ionizing radiation, ara-C, etoposide, cisplatin, mitomycin, methotrexate and ultraviolet (UV) radiation induces cleavage of PKCδ in various cell types [1, 2, 31, 35, 53, 69, 71, 73-77]. TNF-α, Fas ligation and IL-2 deprivation of activated T cells also result in the generation of an active catalytic fragment of PKCδ [1, 78]. Similarly, treatment of the insulinoma cell line INS-1 with IL-1β and streptozotocin and the spontaneous apoptosis of neutrophils are involved in caspase-3-dependent cleavage of PKCδ [34, 66]. In contrast, no cleavage of PKCδ is observed in the apoptosis of LNCaP cells treated with phorbol esters, in CHO cells treated with hydrogen peroxide or in the apoptosis of HaCaT cells in response to UV radiation [30, 79, 80].

The role for caspase-3 in the cleavage of PKCδ has been demonstrated using caspase-3 specific inhibitors. Inhibition of caspase-3 activity by Z-DEVD-fmk and Z-VAD-fmk reduced both the cleavage of PKCδ and apoptosis in response to many apoptotic stimuli [31, 53, 71, 73-75]. Moreover, PKCδ inhibitors and a PKCδ dominant negative mutant inhibited both the activation of caspase-3 and the cleavage of PKCδ suggesting that PKCδ may also act upstream of caspase-3 and pointing to the existence of a positive regulatory loop between PKCδ and caspase-3 [31, 53, 71, 75]. In accordance with this concept, a role for PKCδ in the activation of caspase-9 was demonstrated in response to etoposide, taxol, brefeldin and UV radiation [81]. The mechanism by which PKCδ activates caspase-9 and -3 are not clear; however, phosphorylation of caspases by PKC is a plausible pathway since phosphorylated forms of the caspases have been recently described [82] and human caspase-9 was recently reported to be phosphorylated and modulated by Akt [83].

Recently, a novel spliced variant of mouse PKCδ was identified. This variant, PKCδII, contains a 78 bp fragment inserted in the caspase-3 cleavage site [26]. PKCδII is induced by H7 in mouse thymocytes and, as predicted, does not undergo cleavage by caspase-3, suggesting a role for this splicing variant in the modulation of apoptosis.

PKCδ Phosphorylation

Upon activation, PKCδ phosphorylates serine/threonine residues in specific substrates, although the specific amino acid substrate sequences phosphorylated by PKCδ are not well characterized. Based on a study conducted with synthetic peptides, a general consensus phosphorylation site motif for PKC was identified as RXX(S/T)XRX, where X can be any amino acid [84]. A recent study demonstrated the specific amino acid sequence motifs of the major PKC isozymes and the common and different amino acid substrate sequences among the PKC isozymes [85]. In particular, PKCδ requires a hydrophobic amino acid at position +1 at the C-terminal of the phosphorylation site (Ser), basic amino acids at positions -6, -5, -4 and -2 and glycine at position -1. All the PKC isozymes evaluated require substrates with arginine at position -3, which reconciles with the consensus sequence motif. The results also indicate that the optimal amino acid sequence for PKCδ phosphorylation is A(A/R)**R**(K/A)**RKG**<u>**S**</u>**F**F(Y/F)G**G**, where the underlined serine (S) is the phosphorylation site and the bold letters are particularly important for sequence recognition by PKCδ. Some PKC isozymes share common motifs, indicating that PKC phosphorylation sites may be conserved among isozymes [85].

Regulation of Apoptosis by PKCδ

PKC isozymes have long been implicated in the growth factor signal transduction pathway [8]. In contrast, activation of PKCδ is associated with inhibition of cell cycle progression and down-regulation of PKCδ is linked to tumor promotion, suggesting that PKCδ may have a negative effect on cell survival [86, 87]. In many cases, the growth-inhibitory effects of PKCδ have been linked to changes in the expression of factors that influence cell cycle progression. Furthermore, PKCδ plays an essential role in the genotoxic stress response leading to apoptotic cell death in many cell types.

(1) Interaction of PKCδ with Protein Kinases and Phosphatases

Apoptotic stimuli such as genotoxic and oxidative stress induce activation of serine/threonine and tyrosine kinases [88, 89]. c-Abl is one of the tyrosine kinases activated by DNA damage [90-96] and has been implicated in the phosphorylation of PKCδ in response to genotoxic and oxidative stress [35, 58, 59]. c-Abl is a ubiquitously expressed enzyme that localizes to the nucleus and cytoplasm and participates in cell apoptosis by Rad9- and p73-dependent mechanisms [94, 97-99]. A recent study demonstrated that cytoplasmic c-Abl translocates into the nucleus in response to DNA damage by dissociating from 14-3-3 proteins induced by JNK phosphorylation of 14-3-3 [95, 100]. PKCδ has been reported to interact with c-Abl in response to both genotoxic and oxidative stress [58, 59]. c-Abl is constitutively associated with PKCδ via the SH3 domain of c-Abl. Upon exposure to ionizing radiation, PKCδ is tyrosine phosphorylated, activated and translocated into the nucleus, which are mediated by c-Abl. Similarly, hydrogen peroxide has been reported to enhance the association of PKCδ with c-Abl and the phosphorylation of PKCδ at Tyr512, which increases PKCδ activity. The association between PKCδ and c-Abl is important for the activity of c-Abl as well as PKCδ, since cells overexpressing the regulatory domain of PKCδ exhibit a lower activity of c-Abl in response to hydrogen peroxide [59]. An increase in the association of PKCδ and c-Abl and a PKCδ-dependent increase in the activity of c-Abl were also observed in cells treated with a combination of cisplatin and methylglyoxal [101].

Lyn tyrosine kinase regulates SHP-1 tyrosine phosphatase and has been implicated as a regulator of JNK in response to DNA damage [102, 103]. Furthermore, activation of Lyn leads to tyrosine phosphorylation and activation of PKCδ, resulting in activation of MEKK1 (MEK kinase 1)/MKK7 (MAPK kinase 7)/JNK pathway [35]. These results suggest that Lyn tyrosine kinase is also involved in PKCδ-mediated genotoxic stress-induced apoptosis.

p73 is a p53 homolog that is associated with the cellular response to DNA damage and interacts with c-Abl in this response [97-99]. A recent study showed that the catalytic fragment of PKCδ interacts with p73β and phosphorylates the transactivation and DNA binding domains of p73β. This phosphorylation is associated with accumulation of p73β and induction of p73β transactivation and apoptotic function [104]. In view of the roles for c-Abl and p73 in apoptosis, these results further demonstrate the importance of PKCδ as an essential link in this apoptotic signaling complex.

The signals induced by protein phosphorylation are controlled by the activity of specific tyrosine phosphatases. Indeed, c-Abl tyrosine phosphorylates and activates the tyrosine phosphatase SHP-1 in response to DNA-damaging agents [91]. PKCδ, which activates c-Abl, forms a complex with c-Abl and SHP-1, phosphorylates SHP-1 and decreases its activity [105]. Thus, PKCδ regulates apoptotic signals of the key apoptotic tyrosine kinase c-Abl not only by its activation but also by inhibiting the activity of its tyrosine phosphatase.

In addition to the interaction between PKCδ and c-Abl, there are other signaling pathways that may participate directly or indirectly in the apoptotic effect of PKCδ. One of these pathways involves various members of the MAP kinase superfamily that are activated in response to various extracellular stimuli and stress [106]. The activation and inactivation of Erk1/2, JNK and p38 have been associated with PKCδ signaling during apoptosis. Thus, treatment of salivary gland acinar cells with PMA induces apoptosis via activation of PKCδ and PKCα [107]. PMA also induces biphasic activation of JNK and a decrease in the activation of Erk [107]. Similarly, etoposide induces a sustained activation of JNK in salivary gland acinar cells and this activation is abolished by rottlerin [31]. Furthermore, activation of JNK by PKCδ through MKK7 in response to ionizing radiation has been demonstrated in human thyroid cells [108]. This activation is also inhibited by rottlerin [108]. In another study, PKCδ increased the phosphorylation of Erk1/2 and this effect enhanced the sensitivity of NIH3T3 cells to ionizing radiation and reduced the radiation-induced G2/M arrest [109].

PKCδ also activates nuclear factor κ B (NF-κB). This ubiquitous transcription factor plays a key role in regulating immune and inflammatory responses; it also influences tumorigenesis through the induction of target genes that accelerate transit through the cell cycle, block apoptosis, promote angiogenesis and enhance tumor cell invasiveness. NF-κB is held in the cytoplasm by inhibitor κB (IκB) proteins that mask its nuclear localization signals. There is evidence that PKCδ itself, or its downstream effector PKCμ (or protein kinase D), can act as IkB kinases (IKKs) to phosphorylate IκB [110, 111]. The phosphorylated form of IκB undergoes ubiquitination and degradation by the 26S proteasome, liberating NF-κB, which then translocates to the nucleus and binds to κB-regulatory elements. PKCδ is also reported to increase the transactivation potential of NF-κB via an IKK/IκB-independent pathway [112-114]. Interestingly, recent studies identified reciprocal regulation of PKCδ by NF-κB, by showing that a NF-κB-responsive regulatory element in the PKCδ promoter links TNFα stimulation to increase PKCδ mRNA and protein expression [115]. Finally, our recent studies demonstrated that both PKCδ and IKKα, but not IKKβ, translocate to the nucleus in response to oxidative stress (Yoshida K, et al. unpublished observation). We also found that PKCδ interacts with and activates IKKα. Importantly, our data suggest that, upon exposure to oxidative stress, PKCδ-mediated IKKα activation does not contribute to NF-κB activation; instead, nuclear IKK regulates p53 transcription activity by phosphorylation at Ser20. These findings collectively support a novel mechanism in which the PKCδ→IKKα signaling pathway contributes to ROS-induced p53 activation.

(2) Translocation of PKCδ during Apoptotic Responses

Depending on the cell types and apoptotic stimuli, PKCδ translocates to nearly all subcellular organelles, including the nucleus, mitochondria, the golgi complex, endoplasmic reticulum and plasma membrane [116]. At each subcellular organelle, PKCδ phosphorylates different substrates, including various responses that eventually lead to cell death. Identification of these substrates is critical to understanding the mechanism of PKCδ, but it has been very challenging to identify physiologic substrates in each organelle. In this context, deletion or inactivation of the substrate must lead to at least a partial loss of PKCδ-induced response.

a) Mitochondrial Targeting

One of the main targets of PKCδ is the mitochondria. PKCδ has been reported to translocate into the mitochondria in response to PMA in mouse keratinocytes and U-937 cells

[117, 118]. Similar results were obtained in response to oxidative stress [119]. In addition, several studies have demonstrated that overexpression and activation of PKCδ exert changes in mitochondrial functions such as a decrease in mitochondrial membrane potential and release of cytochrome c. Treatment of HeLa cells with cisplatin induces cleavage of PKCδ and inhibition of PKCδ attenuates the activation of caspases and the late release of cytochrome c by cisplatin [74]. PKCδ also mediates the decrease in membrane potential and activation of caspase-9 in salivary epithelial cells treated with various apoptotic stimuli [81].

PKCδ has been recently reported to enhance ceramide formation via mitochondrial signaling in prostate cancer cells [120]. Activation of PKCδ with etoposide induces ceramide formation and translocation of PKCδ to the mitochondria that is followed by cytochrome c release, caspase-9 activation and a sustained increase in ceramide levels. Furthermore, treatment of cells with rottlerin attenuates the etoposide-induced ceramide formation and apoptosis, suggesting that the generation of ceramide and the mitochondrial translocation of PKCδ act to amplify apoptotic signals via activation of the mitochondrial pathway.

A recent study demonstrated that the phospholipids scramblase 3 (PLS3) is a mitochondrial target of PKCδ [121]. PLS3 is a member of the scramblase family, which is responsible for bidirectional movement of phospholipids in the lipid bilayer. PLS3 is found exclusively in the mitochondria [121, 122]. The function of PLS3 is currently unclear, but it is likely involved in the translocation of cardiolipin from the mitochondrial inner membrane to the outer membrane during apoptosis. Mitochondria with expression of an inactive mutant of PLS3 have a low level of cardiolipin and poor respiration [122]. They also display a unique morphology, being larger in size, fewer in number and with tightly packed cristae, consistent with the notion that PLS3 moves phospholipids from the inner membrane to the outer membrane [122]. Cardiolipin translocation is directly tied to the sensitivity of mitochondria to tBid-induced cytochrome c release. Because tBid targeting to the mitochondria is mediated by cardiolipin [123], the translocation of cardiolipin from the inner membrane to the outer membrane facilitates the recruitment of tBid [122]. This idea was confirmed by the finding that mitochondria overexpressing PLS3 are more sensitive to tBid-induced cytochrome c release, whereas those expressing mutant PLS3 are more resistant [122]. In addition, HeLa cells expressing PLS3 become more sensitive in TUNEL studies when cells were treated with PMA. Moreover, the expression of mitochondria-targeted PKCδ in cells results in apoptosis, while overexpression of PLS3 enhances this effect. In contrast, overexpression of the inactive PLS3 mutant does not generate this response [121]. These findings support the view that PLS3 is a mitochondrial target of PKCδ-induced apoptosis.

b) Nuclear Targeting

Translocation of PKCδ into the nucleus has been demonstrated in various cells [53, 58, 78, 124-126]. Our recent study demonstrated that PKCδ translocates into the nucleus following exposure of cells with ara-C (Figure 2) [125]. Moreover, pretreatment with rottlerin attenuates nuclear targeting of PKCδ, suggesting that its kinase activity is required for nuclear translocation (Figure 2). A putative nuclear localization signal has been identified at the carboxyl terminus of the catalytic domain of PKCδ [124].

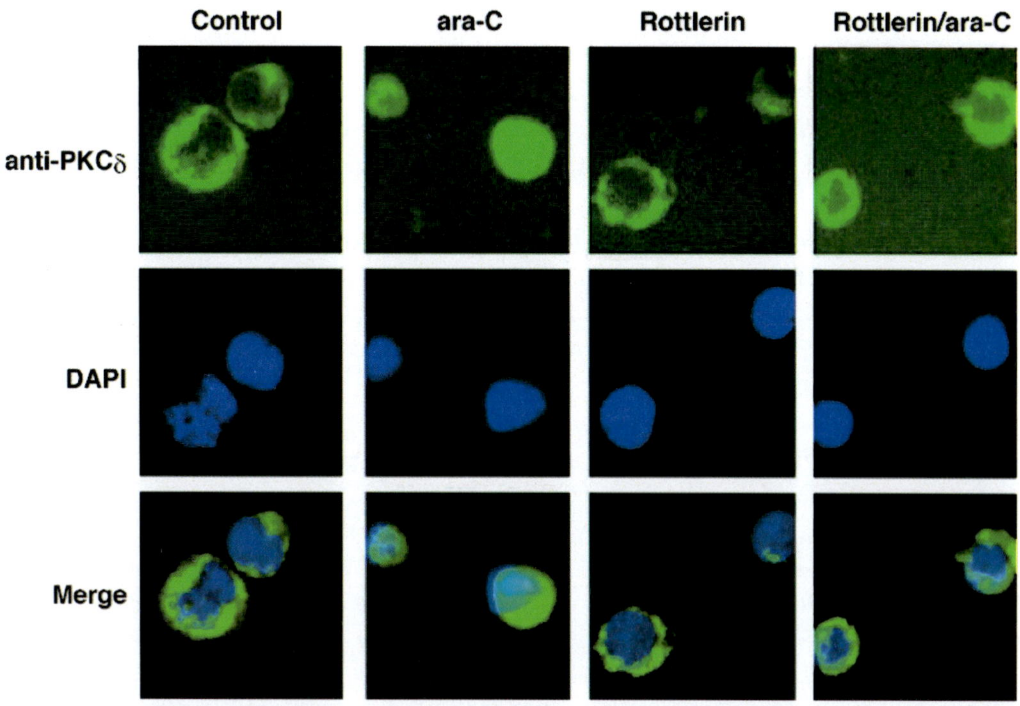

Figure 2. Nuclear translocation of PKCδ in response to DNA damage. U-937 cells were left untreated or treated with 5 mM rottlerin for 30 min followed by treatment with 10 mM ara-C for 2 h. After fixation and blocking, cells were incubated with anti-PKCδ and counterstained with DAPI.

Numerous PKCδ targets and substrates are nuclear proteins that function in apoptotic cell death. One of these proteins is the nuclear DNA-dependent protein kinase (DNA-PK), which plays an essential role in the repair of DNA double strand breaks [127] and interacts with c-Abl during genotoxic stress [128]. Activated PKCδ binds directly to the C-terminus of this enzyme and induces its phosphorylation. This results in the dissociation of DNA-PK from DNA, the inhibition of DNA repair and enhanced DNA fragmentation [129]. Cells lacking DNA-PK are resistant to apoptosis induced by overexpression of the activated catalytic fragment of PKCδ, further suggesting that interaction between PKCδ and DNA-PK participates in apoptosis induced by DNA damage.

Lamin proteins represent another important potential target for the apoptotic function of PKCδ. The disassembly of the nuclear lamina during apoptosis is one of the key processes in the nuclear phase of apoptosis and contributes to the characteristic nuclear appearance of apoptotic cells [130]. The main process in lamina disassembly involves phosphorylation, which is important for the cleavage of the lamin proteins by caspase-6 [131]. Recently, PKCδ was identified as an apoptotic lamin kinase. PKCδ translocates into the nucleus in response to apoptotic stimuli such as ara-C, and co-localizes with and phosphorylates lamin B [72]. Inhibition of PKCδ activity inhibits both the phosphorylation and the caspase-6 dependent cleavage of lamin B.

Our recent study demonstrated that PKCδ is responsible for constitutive and DNA damage-induced phosphorylation of Rad9 [125], a key factor involved in checkpoint regulation of the DNA damage response [132]. The results demonstrated an interaction between PKCδ and Rad9, and showed that PKCδ phosphorylates Rad9 both in vitro and in cells treated with ara-C or ionizing radiation. Nuclear Rad9 forms a critical heterotrimeric complex with Hus1 and Rad1, the 9-1-1 complex that is involved in DNA damage checkpoint control [133]. PKCδ, which translocates to the nucleus during apoptosis, enhances the phosphorylation of Rad9 and the formation of the Rad9-Hus1-Rad1 complex [125]. Interestingly, Rad9 is also phosphorylated by ATM [134] and c-Abl [94]. Using ATM siRNA to down-regulate the level of ATM, a diminished nuclear targeting of PKCδ is observed, suggesting that ATM is required for nuclear targeting of PKCδ and is functionally upstream of PKCδ [125]. These studies provide a direct linkage between PKCδ and DNA damage-induced checkpoint regulation, and form the basis for future studies of the mechanism for PKCδ-mediated apoptosis (Figure 3).

Another downstream effector of PKCδ in the DNA damage response is stress-activated protein kinase (SAPK/JNK) [35]. DNA damage-induced SAPK/JNK activation is attenuated by rottlerin, a dominant negative mutant of PKCδ and PKCδ siRNA. PKCδ does not directly phosphorylates SAPK/JNK, rather SAPK/JNK is indirectly phosphorylated through the SAPK/JNK pathway, PKCδ→MEKK1→MKK7→SAPK/JNK [35]. The finding that SAPK/JNK is a downstream effector of PKCδ provides another mechanism of PKCδ-mediated apoptosis.

The p53 tumor suppressor is activated in the cellular response to genotoxic stress. Transactivation of p53 target genes dictates cell cycle arrest and DNA repair or induction of apoptosis. Recent studies have revealed that phosphorylation of p53 on Ser 46 is associated with the induction of p53AIP1 expression, resulting in the commitment of the cell fate to apoptotic cell death [135, 136]. Moreover, upon exposure to genotoxic stress, p53DINP1 is expressed and then recruits a kinase(s) to p53 that specifically phosphorylates Ser 46 [137]. Our recent data showed that PKCδ is involved in phosphorylation of p53 on Ser 46 [138]. PKCδ-mediated phosphorylation was required for the interaction of PKCδ with p53. The results also demonstrated that p53DINP1 associates with PKCδ upon exposure to genotoxic agents. In concert with these results, PKCδ potentiates p53-dependent apoptosis by Ser 46 phosphorylation in response to genotoxic stress. These findings indicate that PKCδ regulates p53 to induce apoptotic cell death in the cellular response to DNA damage (Figure 3) [138].

Our recent studies have demonstrated that topoisomerase IIα is activated in response to genotoxic stress [139]. Concomitant with the activation, the expression of topoisomerase IIα is increased following DNA damage. The results also demonstrated that PKCδ interacts with topoisomerase IIα. This association was in an S phase-specific manner and was required for stabilization and catalytic activation of topoisomerase IIα in response to DNA damage. Conversely, inhibition of PKCδ kinase activity attenuated DNA damage-induced activation of topoisomerase IIα. Finally, aberrant activation of topoisomerase IIα by PKCδ was associated with induction of apoptosis upon exposure to genotoxic agents. These findings indicate that PKCδ regulates topoisomerase IIα and thereby cell fate in the genotoxic stress response (Figure 3) [139]. A model describing the cellular pathways involved in the effects of PKCδ on apoptosis in response to DNA damage is presented in Figure 3.

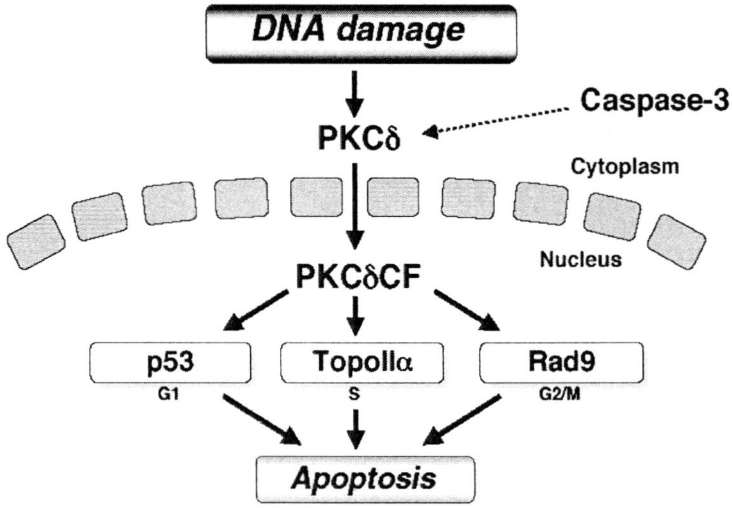

Figure 3. Diagram of PKCδ and its downstream effectors in DNA damage-induced apoptosis. PKCδ translocates to the nucleus upon exposure to genotoxic stress, where it phosphorylates and activates p53, topoisomerase IIα and Rad9 in the G1, S and G2/M phase, respectively.

Conclusions and Future Directions

PKCδ plays an essential role in the regulation of apoptosis in response to a large and diverse array of apoptotic stimuli. PKCδ is thus a pro-apoptotic kinase activated by multiple mechanisms, including translocation, proteolysis and tyrosine phosphorylation. The proteolytic activation of PKCδ is important not only in activating the downstream apoptotic cascade, but also in amplifying upstream caspase signaling.

The phosphorylation of PKCδ on tyrosine residues occurs in response to many of the apoptotic stimuli. Tyrosine phosphorylation represents an early event in the apoptotic pathways, and plays a pivotal role in the pro-apoptotic effects of PKCδ. The precise role for tyrosine phosphorylation in PKCδ functioning is currently unclear, but may involve changes in the catalytic activity, changes in the subcellular localization via binding to specific scaffold proteins or changes in the affinity of PKCδ towards its different substrates.

c-Abl phosphorylates PKCδ during apoptosis; however, the ability of PKCδ to be phosphorylated on multiple tyrosine residues suggests that other tyrosine kinases phosphorylate and associate with PKCδ via distinct tyrosine residues. An important issue is the identity of the additional tyrosine kinases and tyrosine phosphatases that participate in the phosphorylation and dephosphorylation of PKCδ in response to the various apoptotic stimuli.

Most of the studies mentioned above suggest that the role for PKCδ in the induction of apoptosis involves its caspase-dependent cleavage and the presence of a positive loop between PKCδ and caspase-3. The mechanisms involved in the activation of caspase-3 by PKCδ are unclear. PKCδ may act directly on caspase-3 to induce its phosphorylation or indirectly via the activation of upstream caspases such as caspase-8 and -9 or by modulating the activity of other upstream proteins. While deregulation of the PKCδ signaling pathway can contribute to anticancer drug resistance [140], there is little understanding of how the

PKCδ signaling pathway is affected when cancer cells acquire resistance to chemotherapeutic drugs. Considering the importance of PKCδ in DNA damage-induced apoptosis, a thorough understanding of how it regulates apoptosis should benefit cancer therapy. Moreover, novel PKCδ-based therapy may well be used in combination with other agents to create synergism and help prevent the development of drug resistance.

Acknowledgements

This work was supported by grants from the Ministry of Education, Science and Culture of Japan, Kowa Life Science Foundation and the Astellas Foundation for Research on Medical Resources.

References

[1] Emoto, Y., Manome, Y., Meinhardt, G., Kisaki, H., Kharbanda, S., Robertson, M., Ghayur, T., Wong, W. W., Kamen, R., Weichselbaum, R. and Kufe, D. (1995) Proteolytic activation of protein kinase C delta by an ICE-like protease in apoptotic cells. *EMBO J 14*, 6148-6156.

[2] Emoto, Y., Kisaki, H., Manome, Y., Kharbanda, S. and Kufe, D. (1996) Activation of protein kinase Cdelta in human myeloid leukemia cells treated with 1-beta-D-arabinofuranosylcytosine. *Blood 87*, 1990-1996.

[3] Ghayur, T., Hugunin, M., Talanian, R. V., Ratnofsky, S., Quinlan, C., Emoto, Y., Pandey, P., Datta, R., Huang, Y., Kharbanda, S., Allen, H., Kamen, R., Wong, W. and Kufe, D. (1996) Proteolytic activation of protein kinase C delta by an ICE/CED 3-like protease induces characteristics of apoptosis. *J Exp Med 184*, 2399-2404.

[4] Leitges, M., Mayr, M., Braun, U., Mayr, U., Li, C., Pfister, G., Ghaffari-Tabrizi, N., Baier, G., Hu, Y. and Xu, Q. (2001) Exacerbated vein graft arteriosclerosis in protein kinase Cdelta-null mice. *J Clin Invest 108*, 1505-1512.

[5] Takai, Y., Kishimoto, A., Inoue, M. and Nishizuka, Y. (1977) Studies on a cyclic nucleotide-independent protein kinase and its proenzyme in mammalian tissues. I. Purification and characterization of an active enzyme from bovine cerebellum. *J Biol Chem 252*, 7603-7609.

[6] Nishizuka, Y. (1984) The role of protein kinase C in cell surface signal transduction and tumour promotion. *Nature 308*, 693-698.

[7] Nishizuka, Y. (1988) The molecular heterogeneity of protein kinase C and its implications for cellular regulation. *Nature 334*, 661-665.

[8] Nishizuka, Y. (1992) Intracellular signaling by hydrolysis of phospholipids and activation of protein kinase C. *Science 258*, 607-614.

[9] Nishizuka, Y. (1995) Protein kinase C and lipid signaling for sustained cellular responses. *Faseb J 9*, 484-496.

[10] Niedel, J. E., Kuhn, L. J. and Vandenbark, G. R. (1983) Phorbol diester receptor copurifies with protein kinase C. *Proc Natl Acad Sci U S A 80*, 36-40.

[11] Casabona, G. (1997) Intracellular signal modulation: a pivotal role for protein kinase C. *Prog Neuropsychopharmacol Biol Psychiatry 21*, 407-425.

[12] Clemens, M. J., Trayner, I. and Menaya, J. (1992) The role of protein kinase C isoenzymes in the regulation of cell proliferation and differentiation. *J Cell Sci 103 (Pt 4)*, 881-887.

[13] Goodnight, J., Mischak, H. and Mushinski, J. F. (1994) Selective involvement of protein kinase C isozymes in differentiation and neoplastic transformation. *Adv Cancer Res 64*, 159-209.

[14] Hofmann, J. (1997) The potential for isoenzyme-selective modulation of protein kinase C. *FASEB J 11*, 649-669.

[15] Hug, H. and Sarre, T. F. (1993) Protein kinase C isoenzymes: divergence in signal transduction? *Biochem J 291 (Pt 2)*, 329-343.

[16] Gschwendt, M., Kittstein, W. and Marks, F. (1986) A novel type of phorbol ester-dependent protein phosphorylation in the particulate fraction of mouse epidermis. *Biochem Biophys Res Commun 137*, 766-774.

[17] Ono, Y., Fujii, T., Ogita, K., Kikkawa, U., Igarashi, K. and Nishizuka, Y. (1987) Identification of three additional members of rat protein kinase C family: delta-, epsilon- and zeta-subspecies. *FEBS Lett 226*, 125-128.

[18] Mischak, H., Bodenteich, A., Kolch, W., Goodnight, J., Hofer, F. and Mushinski, J. F. (1991) Mouse protein kinase C-delta, the major isoform expressed in mouse hemopoietic cells: sequence of the cDNA, expression patterns, and characterization of the protein. *Biochemistry 30*, 7925-7931.

[19] Mizuno, K., Kubo, K., Saido, T. C., Akita, Y., Osada, S., Kuroki, T., Ohno, S. and Suzuki, K. (1991) Structure and properties of a ubiquitously expressed protein kinase C, nPKC delta. *Eur J Biochem 202*, 931-940.

[20] Aris, J. P., Basta, P. V., Holmes, W. D., Ballas, L. M., Moomaw, C., Rankl, N. B., Blobel, G., Loomis, C. R. and Burns, D. J. (1993) Molecular and biochemical characterization of a recombinant human PKC-delta family member. *Biochim Biophys Acta 1174*, 171-181.

[21] Gschwendt, M. (1999) Protein kinase C delta. *Eur J Biochem 259*, 555-564.

[22] Newton, A. C. (2001) Protein kinase C: structural and spatial regulation by phosphorylation, cofactors, and macromolecular interactions. *Chem Rev 101*, 2353-2364.

[23] Altman, A. and Villalba, M. (2002) Protein kinase C-theta (PKC theta): a key enzyme in T cell life and death. *J Biochem (Tokyo) 132*, 841-846.

[24] Huppi, K., Siwarski, D., Goodnight, J. and Mischak, H. (1994) Assignment of the protein kinase C delta polypeptide gene (PRKCD) to human chromosome 3 and mouse chromosome 14. *Genomics 19*, 161-162.

[25] Kurkinen, K. M., Keinanen, R. A., Karhu, R. and Koistinaho, J. (2000) Genomic structure and chromosomal localization of the rat protein kinase Cdelta-gene. *Gene 242*, 115-123.

[26] Sakurai, Y., Onishi, Y., Tanimoto, Y. and Kizaki, H. (2001) Novel protein kinase C delta isoform insensitive to caspase-3. *Biol Pharm Bull 24*, 973-977.

[27] Ueyama, T., Ren, Y., Ohmori, S., Sakai, K., Tamaki, N. and Saito, N. (2000) cDNA cloning of an alternative splicing variant of protein kinase C delta (PKC deltaIII), a new truncated form of PKCdelta, in rats. *Biochem Biophys Res Commun 269*, 557-563.

[28] Zhang, G., Kazanietz, M. G., Blumberg, P. M. and Hurley, J. H. (1995) Crystal structure of the cys2 activator-binding domain of protein kinase C delta in complex with phorbol ester. *Cell 81*, 917-924.

[29] Pappa, H., Murray-Rust, J., Dekker, L. V., Parker, P. J. and McDonald, N. Q. (1998) Crystal structure of the C2 domain from protein kinase C-delta. *Structure 6*, 885-894.

[30] Konishi, H., Yamauchi, E., Taniguchi, H., Yamamoto, T., Matsuzaki, H., Takemura, Y., Ohmae, K., Kikkawa, U. and Nishizuka, Y. (2001) Phosphorylation sites of protein kinase C delta in H2O2-treated cells and its activation by tyrosine kinase in vitro. *Proc Natl Acad Sci U S A 98*, 6587-6592.

[31] Reyland, M. E., Anderson, S. M., Matassa, A. A., Barzen, K. A. and Quissell, D. O. (1999) Protein kinase C delta is essential for etoposide-induced apoptosis in salivary gland acinar cells. *J Biol Chem 274*, 19115-19123.

[32] Chen, N., Ma, W., Huang, C. and Dong, Z. (1999) Translocation of protein kinase Cepsilon and protein kinase Cdelta to membrane is required for ultraviolet B-induced activation of mitogen-activated protein kinases and apoptosis. *J Biol Chem 274*, 15389-15394.

[33] Denning, M. F., Dlugosz, A. A., Threadgill, D. W., Magnuson, T. and Yuspa, S. H. (1996) Activation of the epidermal growth factor receptor signal transduction pathway stimulates tyrosine phosphorylation of protein kinase C delta. *J Biol Chem 271*, 5325-5331.

[34] Carpenter, L., Cordery, D. and Biden, T. J. (2002) Inhibition of protein kinase C delta protects rat INS-1 cells against interleukin-1beta and streptozotocin-induced apoptosis. *Diabetes 51*, 317-324.

[35] Yoshida, K., Miki, Y. and Kufe, D. (2002) Activation of SAPK/JNK signaling by protein kinase Cdelta in response to DNA damage. *J Biol Chem 277*, 48372-48378.

[36] Li, W., Mischak, H., Yu, J. C., Wang, L. M., Mushinski, J. F., Heidaran, M. A. and Pierce, J. H. (1994) Tyrosine phosphorylation of protein kinase C-delta in response to its activation. *J Biol Chem 269*, 2349-2352.

[37] Song, J. S., Swann, P. G., Szallasi, Z., Blank, U., Blumberg, P. M. and Rivera, J. (1998) Tyrosine phosphorylation-dependent and -independent associations of protein kinase C-delta with Src family kinases in the RBL-2H3 mast cell line: regulation of Src family kinase activity by protein kinase C-delta. *Oncogene 16*, 3357-3368.

[38] Brodie, C., Bogi, K., Acs, P., Lorenzo, P. S., Baskin, L. and Blumberg, P. M. (1998) Protein kinase C delta (PKCdelta) inhibits the expression of glutamine synthetase in glial cells via the PKCdelta regulatory domain and its tyrosine phosphorylation. *J Biol Chem 273*, 30713-30718.

[39] Denning, M. F., Dlugosz, A. A., Howett, M. K. and Yuspa, S. H. (1993) Expression of an oncogenic rasHa gene in murine keratinocytes induces tyrosine phosphorylation and reduced activity of protein kinase C delta. *J Biol Chem 268*, 26079-26081.

[40] Joseloff, E., Cataisson, C., Aamodt, H., Ocheni, H., Blumberg, P., Kraker, A. J. and Yuspa, S. H. (2002) Src family kinases phosphorylate protein kinase C delta on

tyrosine residues and modify the neoplastic phenotype of skin keratinocytes. *J Biol Chem 277*, 12318-12323.

[41] Szallasi, Z., Denning, M. F., Chang, E. Y., Rivera, J., Yuspa, S. H., Lehel, C., Olah, Z., Anderson, W. B. and Blumberg, P. M. (1995) Development of a rapid approach to identification of tyrosine phosphorylation sites: application to PKC delta phosphorylated upon activation of the high affinity receptor for IgE in rat basophilic leukemia cells. *Biochem Biophys Res Commun 214*, 888-894.

[42] Li, W., Yu, J. C., Michieli, P., Beeler, J. F., Ellmore, N., Heidaran, M. A. and Pierce, J. H. (1994) Stimulation of the platelet-derived growth factor beta receptor signaling pathway activates protein kinase C-delta. *Mol Cell Biol 14*, 6727-6735.

[43] Soltoff, S. P. and Toker, A. (1995) Carbachol, substance P, and phorbol ester promote the tyrosine phosphorylation of protein kinase C delta in salivary gland epithelial cells. *J Biol Chem 270*, 13490-13495.

[44] Moussazadeh, M. and Haimovich, B. (1998) Protein kinase C-delta activation and tyrosine phosphorylation in platelets. *FEBS Lett 438*, 225-230.

[45] Li, W., Jiang, Y. X., Zhang, J., Soon, L., Flechner, L., Kapoor, V., Pierce, J. H. and Wang, L. H. (1998) Protein kinase C-delta is an important signaling molecule in insulin-like growth factor I receptor-mediated cell transformation. *Mol Cell Biol 18*, 5888-5898.

[46] Popoff, I. J. and Deans, J. P. (1999) Activation and tyrosine phosphorylation of protein kinase C delta in response to B cell antigen receptor stimulation. *Mol Immunol 36*, 1005-1016.

[47] Barbazuk, S. M. and Gold, M. R. (1999) Protein kinase C-delta is a target of B-cell antigen receptor signaling. *Immunol Lett 69*, 259-267.

[48] Benes, C. and Soltoff, S. P. (2001) Modulation of PKCdelta tyrosine phosphorylation and activity in salivary and PC-12 cells by Src kinases. *Am J Physiol Cell Physiol 280*, C1498-1510.

[49] Haleem-Smith, H., Chang, E. Y., Szallasi, Z., Blumberg, P. M. and Rivera, J. (1995) Tyrosine phosphorylation of protein kinase C-delta in response to the activation of the high-affinity receptor for immunoglobulin E modifies its substrate recognition. *Proc Natl Acad Sci U S A 92*, 9112-9116.

[50] Kronfeld, I., Kazimirsky, G., Lorenzo, P. S., Garfield, S. H., Blumberg, P. M. and Brodie, C. (2000) Phosphorylation of protein kinase Cdelta on distinct tyrosine residues regulates specific cellular functions. *J Biol Chem 275*, 35491-35498.

[51] Acs, P., Beheshti, M., Szallasi, Z., Li, L., Yuspa, S. H. and Blumberg, P. M. (2000) Effect of a tyrosine 155 to phenylalanine mutation of protein kinase cdelta on the proliferative and tumorigenic properties of NIH 3T3 fibroblasts. *Carcinogenesis 21*, 887-891.

[52] Konishi, H., Tanaka, M., Takemura, Y., Matsuzaki, H., Ono, Y., Kikkawa, U. and Nishizuka, Y. (1997) Activation of protein kinase C by tyrosine phosphorylation in response to H2O2. *Proc Natl Acad Sci U S A 94*, 11233-11237.

[53] Blass, M., Kronfeld, I., Kazimirsky, G., Blumberg, P. M. and Brodie, C. (2002) Tyrosine phosphorylation of protein kinase Cdelta is essential for its apoptotic effect in response to etoposide. *Mol Cell Biol 22*, 182-195.

[54] Blake, R. A., Garcia-Paramio, P., Parker, P. J. and Courtneidge, S. A. (1999) Src promotes PKCdelta degradation. *Cell Growth Differ 10*, 231-241.

[55] Gschwendt, M., Kielbassa, K., Kittstein, W. and Marks, F. (1994) Tyrosine phosphorylation and stimulation of protein kinase C delta from porcine spleen by src in vitro. Dependence on the activated state of protein kinase C delta. *FEBS Lett 347*, 85-89.

[56] Shanmugam, M., Krett, N. L., Peters, C. A., Maizels, E. T., Murad, F. M., Kawakatsu, H., Rosen, S. T. and Hunzicker-Dunn, M. (1998) Association of PKC delta and active Src in PMA-treated MCF-7 human breast cancer cells. *Oncogene 16*, 1649-1654.

[57] Zang, Q., Lu, Z., Curto, M., Barile, N., Shalloway, D. and Foster, D. A. (1997) Association between v-Src and protein kinase C delta in v-Src-transformed fibroblasts. *J Biol Chem 272*, 13275-13280.

[58] Yuan, Z. M., Utsugisawa, T., Ishiko, T., Nakada, S., Huang, Y., Kharbanda, S., Weichselbaum, R. and Kufe, D. (1998) Activation of protein kinase C delta by the c-Abl tyrosine kinase in response to ionizing radiation. *Oncogene 16*, 1643-1648.

[59] Sun, X., Wu, F., Datta, R., Kharbanda, S. and Kufe, D. (2000) Interaction between protein kinase C delta and the c-Abl tyrosine kinase in the cellular response to oxidative stress. *J Biol Chem 275*, 7470-7473.

[60] Wrenn, R. W. (2001) Carbachol stimulates TYR phosphorylation and association of PKCdelta and PYK2 in pancreas. *Biochem Biophys Res Commun 282*, 882-886.

[61] Dempsey, E. C., Newton, A. C., Mochly-Rosen, D., Fields, A. P., Reyland, M. E., Insel, P. A. and Messing, R. O. (2000) Protein kinase C isozymes and the regulation of diverse cell responses. *Am J Physiol Lung Cell Mol Physiol 279*, L429-438.

[62] Nakai, M., Hojo, K., Yagi, K., Saito, N., Taniguchi, T., Terashima, A., Kawamata, T., Hashimoto, T., Maeda, K., Gschwendt, M., Yamamoto, H., Miyamoto, E. and Tanaka, C. (1999) Amyloid beta protein (25-35) phosphorylates MARCKS through tyrosine kinase-activated protein kinase C signaling pathway in microglia. *J Neurochem 72*, 1179-1186.

[63] Grandvaux, N., Elsen, S. and Vignais, P. V. (2001) Oxidant-dependent phosphorylation of p40phox in B lymphocytes. *Biochem Biophys Res Commun 287*, 1009-1016.

[64] Deszo, E. L., Brake, D. K., Cengel, K. A., Kelley, K. W. and Freund, G. G. (2001) CD45 negatively regulates monocytic cell differentiation by inhibiting phorbol 12-myristate 13-acetate-dependent activation and tyrosine phosphorylation of protein kinase Cdelta. *J Biol Chem 276*, 10212-10217.

[65] Yamamoto, T., Matsuzaki, H., Konishi, H., Ono, Y. and Kikkawa, U. (2000) H(2)O(2)-induced tyrosine phosphorylation of protein kinase cdelta by a mechanism independent of inhibition of protein-tyrosine phosphatase in CHO and COS-7 cells. *Biochem Biophys Res Commun 273*, 960-966.

[66] Pongracz, J., Webb, P., Wang, K., Deacon, E., Lunn, O. J. and Lord, J. M. (1999) Spontaneous neutrophil apoptosis involves caspase 3-mediated activation of protein kinase C-delta. *J Biol Chem 274*, 37329-37334.

[67] Shizukuda, Y., Reyland, M. E. and Buttrick, P. M. (2002) Protein kinase C-delta modulates apoptosis induced by hyperglycemia in adult ventricular myocytes. *Am J Physiol Heart Circ Physiol 282*, H1625-1634.

[68] Bertho, N., Blancheteau, V. M., Setterblad, N., Laupeze, B., Lord, J. M., Drenou, B., Amiot, L., Charron, D. J., Fauchet, R. and Mooney, N. (2002) MHC class II-mediated apoptosis of mature dendritic cells proceeds by activation of the protein kinase C-delta isoenzyme. *Int Immunol 14*, 935-942.

[69] Datta, R., Banach, D., Kojima, H., Talanian, R. V., Alnemri, E. S., Wong, W. W. and Kufe, D. W. (1996) Activation of the CPP32 protease in apoptosis induced by 1-beta-D-arabinofuranosylcytosine and other DNA-damaging agents. *Blood 88*, 1936-1943.

[70] Denning, M. F., Wang, Y., Tibudan, S., Alkan, S., Nickoloff, B. J. and Qin, J. Z. (2002) Caspase activation and disruption of mitochondrial membrane potential during UV radiation-induced apoptosis of human keratinocytes requires activation of protein kinase C. *Cell Death Differ 9*, 40-52.

[71] Anantharam, V., Kitazawa, M., Wagner, J., Kaul, S. and Kanthasamy, A. G. (2002) Caspase-3-dependent proteolytic cleavage of protein kinase Cdelta is essential for oxidative stress-mediated dopaminergic cell death after exposure to methylcyclopentadienyl manganese tricarbonyl. *J Neurosci 22*, 1738-1751.

[72] Cross, T., Griffiths, G., Deacon, E., Sallis, R., Gough, M., Watters, D. and Lord, J. M. (2000) PKC-delta is an apoptotic lamin kinase. *Oncogene 19*, 2331-2337.

[73] Mandil, R., Ashkenazi, E., Blass, M., Kronfeld, I., Kazimirsky, G., Rosenthal, G., Umansky, F., Lorenzo, P. S., Blumberg, P. M. and Brodie, C. (2001) Protein kinase Calpha and protein kinase Cdelta play opposite roles in the proliferation and apoptosis of glioma cells. *Cancer Res 61*, 4612-4619.

[74] Basu, A., Woolard, M. D. and Johnson, C. L. (2001) Involvement of protein kinase C-delta in DNA damage-induced apoptosis. *Cell Death Differ 8*, 899-908.

[75] Basu, A. and Akkaraju, G. R. (1999) Regulation of caspase activation and cis-diamminedichloroplatinum(II)-induced cell death by protein kinase C. *Biochemistry 38*, 4245-4251.

[76] Park, I. C., Park, M. J., Hwang, C. S., Rhee, C. H., Whang, D. Y., Jang, J. J., Choe, T. B., Hong, S. I. and Lee, S. H. (2000) Mitomycin C induces apoptosis in a caspases-dependent and Fas/CD95-independent manner in human gastric adenocarcinoma cells. *Cancer Lett 158*, 125-132.

[77] Denning, M. F., Wang, Y., Nickoloff, B. J. and Wrone-Smith, T. (1998) Protein kinase Cdelta is activated by caspase-dependent proteolysis during ultraviolet radiation-induced apoptosis of human keratinocytes. *J Biol Chem 273*, 29995-30002.

[78] Scheel-Toellner, D., Pilling, D., Akbar, A. N., Hardie, D., Lombardi, G., Salmon, M. and Lord, J. M. (1999) Inhibition of T cell apoptosis by IFN-beta rapidly reverses nuclear translocation of protein kinase C-delta. *Eur J Immunol 29*, 2603-2612.

[79] Fukunaga, M., Oka, M., Ichihashi, M., Yamamoto, T., Matsuzaki, H. and Kikkawa, U. (2001) UV-induced tyrosine phosphorylation of PKC delta and promotion of apoptosis in the HaCaT cell line. *Biochem Biophys Res Commun 289*, 573-579.

[80] Fujii, T., Garcia-Bermejo, M. L., Bernabo, J. L., Caamano, J., Ohba, M., Kuroki, T., Li, L., Yuspa, S. H. and Kazanietz, M. G. (2000) Involvement of protein kinase C delta (PKCdelta) in phorbol ester-induced apoptosis in LNCaP prostate cancer cells. Lack of proteolytic cleavage of PKCdelta. *J Biol Chem 275*, 7574-7582.

[81] Matassa, A. A., Carpenter, L., Biden, T. J., Humphries, M. J. and Reyland, M. E. (2001) PKCdelta is required for mitochondrial-dependent apoptosis in salivary epithelial cells. *J Biol Chem 276*, 29719-29728.

[82] Martins, L. M., Kottke, T. J., Kaufmann, S. H. and Earnshaw, W. C. (1998) Phosphorylated forms of activated caspases are present in cytosol from HL-60 cells during etoposide-induced apoptosis. *Blood 92*, 3042-3049.

[83] Cardone, M. H., Roy, N., Stennicke, H. R., Salvesen, G. S., Franke, T. F., Stanbridge, E., Frisch, S. and Reed, J. C. (1998) Regulation of cell death protease caspase-9 by phosphorylation. *Science 282*, 1318-1321.

[84] Pearson, R. B. and Kemp, B. E. (1991) Protein kinase phosphorylation site sequences and consensus specificity motifs: tabulations. *Methods Enzymol 200*, 62-81.

[85] Nishikawa, K., Toker, A., Johannes, F. J., Songyang, Z. and Cantley, L. C. (1997) Determination of the specific substrate sequence motifs of protein kinase C isozymes. *J Biol Chem 272*, 952-960.

[86] Watanabe, T., Ono, Y., Taniyama, Y., Hazama, K., Igarashi, K., Ogita, K., Kikkawa, U. and Nishizuka, Y. (1992) Cell division arrest induced by phorbol ester in CHO cells overexpressing protein kinase C-delta subspecies. *Proc Natl Acad Sci U S A 89*, 10159-10163.

[87] Lu, Z., Hornia, A., Jiang, Y. W., Zang, Q., Ohno, S. and Foster, D. A. (1997) Tumor promotion by depleting cells of protein kinase C delta. *Mol Cell Biol 17*, 3418-3428.

[88] Franklin, R. A. and McCubrey, J. A. (2000) Kinases: positive and negative regulators of apoptosis. *Leukemia 14*, 2019-2034.

[89] Cross, T. G., Scheel-Toellner, D., Henriquez, N. V., Deacon, E., Salmon, M. and Lord, J. M. (2000) Serine/threonine protein kinases and apoptosis. *Exp Cell Res 256*, 34-41.

[90] Huang, Y., Yuan, Z. M., Ishiko, T., Nakada, S., Utsugisawa, T., Kato, T., Kharbanda, S. and Kufe, D. W. (1997) Pro-apoptotic effect of the c-Abl tyrosine kinase in the cellular response to 1-beta-D-arabinofuranosylcytosine. *Oncogene 15*, 1947-1952.

[91] Kharbanda, S., Bharti, A., Pei, D., Wang, J., Pandey, P., Ren, R., Weichselbaum, R., Walsh, C. T. and Kufe, D. (1996) The stress response to ionizing radiation involves c-Abl-dependent phosphorylation of SHPTP1. *Proc Natl Acad Sci U S A 93*, 6898-6901.

[92] Kharbanda, S., Ren, R., Pandey, P., Shafman, T. D., Feller, S. M., Weichselbaum, R. R. and Kufe, D. W. (1995) Activation of the c-Abl tyrosine kinase in the stress response to DNA-damaging agents. *Nature 376*, 785-788.

[93] Kharbanda, S., Yuan, Z. M., Weichselbaum, R. and Kufe, D. (1997) Functional role for the c-Abl protein tyrosine kinase in the cellular response to genotoxic stress. *Biochim Biophys Acta 1333*, O1-7.

[94] Yoshida, K., Komatsu, K., Wang, H. G. and Kufe, D. (2002) c-Abl tyrosine kinase regulates the human Rad9 checkpoint protein in response to DNA damage. *Mol Cell Biol 22*, 3292-3300.

[95] Yoshida, K., Yamaguchi, T., Natsume, T., Kufe, D. and Miki, Y. (2005) JNK phosphorylation of 14-3-3 proteins regulates nuclear targeting of c-Abl in the apoptotic response to DNA damage. *Nat Cell Biol 7*, 278-285.

[96] Yuan, Z. M., Huang, Y., Ishiko, T., Kharbanda, S., Weichselbaum, R. and Kufe, D. (1997) Regulation of DNA damage-induced apoptosis by the c-Abl tyrosine kinase. *Proc Natl Acad Sci U S A 94*, 1437-1440.

[97] Agami, R., Blandino, G., Oren, M. and Shaul, Y. (1999) Interaction of c-Abl and p73alpha and their collaboration to induce apoptosis. *Nature 399*, 809-813.

[98] Yuan, Z. M., Shioya, H., Ishiko, T., Sun, X., Gu, J., Huang, Y. Y., Lu, H., Kharbanda, S., Weichselbaum, R. and Kufe, D. (1999) p73 is regulated by tyrosine kinase c-Abl in the apoptotic response to DNA damage. *Nature 399*, 814-817.

[99] Gong, J. G., Costanzo, A., Yang, H. Q., Melino, G., Kaelin, W. G., Jr., Levrero, M. and Wang, J. Y. (1999) The tyrosine kinase c-Abl regulates p73 in apoptotic response to cisplatin-induced DNA damage. *Nature 399*, 806-809.

[100] Yoshida, K. and Miki, Y. (2005) Enabling death by the Abl tyrosine kinase: mechanisms for nuclear shuttling of c-Abl in response to DNA damage. *Cell Cycle 4*, 777-779.

[101] Godbout, J. P., Pesavento, J., Hartman, M. E., Manson, S. R. and Freund, G. G. (2002) Methylglyoxal enhances cisplatin-induced cytotoxicity by activating protein kinase Cdelta. *J Biol Chem 277*, 2554-2561.

[102] Yoshida, K., Weichselbaum, R., Kharbanda, S. and Kufe, D. (2000) Role for Lyn tyrosine kinase as a regulator of stress-activated protein kinase activity in response to DNA damage. *Mol Cell Biol 20*, 5370-5380.

[103] Yoshida, K., Kharbanda, S. and Kufe, D. (1999) Functional interaction between SHPTP1 and the Lyn tyrosine kinase in the apoptotic response to DNA damage. *J Biol Chem 274*, 34663-34668.

[104] Ren, J., Datta, R., Shioya, H., Li, Y., Oki, E., Biedermann, V., Bharti, A. and Kufe, D. (2002) p73beta is regulated by protein kinase Cdelta catalytic fragment generated in the apoptotic response to DNA damage. *J Biol Chem 277*, 33758-33765.

[105] Yoshida, K. and Kufe, D. (2001) Negative regulation of the SHPTP1 protein tyrosine phosphatase by protein kinase C delta in response to DNA damage. *Mol Pharmacol 60*, 1431-1438.

[106] Chen, Z., Gibson, T. B., Robinson, F., Silvestro, L., Pearson, G., Xu, B., Wright, A., Vanderbilt, C. and Cobb, M. H. (2001) MAP kinases. *Chem Rev 101*, 2449-2476.

[107] Reyland, M. E., Barzen, K. A., Anderson, S. M., Quissell, D. O. and Matassa, A. A. (2000) Activation of PKC is sufficient to induce an apoptotic program in salivary gland acinar cells. *Cell Death Differ 7*, 1200-1209.

[108] Mitsutake, N., Namba, H., Shklyaev, S. S., Tsukazaki, T., Ohtsuru, A., Ohba, M., Kuroki, T., Ayabe, H. and Yamashita, S. (2001) PKC delta mediates ionizing radiation-induced activation of c-Jun NH(2)-terminal kinase through MKK7 in human thyroid cells. *Oncogene 20*, 989-996.

[109] Lee, Y. J., Soh, J. W., Dean, N. M., Cho, C. K., Kim, T. H., Lee, S. J. and Lee, Y. S. (2002) Protein kinase Cdelta overexpression enhances radiation sensitivity via extracellular regulated protein kinase 1/2 activation, abolishing the radiation-induced G(2)-M arrest. *Cell Growth Differ 13*, 237-246.

[110] Vancurova, I., Miskolci, V. and Davidson, D. (2001) NF-kappa B activation in tumor necrosis factor alpha-stimulated neutrophils is mediated by protein kinase Cdelta. Correlation to nuclear Ikappa Balpha. *J Biol Chem 276*, 19746-19752.

[111] Storz, P., Doppler, H. and Toker, A. (2004) Protein kinase Cdelta selectively regulates protein kinase D-dependent activation of NF-kappaB in oxidative stress signaling. *Mol Cell Biol 24*, 2614-2626.

[112] Rahman, A., Anwar, K. N., Uddin, S., Xu, N., Ye, R. D., Platanias, L. C. and Malik, A. B. (2001) Protein kinase C-delta regulates thrombin-induced ICAM-1 gene expression in endothelial cells via activation of p38 mitogen-activated protein kinase. *Mol Cell Biol 21*, 5554-5565.

[113] Vanden Berghe, W., Plaisance, S., Boone, E., De Bosscher, K., Schmitz, M. L., Fiers, W. and Haegeman, G. (1998) p38 and extracellular signal-regulated kinase mitogen-activated protein kinase pathways are required for nuclear factor-kappaB p65 transactivation mediated by tumor necrosis factor. *J Biol Chem 273*, 3285-3290.

[114] Rahman, A., True, A. L., Anwar, K. N., Ye, R. D., Voyno-Yasenetskaya, T. A. and Malik, A. B. (2002) Galpha(q) and Gbetagamma regulate PAR-1 signaling of thrombin-induced NF-kappaB activation and ICAM-1 transcription in endothelial cells. *Circ Res 91*, 398-405.

[115] Suh, K. S., Tatunchak, T. T., Crutchley, J. M., Edwards, L. E., Marin, K. G. and Yuspa, S. H. (2003) Genomic structure and promoter analysis of PKC-delta. *Genomics 82*, 57-67.

[116] Brodie, C. and Blumberg, P. M. (2003) Regulation of cell apoptosis by protein kinase c delta. *Apoptosis 8*, 19-27.

[117] Majumder, P. K., Pandey, P., Sun, X., Cheng, K., Datta, R., Saxena, S., Kharbanda, S. and Kufe, D. (2000) Mitochondrial translocation of protein kinase C delta in phorbol ester-induced cytochrome c release and apoptosis. *J Biol Chem 275*, 21793-21796.

[118] Li, L., Lorenzo, P. S., Bogi, K., Blumberg, P. M. and Yuspa, S. H. (1999) Protein kinase Cdelta targets mitochondria, alters mitochondrial membrane potential, and induces apoptosis in normal and neoplastic keratinocytes when overexpressed by an adenoviral vector. *Mol Cell Biol 19*, 8547-8558.

[119] Majumder, P. K., Mishra, N. C., Sun, X., Bharti, A., Kharbanda, S., Saxena, S. and Kufe, D. (2001) Targeting of protein kinase C delta to mitochondria in the oxidative stress response. *Cell Growth Differ 12*, 465-470.

[120] Sumitomo, M., Ohba, M., Asakuma, J., Asano, T., Kuroki, T., Asano, T. and Hayakawa, M. (2002) Protein kinase Cdelta amplifies ceramide formation via mitochondrial signaling in prostate cancer cells. *J Clin Invest 109*, 827-836.

[121] Liu, J., Chen, J., Dai, Q. and Lee, R. M. (2003) Phospholipid scramblase 3 is the mitochondrial target of protein kinase C delta-induced apoptosis. *Cancer Res 63*, 1153-1156.

[122] Liu, J., Dai, Q., Chen, J., Durrant, D., Freeman, A., Liu, T., Grossman, D. and Lee, R. M. (2003) Phospholipid scramblase 3 controls mitochondrial structure, function, and apoptotic response. *Mol Cancer Res 1*, 892-902.

[123] Lutter, M., Fang, M., Luo, X., Nishijima, M., Xie, X. and Wang, X. (2000) Cardiolipin provides specificity for targeting of tBid to mitochondria. *Nat Cell Biol 2*, 754-761.

[124] DeVries, T. A., Neville, M. C. and Reyland, M. E. (2002) Nuclear import of PKCdelta is required for apoptosis: identification of a novel nuclear import sequence. *EMBO J 21*, 6050-6060.

[125] Yoshida, K., Wang, H. G., Miki, Y. and Kufe, D. (2003) Protein kinase Cdelta is responsible for constitutive and DNA damage-induced phosphorylation of Rad9. *EMBO J 22*, 1431-1441.

[126] Eitel, K., Staiger, H., Rieger, J., Mischak, H., Brandhorst, H., Brendel, M. D., Bretzel, R. G., Haring, H. U. and Kellerer, M. (2003) Protein kinase C delta activation and translocation to the nucleus are required for fatty acid-induced apoptosis of insulin-secreting cells. *Diabetes 52*, 991-997.

[127] Smith, G. C. and Jackson, S. P. (1999) The DNA-dependent protein kinase. *Genes Dev 13*, 916-934.

[128] Kharbanda, S., Pandey, P., Jin, S., Inoue, S., Bharti, A., Yuan, Z. M., Weichselbaum, R., Weaver, D. and Kufe, D. (1997) Functional interaction between DNA-PK and c-Abl in response to DNA damage. *Nature 386*, 732-735.

[129] Bharti, A., Kraeft, S. K., Gounder, M., Pandey, P., Jin, S., Yuan, Z. M., Lees-Miller, S. P., Weichselbaum, R., Weaver, D., Chen, L. B., Kufe, D. and Kharbanda, S. (1998) Inactivation of DNA-dependent protein kinase by protein kinase Cdelta: implications for apoptosis. *Mol Cell Biol 18*, 6719-6728.

[130] Kaufmann, S. H. (1989) Induction of endonucleolytic DNA cleavage in human acute myelogenous leukemia cells by etoposide, camptothecin, and other cytotoxic anticancer drugs: a cautionary note. *Cancer Res 49*, 5870-5878.

[131] Peter, M., Nakagawa, J., Doree, M., Labbe, J. C. and Nigg, E. A. (1990) In vitro disassembly of the nuclear lamina and M phase-specific phosphorylation of lamins by cdc2 kinase. *Cell 61*, 591-602.

[132] al-Khodairy, F., Fotou, E., Sheldrick, K. S., Griffiths, D. J., Lehmann, A. R. and Carr, A. M. (1994) Identification and characterization of new elements involved in checkpoint and feedback controls in fission yeast. *Mol Biol Cell 5*, 147-160.

[133] Volkmer, E. and Karnitz, L. M. (1999) Human homologs of Schizosaccharomyces pombe rad1, hus1, and rad9 form a DNA damage-responsive protein complex. *J Biol Chem 274*, 567-570.

[134] Chen, M. J., Lin, Y. T., Lieberman, H. B., Chen, G. and Lee, E. Y. (2001) ATM-dependent phosphorylation of human Rad9 is required for ionizing radiation-induced checkpoint activation. *J Biol Chem 276*, 16580-16586.

[135] Matsuda, K., Yoshida, K., Taya, Y., Nakamura, K., Nakamura, Y. and Arakawa, H. (2002) p53AIP1 regulates the mitochondrial apoptotic pathway. *Cancer Res 62*, 2883-2889.

[136] Oda, K., Arakawa, H., Tanaka, T., Matsuda, K., Tanikawa, C., Mori, T., Nishimori, H., Tamai, K., Tokino, T., Nakamura, Y. and Taya, Y. (2000) p53AIP1, a potential mediator of p53-dependent apoptosis, and its regulation by Ser-46-phosphorylated p53. *Cell 102*, 849-862.

[137] Okamura, S., Arakawa, H., Tanaka, T., Nakanishi, H., Ng, C. C., Taya, Y., Monden, M. and Nakamura, Y. (2001) p53DINP1, a p53-inducible gene, regulates p53-dependent apoptosis. *Mol Cell 8*, 85-94.

[138] Yoshida, K., Liu, H. and Miki, Y. (2006) Protein Kinase C delta regulates Ser46 phosphorylation of p53 tumor suppressor in the apoptotic response to DNA damage. *J Biol Chem 281*, 5734-5740.

[139] Yoshida, K., Yamaguchi, T., Shinagawa, H., Taira, N., Nakayama, K. I. and Miki, Y. (2006) Protein kinase C delta activates topoisomerase IIalpha to induce apoptotic cell death in response to DNA damage. *Mol Cell Biol 26*, 3414-3431.

[140] Meinhardt, G., Roth, J., Totok, G., Auner, H., Emmerich, B. and Hass, R. (1999) Signaling defect in the activation of caspase-3 and PKCdelta in human TUR leukemia cells is associated with resistance to apoptosis. *Exp Cell Res 247*, 534-542.

In: New Developments in Cell Apoptosis Research
Editor: Alan J. Corvin, pp. 121-158
ISBN: 1-60021-453-3

Chapter V

Heat Shock Proteins against Radiation-Induced Apoptosis: Facts, Hypotheses, Prospects

Alexander E. Kabakov
Department of Radiation Biochemistry, Medical Radiology Research Center, Obninsk, Russia

Abstract

The mammalian heat shock proteins Hsp70, Hsp27 and Hsp90 are known as major molecular chaperones which preserve the native structure of cellular proteins and promote refolding or degradation of stress-damaged protein molecules. For a long time, it was generally accepted that these Hsps are involved in cytoprotection from proteotoxic (protein-damaging) exposures only but not from genotoxic (DNA-damaging) ones. If so, it would be hard to imagine that the Hsps are able somehow to protect mammalian cells against severe ionizing radiation that induces numerous breaks in genomic DNA thereby leading to cell death. Nevertheless, more and more data are published about real contribution of Hsp70, Hsp27 and Hsp90 to protection from the cell-killing effects of irradiation with gamma- or X-rays. Taken together these data allow to conclude that the Hsp-conferred radioprotection is mainly based on an ability of certain Hsps to block the irradiation-responsive apoptotic pathway and/or to stimulate alternative 'cell survival' pathways that interfere with apoptotic signal transduction. The present review affords an intriguing insight into a phenomenon of the Hsp-mediated cellular defense against the radiation-induced apoptosis. Implication of this phenomenon in some radiation-related issues of biology and medicine such as DNA damage/repair, genomic instability, oncogenesis and tumor cell resistance to radiotherapy is also considered herein. Moreover, several prospective approaches are discussed toward pharmacological manipulating the *in vivo* Hsp expression/activity to beneficially influence the outcome of anticancer radiotherapy.

Introduction

The so called 'heat stress response' is a universal adaptive mechanism that takes place in all living organisms from bacteria to man. In mammalian cells, this mechanism is triggered by a heat shock transcription factor 1 (HSF1) when it becomes activated in response to the heat stress-provoked denaturation and aggregation of intracellular proteins. The activated HSF1 initiates transcription of certain *heat shock*-genes encoding a set of cytoprotective 'heat shock proteins' (Hsps). As a result, the stressed cell transiently accumulates inducible Hsps which contribute to the post-stress recovery of cellular homeostasis and confer, for some following period, cytoprotection from the stronger, otherwise lethal stress (see for a review [1-3]).

It is well known that not only heat shock but also many other stressful exposures and some pathophysiological states (e.g. oxidative stress, acidification, energy starvation, hypoxia, ischemia/reperfusion, inflammation, etc) are associated with the HSF1 activation-mediated Hsp induction that yields a transient cell tolerance to either stress. Such a phenomenon of 'cross tolerance' is usually explained by a unique ability of Hsps to attenuate the proteotoxic (protein-damaging) effects of stresses, as all major Hsps are molecular chaperones that maintain the native conformation of cellular proteins and preserve nascent polypeptide chains from their misfolding and aggregating [1, 4, 5]. In the unstressed cell, the constitutively expressed Hsp70 and Hsp90, together with their co-chaperones, assist protein maturation and folding, protein transport, proteolysis, and intracellular signal transduction (reviewed in [5-10]). Inside the tolerant (Hsp-enriched) cell undergoing any proteotoxic exposure, the excess chaperones (Hsps) are able to minimize protein denaturation and aggregation resulting from the exposure and then to promote refolding or degradation of stress-damaged proteins [1, 4, 11]. Besides that, the 'small' Hsp (referred to as Hsp27 or Hsp25) is involved in phosphorylation-dependent regulation of the actin filament dynamics; the same activity appears to be responsible for preservation and/or faster restoration of the cytoskeletal framework [12-14]. Moreover, Hsp90, Hsp70 and Hsp27 can interfere with the apoptotic signaling [15-17]. The latter feature of these Hsps can also contribute to cytoprotection because many kinds of stress trigger the apoptotic pathway while the stressed tolerant cell avoiding early apoptosis will have more facilities for its recovery and survival. Being manifested together under a stressful exposure, the chaperone, cytoskeleton-regulating and antiapoptotic activities of the excess Hsps ensure lesser damage and better recovery of cellular structures and metabolic processes thus resulting in the improved survival of the tolerant cells [11, 14, 17].

Although a cytoprotective capacity of inducible Hsps toward any proteotoxic exposures seems quite obvious, it would be hard to expect analogous Hsp-mediated cytoprotection against genotoxic agents, in particular, ionizing radiation. Unlike heat shock, gamma- or X-rays do not induce early alterations in the native structure of cellular proteins but, instead, cause numerous single- and double-strand breaks in nuclear DNA. This DNA breaking can lead to a failure of the genome program and, consequently, to interphase cell death (apoptosis and necrosis) or reproductive cell death (mitotic catastrophe) [18-20]. Besides the DNA breakage due to direct hits of quanta, ionizing radiation stimulates free radical generation inside the exposed cells, so that additional radiation-associated damage of genomic DNA can occur through attack by the chemically aggressive free radicals [21-23]. There is a cellular system of certain 'checkpoint control': in response to severe DNA damage, a pro-apoptotic protein p53 is up-regulated and activated to initiate cell growth arrest and the cytochrome *c*/caspase-9-mediated apoptotic pathway [24-28]. The p53-dependent apoptosis enables to eliminate the affected cells with the severely damaged genome; *in vivo*, apoptotic elimination of such cells appears to decrease a risk of their malignant transformation following genotoxic (e.g. radiation) exposures. In some cases, the radiation-induced DNA damage may evoke another type of apoptosis via p53-independent pathways [29, 30]. As no direct DNA-related activities have been attributed to the major inducible Hsps, it was generally accepted that Hsps can protect cells from the proteotoxic stresses only but not from the genotoxic ionizing radiation that breaks nuclear DNA and kills the exposed cells.

However, more and more publications report about positive correlations between the intracellular Hsp content and cell resistance to ionizing radiation. So, preirradiation induction of Hsps in various normal or tumor cell cultures [31-34] and in murine tissues [35] was shown to attenuate apoptosis induced by gamma-ray exposures. In comparative studies on normal and *HSF1*-knockout mouse embryo fibroblasts, it was recently established that the HSF1-mediated stress response resulting in Hsp accumulation can, in fact, reduce apoptosis in the irradiated cells thus improving their postirradiation survival [34]. In support of that, experiments with plasmid- or viral vector-induced overexpression of individual Hsps have demonstrated attenuation of apoptosis in irradiated murine cells transiently overexpressing Hsp70 [34, 36-38] or Hsp27 [34, 38-40]. Finally, it was found that 'pharmacological' blockage of the Hsp90 chaperone activity with specific inhibitors selectively radiosensitizes cells of some tumors enhancing their apoptosis after X- or gamma-ray irradiation [41-45]. Taken together all these findings allow to consider Hsp70, Hsp25(27) and Hsp90 as endogenous radioprotectors whose activities contribute to better survival of severely irradiated mammalian cells. Below, an involvement of the major Hsps in the cellular radiation response is reviewed in detail and potential mechanisms of the Hsp-mediated radioprotection as well as its significance for clinical radiotherapy are discussed.

Hsp70 against Radiation-induced Apoptosis: Modulation of Signaling and Attenuation of Genotoxicity

The 70 kDa heat shock protein (Hsc70/Hsp70) is a major chaperone of eukaryotes that regulates protein folding in an ATP-dependent manner [1, 5, 7]. Under stressful conditions, expression of this chaperone usually augments so that excess Hsp70, with assistance of its co-chaperones, can contribute to disaggregation, refolding or degradation of stress-damaged cellular proteins [7, 11].

The first convincing publication about the radioprotective properties of Hsp70 was made by Gordon et al in 1997 [31]. Those researchers found marked attenuation of radiation-induced apoptosis in isolated murine thymocytes and T-lymphocyte clones in which Hsp70 was previously induced by heating or arsenite. Down-regulation of two pro-apoptotic proteins, p53 and Bax, was also observed in the Hsp70-enriched cells. To prove the prominent role of Hsp70 among other stress-inducible proteins the researchers used an approach with *Hsp70*-antisense RNA that abolished both the Hsp70 induction and the radioprotection conferred by the pretreatments [31]. Later, similar protection of irradiated thymocytes from apoptosis was achieved by whole body hyperthermia in mice that induced Hsp70 in all murine tissues including thymus [35]; that antiapoptotic effect correlated well with the intracellular Hsp70 content and was abrogated in the mice injected with flavonoid Quercetin, an inhibitor of the Hsp induction [46-48]. The increased cell survival and impaired postirradiation apoptosis were recently revealed in heat-preconditioned cultures of mouse embryo fibroblasts when the expression of inducible Hsp70 peaked in them [34]. All those demonstrations of the radioprotective potential of stress-induced Hsp70 [31-35] are well supported by the data on non-stressful (viral vector- or plasmid-achieved) Hsp70 overexpression that similarly reduced apoptosis in irradiated cultures of mouse embryo fibroblasts [34] and NIH3T3 cells [36, 37]. Finally, the impairment of postirradiation apoptosis was observed following induction (by low gamma-ray doses) of the radioadaptive response in Hsp70-overexpressing splenocytes from *Hspa4*-transgenic mice [49]; this finding also implies a key role for Hsp70 in the inducible mechanisms of radioprotection.

With regard to revealing of analogous effects in tumors, an acquired resistance to irradiation-induced apoptosis was observed in prostate cancer cells when they accumulated Hsp70 after priming heat shock [32]. When prostate cancer-derived cell lines constitutively expressed inducible Hsp70, prolonged down-regulation of this constitutive expression (by means of the antisense RNA-vector) radiosensitized the Hsp70-depleted cells by enhancing their post-irradiation transition to the senescent phenotype [50]. Similarly, transfection of human leukemic THP-1 cells with interfering RNA to silence Hsp70 significantly decreased post-irradiation clonogenicity of the transfectants [51]. In turn, non-stressful overexpression of inducible Hsp70 in mouse fibrosarcoma (RIF) cells transfected with an Hsp70-expressing plasmid was shown to increase their resistance to irradiation-induced apoptosis [38]. In several cell lines of human glioblastoma exposed to irradiation, their different radioresistance seemed to be positively correlated with the intracellular Hsp70 contents [52]. Contrary to that, in lung carcinoma cell lines, artificial up- and down-regulation of the Hsp70 expression did not affect the intensity of apoptosis resulting from ionizing irradiation [53]. The latter discrepancy may mean that the Hsp70-conferred radioprotection is not the absolutely universal phenomenon intrinsic to all types of tumor cells; though, it should be noted that this phenomenon was found in several other samples of the tumor cell lines [32, 38, 50-52]. The constitutive expression of inducible Hsp70 in some tumors and tumor-derived cell lines evokes a particular interest as one of the factors potentially contributing to the enhanced tumor resistance to radiotherapy. *In vivo*, down-regulation of this constitutive expression would increase the efficiency of radiotherapy by analogy with what was achieved on the prostate cancer and leukemia cell lines treated with the *Hsp70*-antisense RNA- or *Hsp70*-RNAi vector [50, 51]. As the antisense and interfering RNA approaches are not applicable in clinic, it would be more useful to develop appropriate 'pharmacological' ways to decrease the Hsp70 content in tumor cells. In this respect, it appears intriguing that avicins (plant-derived triterpenoid electorphiles) were shown to activate ubiquitination and degradation of Hsp70 in Jurkat T leukemia cells and some other leukemia/lymphoma cell lines but not in normal blood lymphocytes [54]. Although the down-regulating effect of avicins was not very strong (~40% Hsp70 still remained) and it was not tested toward cellular radioresistance, further screening of natural and synthetic compounds may yield suitable drugs to overcome the Hsp70-conferred radioresistance of cancer cells.

While the precise mechanism of Hsp70-mediated radioprotection has not yet been established, one of the most popular ideas is that excess Hsp70 can protect the target cells from irradiation-induced apoptosis [31-38]. After studies on models non-related to radiation, inducible Hsp70 has indeed been characterized as a blocker of apoptotic signal transduction (reviewed in [15-17]). Perhaps, the same antiapoptotic activity attributed to inducible Hsp70 may beneficially be manifested in the context of radiation exposure. For example, ionizing radiation may cause activation of c-Jun-N-terminal kinase (JNK) that eventually leads to apoptosis [55-57]. On models with apoptotic stimuli non-related to ionizing radiation, excess Hsp70 was already shown to prevent such 'lethal' JNK activation in stressed cells [58-60]. Extrapolating those findings to a situation with ionizing irradiation, one can imagine similar (Hsp70-mediated) cytoprotection toward irradiated cells, i.e. excess Hsp70 will suppress the irradiation-responsive JNK activation thereby saving the exposed cells from apoptotic death (Figure 1). So far, however, there are no experimentally obtained data confirming this idea; moreover, the ionizing radiation-associated cytotoxicity is not always due to the JNK activation [61].

Another key component of a cellular signaling network, p38 mitogen-activated protein kinase (p38 MAPK), can also be activated in cells exposed to ionizing radiation; then this activation can trigger a cascade of events resulting in apoptosis [40, 55, 62, 63]. As well as in the case with the JNK activation, the fact of inhibition of the stress-responsive p38 MAPK activation in Hsp70-overexpressing cells was previously demonstrated on models non-relevant to radiation exposure [58, 59]. If to imagine overlapping of the two situations described in [40, 55, 62] and in [58, 59], excess Hsp70 will inhibit the p38 MAPK activation in irradiated cells thus blocking the p38 MAPK-dependent apoptotic pathway in them (illustrated in Figure 1). Although such a hypothesis seems rather attractive, this is still only hypothesis because nobody has yet showed that the Hsp70-mediated inhibition of p38 MAPK activation can really protect from irradiation-induced apoptosis. Furthermore, some facts indicate that the p38 MAPK activation/Hsp27 phosphorylation pathway can be one of the cytoprotective reactions (see Figure 1 and below) and the irradiation-responsive p38 MAPK activation can contribute to DNA repair [64]. Perhaps, the *in vivo* p38 MAPK activation, in dependence of its extent and duration, exerts either proapoptotic or cytoprotective effects. It requires to be examined whether the Hsp70-mediated blockage of JNK- and/or p38 MAPK-directed signal cascades takes place in the context of radiation exposure and, if yes, how this blockage influences the postirradiation cell death/survival.

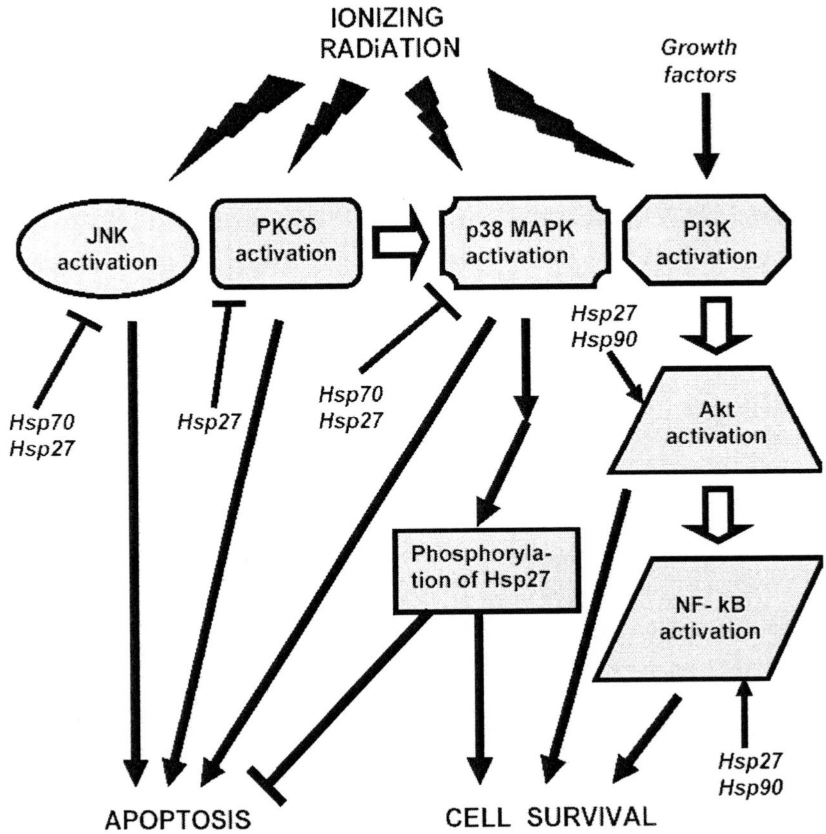

Figure 1. Suggested key points for cytoprotective modulation by Hsp70, Hsp27 and Hsp90 of the major signal transduction pathways leading to either death (apoptosis) or survival of irradiated mammalian cells. In the scheme, all the stimulating effects are designated by arrows, whereas the inhibitory effects are designated by black perpendiculars.

The major irradiation-responsive apoptotic pathway is the p53-dependent one resulting in the apoptosome formation from extramitochondrial (released) cytochrome *c*, Apaf-1 and procaspase-9 followed by 'lethal' activation of caspase-3 [19, 25-28]. It seems likely that excess Hsp70 is able to prevent cytochrome *c* release from mitochondria of irradiated cells (Figure 2) similarly to what was demonstrated in models with apoptotic stimuli non-related to radiation [60, 65-68]. Also, there were several reports that Hsp70 can inhibit apoptosis downstream of the mitochondria by preventing the apoptosome formation and/or caspase activation [69-71]. Theoretically, each irradiated cell, in the case of avoiding its early apoptotic death, will have more time (and chances) to repair the most critical DNA damages and, owing to that, to survive consequences of the exposure. However, at present, there is no evidence that a blockage of the cytochrom *c* release can solely be sufficient to prevent death of irradiated cells. On the contrary, there are some suggestions that the radioprotective (antiapoptotic) potential of Hsp70 is somehow realized just at the level of p53 expression [31, 34, 37]. Indeed, both the impairment of postirradiation up-regulation of p53 and the reduction of postirradiation apoptosis were observed in Hsp70-enriched murine thymocytes and NIH3T3 cells [31, 37] but the authors have not explained why it was so. Although the fact that Hsp70 directly interacts with p53 is well known (see for a review [72, 73]), no attractive hypotheses have been yet proposed to clearly draw a cytoprotective (antiapoptotic) mechanism based on these interactions. In speculating on this issue, it seems logical to suppose that excess Hsp70 somehow ensures accelerated degradation of p53 by proteasomes and, as a consequence, prevents the post-irradiation (p53-dependent) apoptosis. Another speculation may be that in a dependence of its content in the irradiated cell, Hsp70 can act as a switch between the pro-apoptotic function and the DNA repair function of p53. In this case, excess Hsp70 will repress the p53-mediated transcription of genes encoding apoptotic effectors and activate the p53-mediated transcription of other genes contributing to the postirradiation DNA repair thereby reprogramming the cell from suicide to survival (see Figure 2). So far, no findings have been obtained in support of such speculations and, therefore, the role of Hsp70-p53 interactions in the mechanism of cytoprotection from radiation-induced apoptosis remains to be determined.

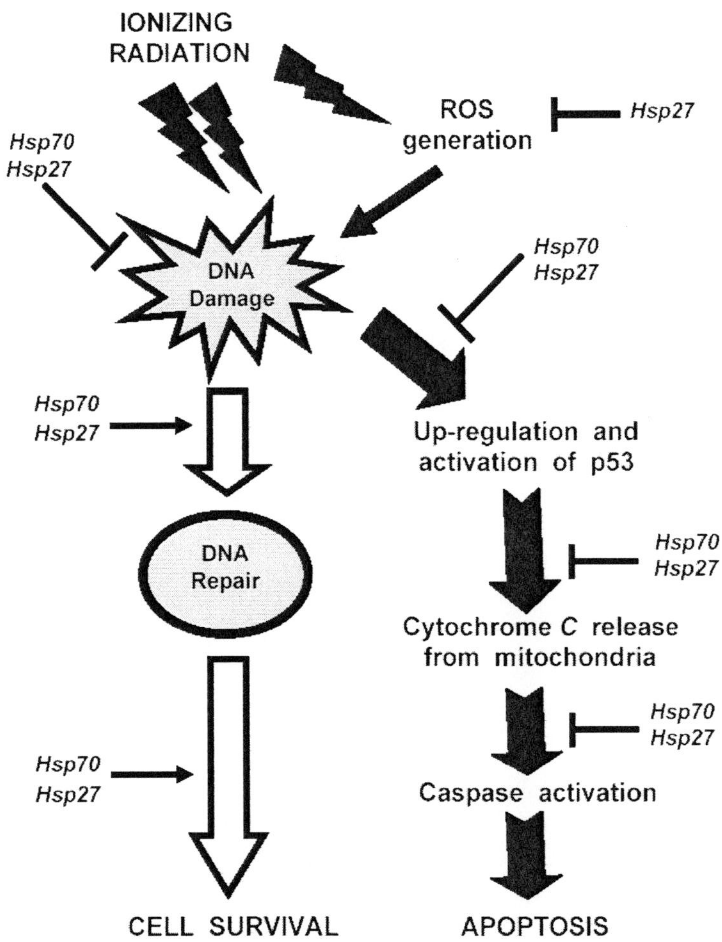

Figure 2. Hypothetical scheme showing key points for the cytoprotective action of Hsp70 and/or Hsp27 against genotoxic and apoptosis-inducing effects of ionizing radiation. In the scheme, the stimulating effects of either Hsp are designated by black arrows, whereas the inhibitory effects are designated by black perpendiculars.

Meanwhile, there are suggestions that the Hsp70-conferred radioprotection is manifested yet at the level of radiation-induced DNA damage. One of such suggestions comes from a paper of Calini et al [33] who did observe attenuation of the gamma-ray-induced DNA breakage in C3H 10T1/2 mouse embryo fibroblasts accumulating Hsp70 after heat preconditioning. Using an approach with targeted gene deletion, Hunt et al. [74] found both the enhanced radiosensitivity and the greater amounts of residual chromosome aberrations in gamma-irradiated fibroblasts from mice deficient on *Hsp70.1*- and *Hsp70.3*-genes (*Hsp70.1/3* $^{-/-}$ mice); this implies the essential genome-stabilizing role for inducible Hsp70 under radiation exposure. Proceeding from the latter finding, it would not be hard to assume that in the reverse situation with overexpressed Hsp70 instead of its deficiency, excess Hsp70 in the irradiation-exposed cell will attenuate the DNA damages resulting from the exposure. Indeed, in another recent study [34], significantly lower amounts of single- and double-strand breaks were determined in nuclear DNA of gamma-irradiated mouse embryo fibroblasts overexpressing Hsp70. Importantly, the lesser DNA breakage correlated with down-regulated expression of p53 and reduction of apoptosis [34]. Since it is the severe DNA damage that provokes the up-regulation and activation of p53 followed by triggering of the p53-dependent apoptotic pathway [19, 24-28], the Hsp70-conferred attenuation of DNA breaking under irradiation appears to define the enhanced radioresistance of the Hsp70-enriched cells Hsp70 content. By the same reason, analogous Hsp70-mediated protection of genomic DNA may rescue irradiated cells from the p53-independent apoptosis that is triggered, in some cases, in response to radiation-induced DNA damage [29, 30]. In any event, if excess Hsp70 is able somehow to minimize DNA breakage in the nuclei of irradiated cells, this may suppress the primary 'cell death signal' and thus prevent triggering of the suicidal (apoptotic) program and trigger, instead, the alternative program aimed at cell survival and recovery (Figure 2).

The most intriguing question herein is: how can Hsp70 protect nuclear DNA from the radiation-induced damage? This does seem enigmatic because no direct DNA-related activities of Hsp70 have been documented so far. Nonetheless, a capability of inducible Hsp70 to confer stabilization/protection of chromosomal DNA under radiation exposure has convincingly been demonstrated on different experimental models [33, 34, 51, 74]. Apparently, the preirradiation accumulation of Hsp70 in the cells can somehow indirectly preserve their nuclear DNA from the genotoxic action of ionizing irradiation. For instance, the Hsp70 overexpression may promote cell phenotype modulation including specific alterations in folding and molecular composition of the chromatin so that chromosomal DNA becomes a less radiosensitive target and, as a result, undergoes lesser primary damage under hits of gamma-quanta (Figure 2). At present, however, nothing is known about any 'radioprotective' changes in the *in situ* chromatin folding that could be associated with overexpressed Hsp70. The other, supposedly more realistic mechanism may act via very early and rapid repair of the radiation-damaged DNA when excess Hsp70 facilitates maturation and nuclear import of certain proteins contributing to the postradiation DNA repair (Figure 2). In support of this hypothesis there is the fact that *in vivo*, Hsp70 can form complexes with DNA topoisomerase I [75], while this enzyme is known to participate in DNA repair [76]. Moreover, there are such supporting data: (i) Hsp70 was associated with base excision repair enzymes in HeLa cells [77], (ii) Hsp70-enriched murine C3H 10T1/2 cells exhibited faster disappearance of residual DNA breaks during the post-irradiation cell recovery [33] and (iii) transfection of human leukemic THP-1 cells with *Hsp70*-RNAi depressed the base excision repair enzyme activity in the transfectants [51]. Additional support comes from the in vitro investigations: in cell-free systems, Hsp70 was shown to stimulate repair of the single-strand DNA breaks by purified DNA polymerase β [78] and purified apurinic/apyrimidinic endonuclease [79]. Recently, Bases [51] found both the enhanced activity of DNA repair enzymes (presumably apurinic/apyrimidinic endonuclease and/or DNA polymerase β) and the improved postirradiation clonogenicity in human leukemic THP-1 cells pre-incubated with exogenous Hsp70. The author associated those effects with entry of the exogenously added chaperone into the target cells where it stimulates the DNA repair enzyme activities, thus contributing to the improved postirradiation DNA repair and cell survival [51]; possibly, the endogenously overexpressed Hsp70 is able to manifest the same activity. Likewise, it has been established that Hsp70 can bind to the catalytic unit of telomerase (TERT) [80], an enzyme maintaining the telomere length and genome stability [81]. A potential significance of these Hsp70-TERT interactions for the DNA repair machinery and cellular radioresistance is actively discussed [74, 82, 83]. Nadin et al [84] found nuclear translocation of inducible Hsp70 and more efficient DNA repair in heat-preconditioned peripheral blood mononuclear cells following treatment with doxorubicin, with a possible participation of Hsp70 in the DNA repair mechanisms being suggested by the authors. Although the latter study [84] was performed on the model of DNA damage non-related to radiation, Hsp70 may similarly participate in the postirradiation DNA repair process. For that matter, implication of Hsp70 in stabilization/repair of genomic DNA seems very intriguing and needs further, more thorough exploration.

Hsp27: The Small Protein with a Huge Cytoprotective Potential

The low-molecular weight (25-27 kDa) heat shock protein is referred as to Hsp27 or Hsp25 (the latter is mainly used for rodent cells) and belongs to a family of stress-inducible chaperones [1, 8, 12]. *In vivo*, this 'small' Hsp can form large oligomers and undergo phosphorylation at serine residues. Interestingly, both the oligomeric status and intracellular localization, and cytoprotective properties of Hsp27 strongly depend on its phosphorylation [12-14].

In the unstressed cell, there is a functional diversity between Hsp27 and Hsp70. So, in the absence of cell-stressing factors, the main function of constitutively expressed Hsp27 appears to lie in phosphorylation-dependent regulation of cell motility based on actin filament dynamics [12, 13]. However, in the case of proteotoxic (e.g. heat) stress, Hsp27 can act as an ATP-independent chaperone assisting Hsp70 to refold stress-denatured proteins [8, 12, 85]. Like Hsp70, Hsp27 possesses an antiapoptotic activity and, when overexpressed in cells, is able to inhibit the cytochrome *c* release/caspase activation induced by apoptosis-inducing stimuli (see Figure 2 and [86]). Thus, under cellular stresses, Hsp27 and Hsp70 may behave as a couple of 'interchangeable players' contributing to cytoprotection in a similar manner.

This similarity is also observed in the context of radiation exposure: Hsp25 (or Hsp27) overexpressed in mammalian cells exerts the radioprotective effects as well as overexpressed Hsp70 does it. So, significantly reduced apoptosis and improved cell survival took place in irradiated murine L929 cell clones stably overexpressing Hsp25 [87-91]. Transient overexpression of Hsp27 in viral vector-infected mouse embryo fibroblasts also weakened irradiation-induced apoptosis and increased the post-irradiation cell survival [34]. Accumulation of Hsp25 in thymi of heat-preconditioned mice temporally correlated with an acquisition of thymocyte resistance to apoptosis following gamma-irradiation [35]; in the same study, analogous correlation was also revealed for stress-inducible Hsp70.

As for the effects in tumor-derived cells, it was established that mouse fibrosarcoma (RIF) cells [38] and human Jurkat T leukemia cell line [40], being transfected with Hsp25-expressing plasmids, became more resistant to caspase-dependent apoptosis resulting from gamma-irradiation. Consistently with that, resistance to irradiation-induced apoptosis in three human lung carcinoma cell lines clearly correlated with the intracellular Hsp25 content, while transfection of NCI-H1299 lung carcinoma cells with siRNA to Hsp27 notably enhanced apoptotic death among the irradiated transfectants [39]. On the contrary, Ekedahl et al [53] found no correlation between the Hsp27 content and the intensity of radiation-induced apoptosis in non-small cell lung carcinoma cells. Likewise, cells of a tumorigenic O23 subclone derived from Chinese hamster lung fibroblasts were not rendered more radioresistant after the Hsp27 overexpression in them [92]; therefore, excess Hsp27 does not confer radioprotection in all types of tumor cells. In spite of those non-supporting data [53, 92], a lot of the other published findings [34, 38-40, 87-91] characterize Hsp27 (or Hsp25) as a powerful radioprotector that may beneficially act at different points of the cellular radiation response (see Figures 1 and 2).

Exact molecular mechanisms of the Hsp27-mediated radioprotection have not yet been unraveled but the antiapoptotic capacity attributed to overexpressed Hsp27 [15-17, 86] can play a prominent role in them. Excess Hsp25(27), like excess Hsp70, appears to modulate major stress-activated signaling pathways thereby enhancing cellular radioresistance (Figure 1). It was shown on a model non-related to irradiation that overexpressed Hsp27 suppresses the stress-induced activation of JNK and p38 MAPK [93]. Comparing the data from ref [93] with the data on JNK activation in irradiated cells [55-57] one can hypothesize that excess Hsp27 will protect from the JNK-mediated apoptosis under radiation exposure (Figure 1); though, this hypothesis should experimentally be examined on adequate models.

Besides the JNK activation, ionizing radiation evokes the activation of protein kinase C delta (PKCδ) and its membrane translocation that can result in apoptotic death of target cells, while Hsp25 is able to prevent such PKCδ-triggered apoptosis [39] (Figure 1). The authors assert that this mechanism is based on direct Hsp25-PKCδ interaction that inhibits both the kinase activity and membrane translocation of the enzyme [39]. Such Hsp25-mediated modulation of the irradiation-responsive PKCδ signaling can additionally involve p38 MAPK that is activated by active PKCδ and then stimulates the mitochondrial ROS generation leading to cytochrome *c*/caspase-dependent apoptosis (see below, [40] and Figure 2). On the other hand, the stress-induced p38 MAPK activation usually yields the Hsp25(27) phosphorylation by the p38 MAPK-activated protein kinase 2/3 and this pathway is associated with cellular defense from the actin skeleton destruction [12-14] and caspase-dependent apoptosis ([94], Figure 1). Moreover, the p38 MAPK activation by double-strand DNA breaks may contribute to cellular radioprotective mechanisms [64]. It seems likely that short-term duration and/or moderate levels of the p38 MAPK activation can confer basically cytoprotective (antiapoptotic) effects, whereas hyperactivation of this kinase by severe and sustained stresses ultimately kills the exposed cells. The fact that Hsp25(27), being a terminal substrate in the p38 MAPK cascade [95], is able to down-regulate the p38 MAPK activity [40, 93] implies an existence of a regulatory loop in the p38 MAPK/Hsp27 pathway. Correspondingly, this Hsp27-mediated attenuation of the p38 MAPK activation may contribute to the enhanced radioresistance of Hsp27-enriched cells (Figure 1).

Certain growth factors (e.g. EGF, VEGF), upon binding to their receptors, yield the activation of phosphoinositide 3-kinase (PI3K) that, in its turn, activates the serine-threonine kinase Akt; this PI3K/Akt pathway is known to be implicated in desensitizing cells to a number of apoptotic stimuli including radiation [96, 97]. Moreover, ionizing radiation may stimulate the PI3K/Akt pathway in some types of tumors [98-100]. It was shown on a model non-relevant to radiation that Hsp27 can activate Akt so that this Akt activation deters apoptosis [101, 102]. If analogous regulation of Akt by Hsp27 takes place in irradiated cells, this could be considered as one more mechanism of the Hsp27-mediated radioprotection (Figure 1) but, certainly, such a supposition requires experimental examination to be either confirmed or rejected.

The antiapoptotic effects of the Akt activation are partly realized via the Akt-induced activation of NF-κB that triggers the multiple transcriptional response contributing to cytoprotection [103, 104]. In addition to the suggested pathway when Hsp27 can promote the NF-κB activation by activating Akt [101, 102], there is the other pathway in which the NF-κB activation occurs through the Hsp25(27)-mediated degradation of the main inhibitor of

NF-κB, IκB [91, 105]. According to Parcellier et al [105], the Hsp27 overexpression did enhance ubiquitin-dependent proteasomal degradation of IκB and, as a consequence, nuclear translocation, DNA-binding and transcriptional activity of NF-κB became enhanced as well. Importantly, both the NF-κB activation and the increased degradation of IκB were found in murine L929 cells overexpressing Hsp25 and exhibiting enhanced radioresistance [91]. So, the revealed ability of excess Hsp25(27) to promote the NF-κB activation appears to be a real component of the Hsp-conferred radioprotection (Figure 1). Apparently, some products of the NF-κB-triggered gene transcription are able to increase cellular resistance to the cytotoxic impact of ionizing radiation; in particular, manganese superoxide dismutase (SOD2) can be one of such products (see below and [91]).

A number of publications indicate that the antiapoptotic activity of Hsp27 can effectually be manifested at the stage of blocking the cytochrome *c* release from mitochondria [86] or downstream of the mitochondria, at the stage of apoptosome formation and/or activation of caspase-3 [106-108]. Although all those findings were obtained in studies non-related to radiation, analogous mechanism(s) may act in the irradiated, Hsp27-enriched cells to prevent the cytochrome *c*/caspase-mediated apoptosis (Figure 2). Interestingly, the stable Hsp25 overexpression in murine L929 cells was associated with substantially increased expression of another antiapoptotic protein, Bcl-2; this up-regulation of Bcl-2 was suggested to be responsible for the radioprotection conferred by excess Hsp25 [88]. With regard to a mechanism by which excess Hsp27 up-regulates Bcl-2, Cho et al [90] believed that it is due to down-regulation of extracellular signal-regulated kinase 2 (ERK2); such down-regulation of ERK2 did occur in Hsp25-overexpressing L929 cells and seem essential for both the increase in Bcl-2 and reduction of irradiation-induced apoptosis. The Hsp25(27)-conferred enhancement of Bcl-2 expression could really exert the radioprotective effect because excess Bcl-2 is really able to block the cytochrome *c* release from mitochondria of irradiated cells and, thus, to save them from caspase-dependent apoptosis [109]. Meantime, other researchers failed to find analogous up-regulation of Bcl-2 in Hsp27-enriched mouse embryo fibroblasts [34]; therefore, in various types of cells, the Hsp27-mediated radioprotection (if takes place) can be based on different pathways.

The mouse embryo fibroblasts transiently overexpressing Hsp27 exhibited the significantly impaired up-regulation of the post-irradiation p53 expression that correlated with reduction of irradiation-induced apoptosis [34]. The latter finding suggests that preirradiation accumulation of Hsp27 in the target cells can somehow suppress the irradiation-provoked increase in intracellular p53 content, thus preventing the forthcoming p53-dependent apoptosis (Figure 2). To accept such a scheme of the Hsp27-mediated radioprotection it is necessary to answer the question: by which mean overexpressed Hsp27 down-regulates the p53 content in irradiated cells? There is a rather attractive hypothesis that excess Hsp27 is able to attenuate the DNA damage resulting from radiation exposure [34]; such a presumptive activity may impair both the DNA damage-responsive up-regulation/activation of p53 and the p53-dependent apoptosis. This hypothesis is supported by the fact that the Hsp27 overexpression can attenuate the DNA damage in the gamma-ray-exposed cells (see below and ref [34]).

The radioprotective effect of Hsp27 on nuclear DNA seems very similar to that of Hsp70. In particular, Hsp27 overexpressed in cultures of mouse embryo fibroblasts significantly diminished amounts of single- and double-strand breaks induced in nuclear DNA by gamma-irradiation; the better preservation of DNA correlated well with the down-regulation of p53 and the lower intensity of apoptosis [34]. In the same work, the almost identical effects were found following the Hsp70 overexpression and it is the attenuation of DNA breakage that was suggested to define the radioprotection conferred by excess Hsp27 or Hsp70 [34]. Actually, if excess Hsp27 was able to protect genomic DNA from the radiation-induced breaking, such a functional activity of the protein could be responsible for the enhanced cellular resistance to postirradiation apoptosis. At present, it is still poorly understood how Hsp27 can prevent the DNA breakage resulting from direct hits of gamma-quanta. As well as it was already discussed for Hsp70 (see above), the revealed attenuation of DNA breakage may be a result of action of yet unknown Hsp-mediated mechanisms that ensure (e.g. by altering the chromatin structure) greater resistance of DNA strands to primary damage induced by ionizing radiation (Figure 1). On the other hand, the lesser DNA damage may be detected following very early and fast repair of the DNA breaks that is stimulated by excess Hsp27 (Figure 2). The latter idea is supported by some preliminary data on the accelerated post-irradiation DNA repair in Hsp27-enriched cells (A.E. Kabakov et al., unpublished observations). Moreover, it was recently found that gamma-irradiation evokes rapid re-localization of Hsp27 from the cytoplasm to the nucleus; this irradiation-responsive nuclear translocation of Hsp27 requires its phosphorylation and can be abolished by an inhibitor of p38 MAPK, SB203580 (A.E. Kabakov et al., unpublished data). Such concentrating of phosphorylated Hsp27 into the nuclei of irradiated cells appears to contribute to the faster post-irradiation DNA repair but the entire molecular machinery of Hsp27-mediated protection from the DNA breakage remains to be clarified.

It should be noted that Hsp27 and Hsp70 being overexpressed together in the same cells confer more pronounced protection from the irradiation-induced DNA breakage and apoptosis than the overexpression of only Hsp27 or Hsp70 alone [34]. Taking into consideration that Hsp27 can assist Hsp70 in refolding of denatured proteins [85], one can suppose that Hsp27 similarly assists Hsp70 in defensive and repair processes under radiation exposure. The stronger radioprotection in the case of simultaneous overexpression of Hsp27 and Hsp70 [34] may mean tight cooperation between these Hsps that consequently yields the better (additive) effects. On the other hand, Hsp27 and Hsp70 may independently act in parallel radioprotective mechanisms; if so, the combined overexpression of both Hsp27 and Hsp70 will also confer the more prominent radioprotection as compared with the individual Hsp overexpression. Thus, in the context of irradiation, an excess of either Hsp seems to be able to partially compensate for deficiency or dysfunction of the other.

It is well known that ionizing radiation generates free radicals inside the target cells. Some of those radicals appear to be formed under direct radiolysis of molecules; in addition, abnormal overproduction of reactive oxygen species (ROS) is stimulated in irradiated cells [21-23]. Both the radiolysis-generated free radicals and ROS can break or chemically modify genomic DNA, thus aggravating the radiation-associated genotoxicity and triggering the DNA damage-responsive apoptosis. Moreover, irrespective of the DNA state, an intracellular burst of ROS *per se* can trigger the apoptotic pathway (see for a review [110, 111]). In cellular models with oxidative stress, overexpressed Hsp25 was shown to attenuate oxidative modification of DNA [112] and protect from ROS-induced apoptosis [113]. Hence, it would be logical to expect a similar (protective) action of excess Hsp25(27) toward apoptosis-inducing free radicals generated under irradiation (Figure 1). Baek et al [87] suggested that the enhanced radioresistance of murine L929 cells stably overexpressing Hsp25 can be due to faster replenishment of reduced glutathione (GSH) from its oxidized form (GSSG). Those authors did find the increase in a GSH/GSSG ratio as well as the greater activities of glutathione reductase and glutathione peroxidase in the radioresistant (Hsp25-enriched) cells [87]. According to the authors' idea, such an influence of overexpressed Hsp25 on the glutathione-redox cycle beneficially up-regulates the intracellular level of GSH that protects against detrimental free radicals produced by irradiation. Later, it was shown on the same model with L929 cells that excess Hsp25 up-regulates the manganese superoxide dismutase (SOD2) expression through activation of NF-κB [91]. If so, the up-regulated SOD2 can also contribute to detoxifying of ROS generated under gamma-irradiation. It should be noticed that, besides the described up-regulation of SOD2 [91], activated NF-κB triggers the multiple transcriptional response and some products of this multi-gene transcription will contribute to cell survival and radioresistance (see above, refs [103, 104] and Figure 1). Finally, in a recent study performed on Jurkat T cells [40], it was shown that ionizing irradiation activates PKCδ followed by PKCδ-mediated activation of p38 MAPK; in its turn, activated p38 MAPK stimulates ROS generation by mitochondria thus promoting caspase-dependent apoptosis in the irradiated cells. The Hsp25 overexpression suppressed all those effects of irradiation including the consecutive activation of PKCδ and p38 MAPK, mitochondrial production of

ROS and ROS-induced apoptotic cell death (see [40] and Figures 1, 2). The authors have suggested that it is the Hsp25-mediated inhibition of the PKCδ activation that confers cytoprotection against apoptosis resulting from the irradiation-stimulated ROS generation [40].

At first glance, it seems rather confusing that there are so many different versions describing potential mechanisms of Hsp25(27)-conferred radioprotection [34, 39, 40, 87-91]. At least five diverse mechanisms were successively proposed after investigations on the same model with Hsp25-overexpressing L929 cells [87-91]. However, it should be noted that all those proposed mechanisms are not incompatible, quite the contrary, they may be interconnected with each other. For example, in the context of radiation exposure, the up-regulation of SOD2 revealed in the Hsp25-enriched cells [91] may contribute to the delay in GSH depletion that was described earlier [87]. Also, the lesser DNA breaking found by one research group [34] may partly be due to the suppression of ROS generation discovered by the other group [40]. Apparently, excess Hsp25(27) does possess a powerful cytoprotective potential allowing this protein to beneficially modulate the intracellular redox balance and signaling cascades so that the genotoxic impact of irradiation is attenuated and the apoptotic pathway is blocked (see Figures 1 and 2). Meanwhile, fine machinery and all cofactors of the Hsp27-mediated radioprotection still need to be elucidated.

Hsp90 as an 'Undesirable' Radioprotector: A Target for Enhancing Tumor Radiosensitivity

Being one of the major chaperones of eukaryotes, the 90 kDa heat shock protein (Hsp90) functions in the vital machinery of protein folding and trafficking [5, 6, 8, 10]. Importantly, Hsp90 regulates conformation, functional activity and stability of its client proteins which are key mediators in signal transduction pathways related to cell cycle, stress response and post-stress cell survival, apoptosis, etc (see for a review [6, 8-10, 114, 115]). The constitutively expressed Hsp90 is a very abundant cytosolic component whose percentage can account for 2-5% of total cellular protein [116]. Therefore, it would be hard to substantively alter such a high intracellular content of Hsp90 using traditional methods with introduction of artificial gene vectors or oligonucleotides. Fortunately, a few specific drugs are known that selectively inhibit the chaperone activity of Hsp90. *In vivo*, action of such drugs can lead to inactivation and faster ubiquitin/proteasome-mediated degradation of the Hsp90 client proteins including EGF receptor, steroid hormone receptor, Raf-1, HER2, v-Srk, ErbB2, Akt and others [41-45, 115, 116]. As some of those client proteins contribute to expansive progression and improved survival of malignant cells, the artificially stimulated inactivation and degradation of these proteins through the drug-induced Hsp90 inhibition may be synergistic with chemotherapy and radiotherapy. That is why targeting tumor cells with the pharmacological inhibitors of Hsp90 is now actively studied on various experimental models, tested in clinical trials and considered as a promising strategy for treatment of some kinds of human cancer [97, 116-118].

For instance, macrocyclic antibiotic radicicol strongly inhibits the chaperone-substrate complex formation between Hsp90 and its client proteins thus promoting degradation of these proteins by proteasomes [119]. Accordingly, radicicol was shown to markedly potentiate the radiation-induced cytotoxicity in a human hormone-sensitive prostate cancer cell line and the achieved radiosensitizing effect correlated clearly with the drug-induced enhancement of proteasomal degradation of androgen receptor and Raf-1 [120].

Ansamycin antibiotic geldanamycin as well as its less toxic derivatives 17-N-allilamino-17-demethoxygeldanamycin (17AAG) and 17-dimethylaminoethylamino-17-demethoxy geldanamycin (17DMAG) are potent inhibitors of the Hsp90 chaperone activity and can exert prominent anticancer effects. At the molecular level, these drugs inhibit the ATPase activity of Hsp90 thus depressing its ATP-dependent chaperone function [115-118]. There are publications demonstrating that such pharmacological inhibition of Hsp90 can lead to radiosensitization of tumor cells. So, Russell et al [121] have described that 17AAG enhances sensitivity of four human tumor cell lines (two gliomas and two prostate carcinomas) to X-rays; moreover, the drug had no effect on the radiosensitivity of normal human fibroblasts. In other studies, 17AAG was shown to enhance irradiation-induced apoptosis and/or growth suppression in human prostate carcinoma cells [42, 122], squamous cell carcinoma monolayer cultures or spheroids [43] and colon carcinoma cells [123]. In its turn, 17DMAG also enhanced the radiosensitivity of cells of three human solid tumors growing in culture or as xenografts [124, 125], with a suggestion being made that the drug-stimulated degradation of ErbB2 plays a role in the phenomenon discovered. In the later work, 17DMAG was characterized as a radiosensitizer for only those human tumor cell lines which do not express ErbB3 [125]. Bisht et al [44] found that both geldanamycin and 17AAG can intensify apoptosis and necrosis in X-ray-irradiated human cervical tumor cells; in addition, these drugs notably radiosensitized cells of several transformed and immortalized fibroblast lines but not of the normal ones. Similarly, geldanamycin significantly sensitized human SQ-5 lung carcinoma, T24 bladder carcinoma and HMV-I melanoma cells to X-ray-induced apoptosis [41, 45], whereas no radiosensitization was observed in a normal human NB1RGB fibroblast line treated with the drug [45]. It was also revealed that the geldanamycin-conferred radiosensitization took place in methylcholantrene-transformed mouse cells without increasing the radiosensitivity of the parental cells of a normal C3H10T1/2 fibroblast line [45].

It seems to be of practical importance that some above cited papers [44, 45, 121] report about a selectivity of the radiosensitizing action of 17AAG and geldanamycin toward tumor cells. There is rather convincing explanation for that: in tumor cells, all their Hsp90 is usually sequestered in multiprotein complexes where it sojourns in its specific conformation that ensures high binding affinity and high ATP-ase activity of the chaperone [126]. In contrast, Hsp90 in normal cells is mainly in a latent uncomplexed state when its 'inactive' conformation defines the low binding affinity. This conformational and functional diversity clarifies why Hsp90 in tumor cells has a 100-fold higher binding affinity for 17AAG than Hsp90 in normal cells [126] and why the Hsp90 inhibitors enhance the radiosensitivity of tumor-derived cell lines in preference to normal fibroblasts [44, 45, 121]. Such encouraging data allow to consider the Hsp90 inhibitors as potential remedies to increase the efficiency of clinical radiotherapy: it is expected that these inhibitors being pretty selective radiosensitizers will strengthen elimination of tumor cells by irradiation without aggravating radiation lesions in normal tissues of patients. Correspondingly, the constitutively expressed Hsp90 plays here a role of the 'undesirable' radioprotector that assists tumor cells to survive the cytotoxic exposure of radiotherapy.

With regard to molecular mechanisms, the enhanced radiosensitivity of tumor cells treated with Hsp90 inhibitors appears to be due to essential contribution of the Hsp90 chaperone activity to activation, functioning and stability of several key proteins responsible for cellular radioresistance and escaping apoptotic elimination (Figure 3). The drug-induced Hsp90 inhibition did lead to down-regulation of some of those key proteins such as EGF receptor, androgen receptor, ErbB2, Raf-1, Akt and focal adhesion kinase (FAK) by stimulating their ubiquitin-dependent degradation at proteasomes [41-45, 120-122]. Likewise, the *in vivo* effects of Hsp90 inhibition indicate that Hsp90 is necessary for the activity of tyrosine kinases involved in transduction of the 'cell survival signal' from the growth factor receptors to the Akt- and NF-κB-mediated cytoprotective pathways ([97, 115, 116], Figure 1).

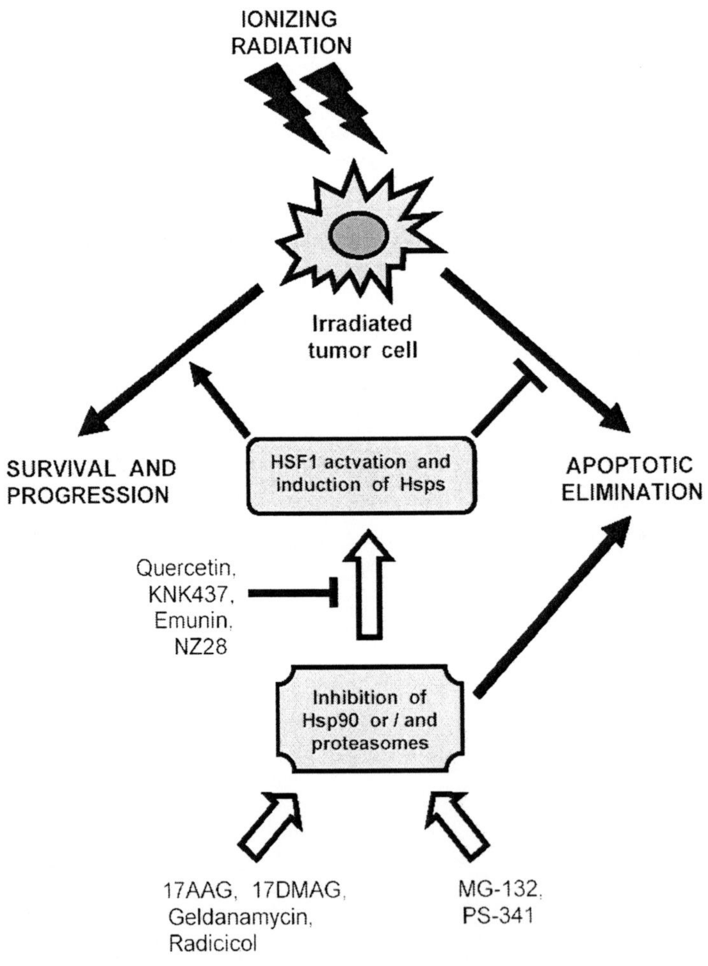

Figure 3. Schematic illustration showing that beneficial synergy expected under combining of radiotherapy with the inhibitors of Hsp90 or the proteasome inhibitors may be impaired through the activation of HSF1 and induction of 'radioprotective' Hsps. To escape such complications it would be rational to use blockers of the HSF1 activation/Hsp induction pathway (see in text).

Besides tumor cells, endothelial cells of the tumor vasculature are important targets for anticancer radiotherapy. Irradiation of vascular endothelial cells with gamma- or X-rays triggers the PI3K/Akt signaling pathway aimed at suppression of apoptosis and improved postirradiation cell survival [62, 127, 128]. Moreover, some tumor-derived growth factors, particularly, vascular endothelial growth factor (VEGF) and basic fibroblast growth factor (bFGF) can enhance radioresistance of vascular endothelial cells via stimulation of the PI3K/Akt pathway [62, 128, 129]. Recent investigations have revealed that clinically accessible (nanomolar) concentrations of 17AAG sensitize human vascular endothelial cells to gamma-irradiation-induced apoptosis and, moreover, abrogate the radioprotective effects of VEGF and bFGF (A.E. Kabakov et al., a manuscript in preparation). This radiosensitization of 17AAG-treated vascular endothelial cells seems to be due to the drug-induced blockage of the Hsp90-dependent PI3K/Akt signaling pathway, while the latter is aimed at prevention of the post-irradiation apoptosis (see refs [62, 127, 128] and Figure 1). Such data obtained on vascular endothelial cells indicate that the radiosensitizing effects attributed to 17AAG may not be exclusively selective toward tumor cells as it was previously stated in some papers [44, 45, 121]. However, if 17AAG radiosensitizes not only tumor cells but endothelial cells of the tumor vasculature as well, this drug would be even more valuable for clinical radiotherapy. In fact, the tumor vasculature supplies nutrients, oxygen and growth factors to the malignant cells thus promoting both expansive progression and enhanced radioresistance of the tumor; therefore, development of clinically applicable methods of radiosensitization of the tumor vasculature cells is one of the top priority tasks for modern radiotherapy of cancer [130, 131].

So, from the point-of-view of radiotherapists, the constitutively expressed Hsp90 behaves as the ‘undesirable’ radioprotector whose chaperone activity can really decrease the efficiency of anticancer radiotherapy. In this respect, the discovered capability of Hsp90 inhibitors to enhance radiation response in the tumor (and the tumor vasculature) cells seems pretty hopeful and may be a basis for a new strategy in radiotherapy. It should be noted, though, that the Hsp90 inhibition was suggested to synergize with chemotherapeutic agents only in the case of wild type p53 expression in target (tumor) cells [132]. By the analogy, the effect of drug-induced Hsp90 inhibition on the *in vivo* radiosensitivity of tumor cells may also depend on the type (wild or mutant) of p53 they express; obviously, this issue must be clarified before the method of tumor radiosensitization by means of the Hsp90 inhibitors will widely be adopted in a clinical setting.

Hsp Induction by the Tumor Radiosensitizers: A Challenge for Radiotherapy?

It should be taken into consideration that the Hsp90 inhibition-potentiated radiosensitivity of tumor cells may be complicated by the HSF1-mediated stress response that is usually triggered by the Hsp90 inhibitors (see Figure 3). Actually, the constitutively expressed Hsp90 (as well as Hsc70) is known to be a natural down-regulator of HSF1: in the non-stressed cell, Hsp90 binds HSF1 and retains it in an inactive state [3, 133]. Geldanamycin, 17AAG and radicicol can disrupt the complex between Hsp90 and HSF1, thereby promoting activation of HSF1 followed by initiating the heat stress transcriptional response and Hsp induction; the latter was associated with partial attenuation of the cytotoxic action of the Hsp90 inhibitors used [134-136]. Probably, 17DMAG exerts the same (Hsp-inducing) effect. If Hsp70 and Hsp27 are induced in the Hsp90 inhibitor-treated tumor cells, this concomitant induction of the 'radioprotective' Hsps can impair the radiosensitizing effect of the Hsp90 inhibition (see Figure 3). Some researchers already proposed approaches to abrogation of this Hsp induction as a real strategy to enhance the anticancer action of 17AAG [135, 136]. For instance, Guo et al [135] have demonstrated that the 17AAG-induced Hsp70 attenuates the drug-stimulated apoptosis in 17AAG-treated human leukemia cells, while abrogation of the Hsp70 induction with siRNA or co-treatment with KNK437 (a benzylidine lactam inhibitor of the Hsp70 induction [137]) can enhance the apoptotic effect of 17AAG. The other research group, after focused screening of 20000 compounds, has characterized the two ones called as NZ28 and emunin that inhibit the HSF1-mediated induction of Hsp70 and Hsp27 in cancer cell cultures treated with 17AAG or radicicol, thereby intensifying the apoptotic cell death provoked by the Hsp90 inhibition [136]. Quercetin and some related flavonoids were also characterized as inhibitors of the Hsp induction in mammalian cells including the tumor ones [46-48, 135]. However, the range of concentrations in which Quercetin represses the expression of inducible Hsp70 and Hsp27 are much higher than those of KNK437 [135] or NZ28 and emunin [136]; therefore, as compared with Quercetin (and the other similarly acting flavonoids), the latter three compounds seem more effective in preventing the undesirable Hsp induction in tumor cells treated with the Hsp90 inhibitors. Taken together, all these 'proof-of-principle' results indicate a real possibility of pharmacological inhibition of the undesirable Hsp induction in treated tumor cells; in the future, similar approaches would be used to enhance the efficiency and selectivity of anticancer radiotherapy.

Besides the Hsp90 inhibition, the 'pharmacological' inhibition of proteasomes is also considered as a prospective approach to cancer treatment and radiosensitization of human tumors (reviewed in [138-140]). Inhibition of the proteasomal machinery of protein degradation in cells is known to lead to their death via apoptosis; hence, the apoptosis-promoting effect of proteasome inhibitors is supposed to be synergistic with the cytotoxic action of chemotherapy and radiotherapy (see Figure 3 and [138-140]). Indeed, such a proteasome inhibitor as MG-132 was shown to sensitize cells of Hodgkin lymphoma [141], human melanoma [142], non-small cell lung cancer [143] and prostate cancer [144] to radiation-induced apoptosis. This radiosensitizing effect of MG-132 appears to be due to inhibition of NF-kB in the drug-treated tumor cells [142-144]. The other well-known proteasome inhibitor, PS-341 (often referred to as velcade or bortezomib), is already used in clinical therapy of cancer [140]. PS-341 induces apoptosis in cancer cells [138-140] and is synergistic with the antitumor activity of 17AAG [145]. Similarly to the effects of MG-132, the PS-341 administration enhanced radiosensitivity of some tumors including colorectal cancer [146], non-small lung cancer [147] and multiple myeloma [148, 149]; this radiosensitization seems to be associated with inhibition of NF-kappaB and enhanced apoptosis following irradiation [146, 148]. However, like the Hsp90 inhibitors, both MG-132 and PS-341 up-regulate Hsp70, Hsp90 and Hsp27 in the treated tumor cells so that this concomitant up-regulation somewhat attenuates the antitumor action of the drug treatment

[136, 145, 150, 151]. As regards a mechanism by which the proteasome inhibitors increase the Hsp levels, apparently, the drug-induced dysfunction of the protein degradation machinery leads to intranuclear accumulation of non-degraded (aberrant) proteins that, in turn, triggers the HSF1 activation/Hsp induction pathway [1-4, 8]. In addition, the p38 MAPK activation occurring in the proteasome inhibitor-treated tumor cells can contribute to the Hsp27 up-regulation [150, 151]. Consequently, according to the above reasoning, the 'radioprotective' Hsps up-regulated by the proteasome inhibitors can impair the radiosensitizing effect attributed to these inhibitors [141-144, 146-149] (see Figure 3). In this respect, it seems of importance that both NZ28 and emunin were shown to inhibit the Hsp70 induction in MM.1S myeloma cells and PC-3 prostate cancer cells treated with MG-132 or PS-341; the inhibition of Hsp induction correlated with the enhancement of apoptosis [136].

Although neither KNK437 nor NZ28, nor emunin were yet tested in combination with the Hsp90 inhibitors and irradiation or the proteasome inhibitors and irradiation, it is expected that such combining may additionally enhance the radiation-induced killing of the target tumor cells. Actually, it seems likely that the radiotherapeutic exposure combined with simultaneous administration of the Hsp90 inhibitor (e.g. 17AAG or 17DMAG) or/and the proteasome inhibitor (e.g. MG-132 or PS-341) and an appropriate non-toxic blocker of the Hsp induction (unknown yet) will strengthen the radiosensitization by preventing the concomitant induction of the radioprotective Hsps (see Figure 3). It should be noted that besides the considered here two situations with the proteasome or Hsp90 inhibitors, some other antitumor agents may also up-regulate Hsps in the target (tumor) cells. While chemotherapy is often combined with radiotherapy, the concomitant Hsp induction in response to cell-stressing chemotherapeutic drugs may decrease an efficiency of both chemotherapy and radiotherapy. To overcome this problem the Hsp-inducing stress response pathway should be blocked by a suitable inhibitor (Figure 3). Overall, the idea of 'pharmacological' down-regulation of the Hsp expression in human tumor cells seems quite atractive and is seriously discussed as one of hopeful trends in future anticancer therapy [152-154]; as for the real 'drug discovery', the first practical steps are already made in this direction [47, 54, 135-137]. Unfortunately, at present, no clinically applicable down-regulators of the *in vivo* Hsp expression are yet known to be beneficially used for cancer treatment. Hence, further active search for such down-regulators and clinical trials of all presumably suitable compounds are of paramount importance for successful pharmacological intervention in radiotherapy of human tumors.

In the general strategy aimed at overcoming the Hsp-mediated tumor cell resistance to antitumor agents, besides the down-regulation of Hsp expression, there is another hopeful approach, namely 'pharmacological' inhibition of functional (e.g. antiapoptotic or chaperone) activities of cytoprotective Hsps. While the clinically applicable inhibitors of Hsp90 (17AAG, 17DMAG) have already been developed and introduced in clinic practice, no pharmacological inhibitors of cytoprotective functions of Hsp70 or Hsp27 have been documented so far. However, some encouraging data come from experimental studies on artificial inhibition of the antiapototic activities of these Hsps [155, 156]. Sakamoto et al [155] have demonstrated that the antiapoptotic activity of Hsp27 is up-regulated through its *in vivo* modification by methylglyoxal (a side product of glycolysis); importantly, if this modification was inhibited by OPB-9195, the enhanced apoptotic cell death under antitumor drug treatments was observed. In the other investigation [156], the antiapoptotic activity of Hsp70 in tumor cells was neutralized by adding an apoptosis-inducing factor (AIF)-derived peptide that selectively bound Hsp70 and, thus, sensitized the treated cells to antitumor drugs stimulating apoptosis. Although both those studies [155, 156] was performed on models non-related to radiation, it seems possible to use analogous approaches for enhancing the apoptotic cell death in irradiated tumors. Herein, combining of the down-regulation of the Hsp induction with the inhibition of antiapoptotic activity of the constitutively expressed Hsp90, Hsp70 and Hsp27 could exert the strongest antitumor effect.

Radiation, Genomic Instability and Oncogenesis: A Role for Hsps

The carcinogenic effects of ionizing radiation are well-known. It is generally accepted that the radiation-induced genomic instability and, particularly, some specific mutations (e.g. in the *p53*-gene) can promote oncogenic transformation of irradiated mammalian cells (reviewed in [18, 157-160]). Apparently, inducible Hsp70 somehow contributes to stabilizing genomic DNA because a deficiency of *Hsp70.1*- and *Hsp70.3*-genes was shown to be causally associated with the increased level of chromosome aberrations following gamma-irradiation [74]. Herein, it seems likely that excess Hsp70 somehow ensures the maintenance of telomere stability; if it does happen, such an Hsp70-mediated mechanism can allow the irradiated cell to avoid its replicative senescence [82, 83]. Moreover, there are data suggesting an indirect (but effectual) involvement of excess Hsp70 in protection/repair of DNA from gamma-irradiation-induced breaks [33, 34, 51]; this activity of Hsp70 appears to be responsible (at least partly) for the reduction of post-irradiation apoptosis in Hsp70-enriched cells (see above and [34]). In its turn, overexpressed Hsp25(27) is also able to attenuate the nuclear DNA breakage resulting from gamma-irradiation [34]. In addition, Hsp25 can suppress the radiation-stimulated ROS generation and protect the irradiated cell from ROS-mediated apoptosis [40]. Finally, among vital functions of Hsp90 there is ensuring of the PI3 kinase/Akt signaling pathway that can contribute to repair of radiation-damaged DNA [98, 161]. Collectively, all those data prove the prominent role of the major Hsps in guarding the genome of mammalian cells undergoing radiation exposure; in fact, it seems so that the up-regulated Hsp(s) can preserve nuclear DNA from the lethal damages induced by ionizing radiation (see refs [33, 34] and Figure 2).

On the other hand, when the increased Hsp content inside the proliferating cells enables them to survive radiation exposure and escape replicative senescence, this will ensure further propagation of a fraction of the radiation-affected (instable) genomes that probably carry a lot of non-lethal mutations, particularly, random 'point mutations'. The latter often yield protein products with substitutions of one amino acid residue by another. In some cases, the genotoxic radiation exposure by itself may be a stimulus for the up-regulation of Hsp expression, as the multiple point mutations arise stochastically in different regions of the genome and may thereafter result in expression of mutant proteins with amino acid substitutions and abnormal folding. If so, the appearance of such misfolded proteins inside the exposed cells will sequester Hsp70 and Hsp90 thus triggering the HSF1 activation/Hsp induction pathway [1-4]. Indeed, it was already shown that both the low doses and high doses of ionizing radiation may up-regulate the Hsp expression in mammalian cells [33, 162-164]. Additionally, mutation of p53 (e.g. following radiation exposure) leads to the enhancement of *Hsp70*-gene transcription through loss of repression of the *Hsp70* promoter [165, 166]. Thus, although the three major Hsps were declared to be the genome-stabilizing factors that protect from the radiation-induced DNA breakage or chromosome aberrations and shortening of telomeres (see the previous paragraph), the other facts and speculations allow to suggest that the excess Hsps can promote maintenance and propagation of the radiation-destabilized genomes carrying various radiogenic mutations. It seems so that, under radiation exposure, the major (inducible) Hsps simultaneously contribute to both the stability and instability of the exposed cell's genome. This looks somewhat paradoxical but only at the first glance because, as the matter of fact, it is the Hsp-conferred escaping of lethal DNA damage and telomere loss that ensures post-irradiation survival of some populations of cells with the mutated (instable) genomes and further proliferation (or even expansion) of such cells. The suggested influence of Hsps on the state and fate of radiation-affected genome could play a role in early bio-evolution and appears to be involved in processes of the radiation-associated oncogenesis (see Figure 4).

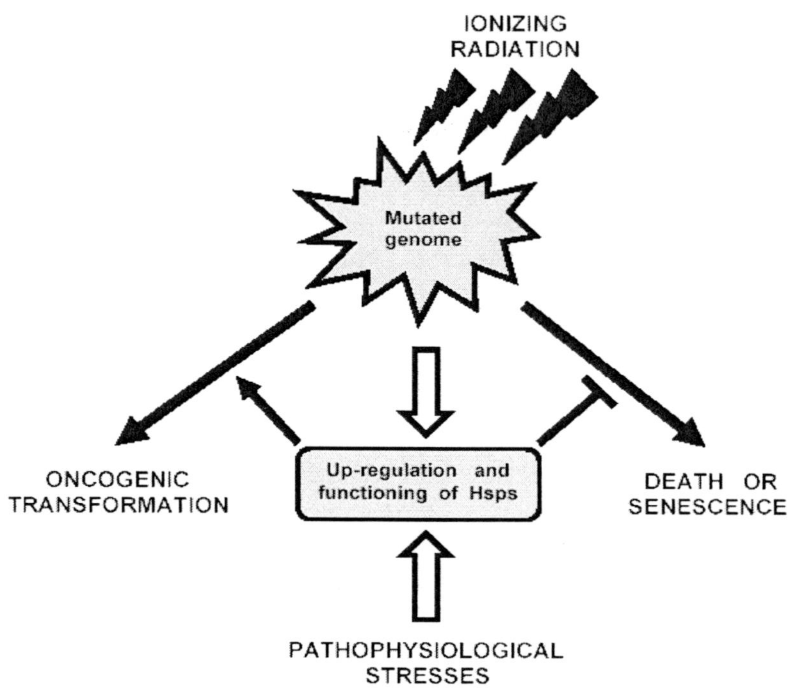

Figure 4. Scheme illustrating potential contribution of the Hsp induction/functioning to oncogenic transformation of irradiated cells. Briefly: cytoprotective Hsps up-regulated owing to radiation-induced mutagenesis and/or concomitant pathophysiological stresses enable the irradiated cell to escape death (apoptosis, necrosis, mitotic catastrophe) or replicative senescence, thus promoting propagation of the radiation-affected genomes and further accumulation and masking of random mutations until fatal appearance of the oncogenic mutations. Then the excess Hsps can promote selection among populations of the transformed cells toward formation of the entirely malignant phenotype (see in text).

Intriguingly, S. Lindquist with coworkers [167, 168] discussed that the known capacity of Hsp90 to stabilize mutated, quasi-unstable protein conformations could be realized in advancing the evolution of variant organisms and appearance of new biological species. So, while various mutations appeared and accumulated in genomes, Hsp90 could function as a 'capacitor for evolution' that promoted formation and maintenance of wide phenotypic variations [167, 168]. It should be noted, however, that the other major and inducible chaperones could also contribute to this evolutionary process. *In vivo*, some mutant (misfolded) proteins appear to become targets for Hsp70 or Hsp27; these chaperones can bind and retain such aberrant protein molecules to preserve them from aggregation and 'correct' their abnormal conformations or facilitate their degradation [5, 7, 8, 11]. Thus by masking the mutant protein expression and securing against its dangerous consequences (e.g. protein aggregation), the excess Hsps enabled a huge diversity of new phenotypes to become variant objects for the acute competition in terms of natural selection. Probably, primary single- and multi-cell organisms, existing in the context of extremely aggressive environment, frequently underwent various proteotoxic and genotoxic exposures including environmental radiation. Under those conditions of the alternate stresses, the radioprotection episodically conferred by the stress-induced Hsps might protect some irradiated cells from the lethal DNA damage (discussed in [34]). If it was so, the inducible (radioprotective) Hsps did contribute to the evolution by promoting survival of some fractions of radiation-affected and genetically changed cells and thus supplying new heritable material (the viable cells carrying radiogenic mutations) for the natural selection.

By analogy, under carcinogenic exposures, the major inducible chaperones (Hsp90, Hsp70 and Hsp27) are supposed to promote processes of oncogenic cell transformation and cancer progression that to some extent resemble Darwin's evolution with its natural selection [115, 169, 170]. In fact, oncogenesis can be considered as accelerated evolution selecting from cell phenotype variations the most adapted and resistant one whose clones will get apparent advantages in competition with all other cell phenotypes within the present organism. During transformation, even appearance of oncogenic mutations in the cell's genome may not yet be sufficient to result in fatal development of malignancy; just the selection of primary populations of the diversely transformed (potentially tumorigenic) cells can generate the quickly growing and highly resistant clones that eventually evolve into the invasive tumor. This 'evolution into tumor' may occur within one tissue microlocus: when a fraction of cells is influenced by carcinogenic factors (e.g. by ionizing radiation), such an influence usually yields a diversity of mutated genomes and a wide spectrum of variable cell phenotypes. Then these different cell phenotypes undergo selection based on the acute inter-cell competition involving both the transformed cells and the surrounding normal cells. Among the phenotype-dependent determinants for such selection may be the rate of cell proliferation, self-sufficiency in growth signals, resistance to apoptosis, capacity to anchorage-independent growth, capability to stimulate angiogenesis, etc. It has been suggested that the excess Hsps contribute to this selection process as 'facilitators of onco-evolution' that enable the involved cells to accumulate mutations and escape apoptosis or replicative senescence. Indeed, many key proteins (hormone receptors, protein kinases, telomerase, Akt, HIF1α, and others) responsible for uncontrolled cell proliferation, immortalization, stimulation of angiogenesis, suppression of apoptosis, etc are clients of

Hsp90 (see refs [114, 115, 170]). In turn, excess Hsp70 and Hsp27 are able to inhibit major pathways leading to the apoptotic cell death or senescence; such activities of these Hsps can also promote tumorigenesis (reviewed in [17, 170-172]).

Probably, the same mechanisms act under oncogenesis provoked by ionizing radiation (Figure 4). High dose irradiation of the entire human organism (e.g. as a result of technogenic catastrophes associated with radiation fallout) can cause appearance of massive populations of mutated cells displaying a rich variation of phenotypes; some of such cells may (via phenotypic selection) evolve into the tumor-forming clones. Similar tumorigenic processes in human tissues may be triggered following occupational irradiation or even medicine-related irradiation including radiotherapy [173, 174]. In turn, chronic low dose irradiation leads to a gradual accumulation of random (radiogenic) mutations in the genomes of target cells that may also become material for the subsequent phenotypic selection in terms of onco-evolution. The latter situation is particularly characteristic for local internal irradiation by radionuclides incorporated in human tissues and organs. For example, the skeleton-seeking radionuclides strontium, radium and plutonium being incorporated into bone tissue can irradiate all adjacent cells including stem hematopoietic cells of bone marrow; the carcinogenic effects of the internally deposited radionuclides have been documented (see for a review [175, 176]). The other typical example is tissue-specific accumulation of radioactive iodine isotopes (^{131}I and ^{125}I) in thyroid gland: it is generally accepted that local exogenous internal resulting from the incorporated iodine radioisotopes is responsible for development of thyroid cancer in humans [177, 178]. In both these situations, limited fractions of target cells undergo long-term irradiation that is accompanied by accumulation of random point mutations in their genomes and selection of newly formed cell phenotypes toward the oncogenic cell transformation. The frequency of radiogenic mutations and the radiation-induced genomic instability are known to causally correlate with a risk of cancer [157-160, 179, 180]. In this respect, p53-mediated apoptosis is the natural mechanism ensuring elimination of irradiated and mutated cells to avoid their oncogenic transformation [24-28, 180]. According to the above described data and speculations, the up-regulated Hsps (Hsp90, Hsp70 and Hsp27) are able to save the irradiated cells from the apoptotic death; if so, the excess Hsps can, in fact, promote oncogenesis in the context of radiation exposure (Figure 4). By the way, this intriguing situation can be considered as conflict of two alternative programs: the Hsp-involving program of 'cell survival' tries to combat the p53-involving program of 'cell suicide' (apoptosis), so that a 'victory' of the former over the latter may result in the oncogenic transformation of the irradiated cells (see Figure 4).

Another intriguing question is why the Hsp levels may be up-regulated in irradiated (mutated) cells during the phenotypic selection toward oncogenic transformation? Some biological effects of ionizing radiation (e.g. expression of mutant and misfolded proteins and/or generation of free radicals) may lead to the HSF1 activation followed by induction of Hsps. Up-regulation of inducible Hsps and their mRNA were indeed found in irradiated mammalian cells [33, 162-164]. On the other hand, some kinds of local pathophysiological stresses such as ischemia and reperfusion, inflammation, edema, hemorrhage, ionic imbalance, etc can activate HSF1 and be responsible for the increased Hsp levels in the involved cells [1, 3, 4]. In that case, local and temporal superposition of the Hsp up-regulation induced by pathophysiological stress and the genotoxic consequences of radiation exposure may augment a risk of the oncogenic cell transformation following the exposure (see Figure 4). Without any stress, the tumorigenic factor heregulin β1 can activate HSF1 and induce Hsps in tumor cells [181]. As regards Hsp70, the enhancement of its expression during the oncogenic cell transformation can be due to radiation-induced alterations in the tumor repressor p53 and the related protein p63 that both repress the *Hsp70*-gene transcription [182, 183]. Mutations of p53 (that are typical for over 45% human tumors) can nullify its capability to repress the *Hsp70* promoter thereby resulting in the enhanced transcription of *Hsp70* gene [165, 166]. In some kinds of cancer, Hsp70 and its co-shaperone Hsp40 can be up-regulated through expression of the certain isoform of p63, Δp63α [183]. Besides that, PTEN mutation, induction of c-Myc and PI3K activation may also contribute to the increased Hsp expression in cells undergoing oncogenic transformation and in cancer cells (reviewed in [170]). Protooncogene c-Myc does activate the Hsp90A promoter, while inhibition of this process can impair the cell-transforming effect of c-Myc expression [184]. Finally, the cell survival- and cancer-associated up-regulation of Hsp27 may occur via involvement of the POU domain-containing transcriptional factor Brn3a(b) [8, 185, 186]. So, there are two major causes defining the enhancement of Hsp expression under the oncogenic cell transformation: (i) dysfunction of p53 through its mutation that leads to de-repression of the *Hsp* gene promoters and (ii) stimulation of oncogenic signaling pathway that result in the activation of *Hsp* gene promoters. At least the former cause seems to play a role in the Hsp-involving tumorigenesis provoked by ionizing radiation (Figure 4), as radiogenic mutations of p53 are really known and their direct relation to oncogenesis is doubtless [157, 179, 180]. Thus, on one hand, the up-regulated Hsp expression in the transformed or cancer cells may be a result of gene mutations and reflect the high degree of genomic instability; on the other hand, the excess Hsps are one of the endogenous factors promoting this instability. All this looks as intervention of the radiation-induced effects in balance between two alternative cellular programs: radiogenic mutation of p53 yields both the failure in the p53-dependent pathway of apoptotic elimination of genetically changed (i.e. potentially oncogenic) cells [24-28, 180] and the enhanced Hsp expression [165, 166, 170]; the latter event, in turn, contributes to the performance of 'cell survival' (antiapoptotic) program and thereby promotes the oncogenic cell transformation (Figure 4). Overall, the suggested coupling of oncogenesis with up-regulated Hsp expression in the radiation-transformed cells is one more example of the Hsp-conferred cytoprotection against postirradiation apoptosis.

Conclusion

The author of the present review hopes that the facts and hypotheses described here will attract an attention of radiobiologists, radiotherapists and pharmacologists to such a novel class of endogenous radioprotectors as Hsps. The pretty numerous data presented in this review do allow to consider mammalian Hsp70, Hsp27 and Hsp90 as inducible radioprotectors which are able to prevent or to a considerable extent attenuate cell death or senescence resulting from severe irradiation. As it follows from the many above findings, the radioprotective properties of all the three major Hsps are particularly manifested in their ability to attenuate apoptosis in normal and tumor cells exposed to ionizing radiation. The precise molecular machinery of Hsp-mediated cytoprotection against radiation-induced apoptosis has not yet been clarified completely; all the more, this seems to be a multi-factor and multi-stage process involving various pathways and cascades of reactions. Nevertheless, some of the reviewed data already enable to suggest three main mechanisms of the radioprotection conferred by the excess Hsps: (i) modulation of the intracellular signaling so that the apoptotic signal transduction is blocked, whereas the 'cell survival' signal transduction is stimulated; (ii) suppression of the radiation-associated free radical (or ROS) generation and ROS-induced apoptosis; (iii) attenuation of the genotoxic impact of ionizing radiation. The latter suggested mechainsm seems particularly intriguing and implies that the excess Hsps can somehow contribute to protection/repair of genomic DNA from radiation-induced damage. If it is confirmed in the future, the generally accepted dogma declaring that 'Hsps are guardians of the proteome but not of the genome' should be revised to take into consideration such a vital function of Hsps as guarding DNA from the environmental genotoxic exposures. In any event, the enigmatic phenomenon of Hsp-conferred radioprotection requires further thorough exploration to achieve greater progress in understanding of molecular mechanisms of apoptosis, senescence and oncogenesis induced by ionizing radiation.

Moreover, this phenomenon is of paramount importance for medicine, as the strategy of modern radiotherapy is basically aimed at elimination of cancer cells via radiation-induced apoptosis. While radiotherapy is one of the most effectual tools in clinical oncologists' store, the enhanced Hsp expression typical for many human tumors can, at least partly, be responsible for the high radioresistance of tumor cells and, therefore, for the low efficiency or the poor outcome of radiotherapeutic procedures. Some antitumor agents and radiosensitizers (e.g. 17AAG or proteasome inhibitors) can additionally up-regulate the Hsp expression in tumor cell, thereby enhancing their resistance to chemotherapy and radiotherapy. In this respect, administration of certain drugs down-regulating the constitutively enhanced Hsp expression in tumors and/or the concomitant Hsp induction in treated tumors could improve the outcome of combined anticancer therapy. Likewise, development of pharmacological inhibitors of the antiapoptotic activities of Hsp70 and Hsp27 could serve for the same purpose. At present, unfortunately, there are not yet such Hsp-down-regulating drugs really applicable for treatment of patients' tumors; though, some recent findings [54, 135, 136, 155, 156] seem quite encouraging and raise optimism toward a possibility of successful resolving this problem in the future. The author hopes that further creative research in this field will help to discover new ways for pharmacological inhibiting the *in vivo* expression and antiapoptotic activities of Hsps in the target cells of human tumors.

References

[1] Welch, W.J. (1992) Mammalian stress response: cell physiology, structure/function of stress proteins, and implication for medicine and disease. *Physiol. Rev. 72,* 1063-1081.

[2] Morimoto, R.I. (1998) Regulation of the heat shock transcriptional response: cross talk between a family of heat shock factors, molecular chaperones, and negative regulators. *Genes Dev. 12,* 3788-3796.

[3] Santoro, M.G. (2000) Heat shock factors and the control of stress response. *Biochem. Pharmacol. 59,* 55-63.

[4] Kabakov, A.E. and Gabai, V.L. (1997) *Heat shock protein and cytoprotection: ATP-deprived mammalian cells (MBIU series).* Austin, TX: Landes Company.

[5] Georgopoulos, C. and Welch, W.J. (1993) Role of the major heat shock proteins as molecular chaperones. *Annu. Rev. Cell Biol. 9,* 601-634.

[6] Wegele, H., Muller, L. and Buchner, J. (2004) Hsp70 and Hsp90 – a relay team for protein folding. *Rev. Physiol. Biochem. Pharmacol. 151,* 1-44.

[7] Mayer, M.P. and Bukau, B. (2005) Hsp70 chaperones: cellular functions and molecular mechanism. *Cell Mol. Life Sci. 62,* 670-684.

[8] Arrigo, A.P. (2005) Heat shock proteins as molecular chaperones. *Med. Sci. (Paris) 21,* 619-625.

[9] Nollen, E.A.A. and Morimoto, R.I. (2002) Chaperoning signaling pathways: molecular chaperones as stress-sensing 'heat shock' proteins. *J. Cell Sci. 115,* 2809-2816.

[10] Pratt, W.B. and Toft, D.O. (2003) Regulation of signaling protein function and trafficking by the hsp90/hsp70-based chaperone machinery. *Exp. Biol. Med. 228,* 111-133.

[11] Kampinga, H.H. (2006) Chaperones in preventing protein denaturation in living cells and protecting against cellular stress. *Handb. Exp. Pharmacol. 172,* 1-42.

[12] Arrigo, A.P. and Landry, J. (1994) Expression and function of the low-molecular-weight heat shock proteins. In R.I. Morimoto, A. Tissieres and C. Georgopoulos (Eds.), *The biology of heat shock proteins and molecular chaperones* (pp. 335-373). Cold Spring Harbor, NY: Cold Spring Harbor Laboratory Press.

[13] Guay, J., Lambert, H., Gingras-Breton, G., Lavoie, J.N., Huot J. and Landry, J. (1997) Regulation of actin filament dynamics by p38 map kinase-mediated phosphorylation of heat shock protein 27. *J. Cell Sci. 110,* 357-368.

[14] Bryantsev, A.L., Loktionova, S.A., Ilyinskaya, O.P., Tararak, E.M., Kampinga, H.H. and Kabakov, A.E. (2002) Distribution, phosphorylation and activities of Hsp25 in heat- stressed H9c2 myoblasts: a functional link to cytoprotection. *Cell Stress Chaperones 7,* 146-155.

[15] Takayama, S., Reed, J.C. and Homma, S. (2003) Heat-shock proteins as regulators of apoptosis. *Oncogene 22,* 9041-9047.

[16] Beere, H.M. (2004) "The stress of dying": the role of heat shock proteins in the regulation of apoptosis. *J. Cell Sci. 117,* 2641-2651.

[17] Didelot, C., Schmitt, E., Brunet, M., Maingret, L., Parcellier, A. and Garrido, C. (2006) Heat shock proteins: endogenous modulators of apoptotic cell death. *Handb. Exp. Pharmacol. 172,* 171-198.

[18] Khanna, K.K. and Jackson, S.P. (2001) DNA double-strand breaks: signaling, repair, and the cancer connection. *Nat. Genet. 27,* 247-254.

[19] Sancar, A., Lindsey-Boltz, L.A., Unsal-Kacmaz, K. and Linn, S. (2004) Molecular mechanisms of mammalian DNA repair and the DNA damage checkpoints. *Annu. Rev. Biochem. 73,* 39-85.

[20] Morrison, C. and Rieder, C.L. (2004) Chromosome damage and progression into and through mitosis in vertebrates. *DNA Repair (Amst.) 3,* 1133-1139.

[21] Lomax, M.E., Gulston, M.K. and O'Neill, P. (2002) Chemical aspects of clustered DNA damage induction by ionizing radiation. Radiat. Prot. Dosimetry 99, 63-68.

[22] Kale, R.K. (2003) Post-irradiation free radical generation: evidence from the conversion of xanthine dehydrogenase into xanthine oxidase. *Indian J. Exp. Biol. 41,* 105-111.

[23] Spitz, D.R., Azzam, E.L., Li, J.J. and Gius, D. (2004) Metabolic oxidation/reduction reactions and cellular responses to ionizing radiation: a unifying concept in stress response biology. *Cancer Metastasis Rev. 23,* 311-322.

[24] Wahl, G.M., Linke, S.P., Paulson, T.G. and Huang, L.C. (1997) Maintaining genetic stability through TP53 mediated checkpoint control. *Cancer Surv. 29,* 183-219.

[25] Pruschy, M., Rocha, S., Zaugg, K., Tenzer, A., Hess, C. Fisher, D.E., Glanzmann, C. and Bodis, S. (2001) Key targets for the execution of radiation-induced tumor cell apoptosis: the role of p53 and caspases. *Int. J. Radiat. Oncol. Biol. Phys. 49,* 561-567.

[26] Fei, P. and El-Deiry, W.S. (2003) P53 and radiation responses. *Oncogene 22,* 5774-5783.

[27] Weber, K.J. and Wenz, F. (2003) p53, apoptosis and radiosensitivity – experimental and clinical data. *Onkologie 25,* 136-141.

[28] Niida, H. and Nakanishi, M. (2006) DNA damage checkpoints in mammals. *Mutagenesis 21,* 3-9.

[29] Yang, S., Kuo, C., Bisi, J.E. and Kim, M.K. (2002) PML-dependent apoptosis after DNA damage is regulated by the checkpoint kinase hCds1/Chk2. *Nat. Cell Biol. 4,* 865-870.

[30] Afshar, G., Jelluma, N., Yang, X., Basila, D., Arvold, N.D., Karlsson, A., Yount, G.L., Dansen, T.B., Koller, E. and Haas-Kogan, D.A. (2006) Radiation-induced caspase-8 mediates p53-independent apoptosis in glioma cells. *Cancer Res. 66,* 4223-4232.

[31] Gordon, S.A, Hoffman, R.A., Simmons, R.L. and Ford, H.R. (1997) Induction of heat shock protein 70 protects thymocytes against radiation-induced apoptosis. *Arch. Surg. 132,* 1277-1282.

[32] Gibbons, N.B., Watson, R.W., Coffey, R.N., Brady, H.P. and Fitzpatrick, J.M. (2000) Heat-shock proteins inhibit induction of prostate cancer cell apoptosis. *Prostate 45,* 58-65.

[33] Calini, V., Urani, C. and Camatini, M. Overexpression of HSP70 is induced by ionizing radiation in 10T1/2 cells and protect from DNA damage. *Toxicol. In Vitro 17,* 561-566.

[34] Kabakov, A.E., Malyutina, Y.V. and Latchman, D.S. (2006) Hsf1-mediated stress response can transiently enhance cellular radioresistance. *Radiat. Res. 165,* 410-423.

[35] Malyutina, Y.V., Makarova, Y.M., Semenets, T.N., Semina, O.V., Mosin, A.F. and Kabakov, A.E. (2005) Whole body hyperthermia in mice confers heat shock protein-dependent radioresistance of their bone marrow and thymocytes. *J. Thermal Biol. 30,* 511-517.

[36] Park, S.-H., Lee, S.-J., Chung, H.-Y., Kim, T.-H., Cho, C.-K., Yoo, S.-Y. and Lee, Y.-S. (2000) Inducible heat shock protein 70 is involved in the radioadaptive response. *Radiat. Res. 153,* 318-326.

[37] Lee, S.-J., Choi, S.A., Lee, K.-H., Chung, H.-Y., Kim, T.-H., Cho, C.-K. and Lee, Y.-S. (2001) Role of inducible heat shock protein 70 in radiation-induced cell death. *Cell Stress Chaperones 6,* 273-281.

[38] Lee, Y.-J., Park, G.-H., Cho, H.-N., Cho, C.-K., Park, Y.-M., Lee, S.-J. and Lee, Y.-S. (2002) Induction of adaptive response by low-dose radiation in RIF cells transfected with Hspb1 (Hsp25) or inducible Hspa (Hsp70). *Radiat. Res. 157,* 371-377.

[39] Lee, Y.-J., Lee, D.-H., Cho, C.-K., Bae, S., Jhon, G.-J., Lee, S.-J., Soh, J.-W. and Lee, Y.-S. (2005) HSP25 inhibits protein kinase Cδ–mediated cell death through direct interaction. *J. Biol. Chem. 280,* 18108-18119.

[40] Lee, Y.-J., Lee, D.-H., Cho, C.-K., Chung, H.-Y., Bae, S., Jhon, G.-J., Soh, J.-W., Jeoung, D.-I., Lee, S.-J. and Lee, Y.-S. (2005) HSP25 inhibits radiation-induced apoptosis through reduction of PKCδ–mediated ROS production. *Oncogene 24,* 3715-3725.

[41] Machida, H., Matsumoto, Y., Shirai, M. and Kubota, N. (2003) Geldanamycin, an inhibitor of Hsp90, sensitizes human tumour cells to radiation. *Int. J. Radiat. Biol. 79,* 973-980.

[42] Enmon, R., Yang, W.-H., Ballangrud, A.M., Solit, D.B., Heller, G., Rosen, N., Scher, H.I. and Sgouros, G. (2003) Combination treatment with 17-*N*-Allylamino-17-Demethoxy Geldanamycin and acute irradiation produces supra-additive growth suppression in human prostate carcinoma spheroids. *Cancer Res. 63,* 8393-8399.

[43] Machida, H., Nakajima, S., Shikano, N., Nishio, J., Okada, S., Asayama, M., Shirai, M. and Kubota, N. (2005) Heat shock protein 90 inhibitor 17-allulamino-17-demethoxygeldanamycin potentiates the radiation response of tumor cells growth as monolayer cultures and spheroids by inducing apoptosis. *Cancer Sci. 96,* 911-917.

[44] Bisht, K.S., Bradbury, C.M., Mattson, D., Kaushal, A., Sowers, A., Markovina, S., Ortiz, K.L., Sieck, L.K., Isaacs, J.S., Brechbiel, M.W., Mitchell, J.B., Neckers, L.M. and Gius, D. (2003) Geldanamycin and 17-allylamino-17-demethoxygeldanamycicn potentiate the *in vitro* and *in vivo* radiation response of cervical tumor cells via the heat shock protein 90-mediated intracellular signaling and cytotoxicity. *Cancer Res. 63,* 8984-8995.

[45] Matsumoto, Y., Machida, H. and Kubota, N. (2005) Preferential sensitization of tumor cells to radiation by heat shock protein 90 inhibitor geldanamycin. *J. Radiat. Res. (Tokyo) 46,* 215-221.

[46] Hosokawa, N., Hirayoshi, K., Kudo, H., Takeshi, H., Aoike, A., Kawai, A. and Nagata, K. (1992) Inhibition of the activation of heat shock factor *in vivo* and *in vitro* by flavonoids. *Mol. Cell. Biol. 12,* 3490-3498.

[47] Asea, A., Ara, G., Teicher, B.A., Stevenson, M.A. and Calderwood, S.K. (2001) Effects of the flavonoid drug quercetin on the response of human prostate tumors to hyperthermia *in vitro* and *in vivo*. *Int. J. Hyperthermia 17,* 347-356.

[48] Rusak, G., Gutzeit, H.O. and Ludwig-Muller, J. (2002) Effects of structurally related flavonoids of hsp gene expression in human promyeloid leukaemia cells. *Food Technol. Biotechnol. 40,* 267-273.

[49] Kang, C.-M., Park, K.-P., Cho, C.-K., Seo, J.-S., Park, W.-Y., Lee, S.-J. and Lee, Y.-S. (2002) Hsps4 (HSP70) is involved in the radioadaptive response: results from mouse splenocytes. *Radiat. Res. 157,* 650-655.

[50] Gabai, V.L., Budagova, K.R. and Sherman, M.Y. (2005) Increased expression of the major heat shock protein Hsp72 in human prostate carcinoma cells is dispensable for their viability but confers resistance to a variety of anticancer agents. *Oncogene 24,* 3328-3338.

[51] Bases, R. (2005) Clonogenicity of human leukemic cells protected from cell-lethal agents by heat shock protein 70. *Cell Stress Chaperones 10,* 37-45.

[52] Brondani Da Rocha, A., Regner, A., Grivicich, I., Pretto Schunemann, D., Diel, C., Kovaleski, G., Brunetto De Farias, C., Mondatori, E., Almeida, L., Braga Filho, A. and Schwartsmann, G. (2004) Radioresistance is associated to increased Hsp70 content in human glioblastoma cell lines. *Int. J. Oncol. 25,* 777-785.

[53] Ekedahl, J., Joseph, B., Marchetti, P., Fauvel, H., Formstecher, P., Lewensohn R. and Zhivotovsky, B. (2003) Heat shock protein 72 does not modulate ionizing radiation-induced apoptosis in U1810 non-small cell lung carcinoma cells. *Cancer Biol. Ther. 2,* 663-669.

[54] Gaikwad, A., Poblenz, A., Haridas, V., Zhang, C., Duvic, M. and Gutterman, J. (2005) Triterpenoid electrophiles (avicins) suppress heat shock protein-70 and X-linked inhibitor of apoptosis proteins in malignant cells by activation of ubiquitin machinery: Implications for proapoptotic activity. *Clin. Cancer Res. 11,* 1953-1962.

[55] Dent, P., Yacoub, A., Contessa, J., Caron, R., Amorino, G., Valerie, K., Hagan, M.P., Grant, S. and Schmidt-Ullrich, R. (2003) Stress and radiation-induced activation of multiple intracellular signaling pathways. *Radiat. Res. 159,* 283-300.

[56] Eliseev, R.A., Zuscik, M.J., Schwarz, E.M., O'Keefe, R.J., Drissi, H. and Rosier, R.N. (2005) Increased radiation-induced apoptosis of Saos2 cells via inhibition of NFkappaB: a role for c-Jun N-terminal kinase. *J. Cell. Biochem. 96,* 1262-1273.

[57] Kanzawa, T., Iwado, E., Aoki, H., Iwamaru, A., Hollingsworth, E.F., Sawaya, R., Kondo, S. and Kondo, Y. (2006) Ionizing radiation induces apoptosis and inhibits neuronal differentiation in rat neural stem cells via the c-Jun NH(2)-terminal kinase (JNK) pathway. *Oncogene 25,* 3638-3648.

[58] Gabai, V.L., Meriin, A.B., Mosser, D.D., Caron, A.W., Rits, S., Shifrin, V.I. and Sherman, M.Y. (1997) Hsp70 prevents activation of stress kinases. A novel pathway of cellular thermotolerance. *J. Biol. Chem. 272,* 18033-18037.

[59] Gabai, V.L., Meriin, A.B., Yaglom, J.A., Wei, J.Y., Mosser, D.D. and Sherman, M.Y. (2000) Suppression of stress kinase JNK is involved in Hsp72-mediated protection of myogenic cells from transient energy deprivation. *J. Biol. Chem. 275,* 38088-38094.

[60] Gabai, V.L., Mabuchi, K., Mosser, D.D. and Sherman, M.Y. (2002) Hsp72 and stress kinase c-jun N-terminal kinase regulate the Bid-dependent pathway in tumor necrosis factor-induced apoptosis. *Mol. Cell. Biol. 22,* 3415-3424.

[61] Wang, T., Hu, Y.C., Dong, S., Fan, M., Tamae, D., Ozeki, M., Gao, Q., Gius, D. and Li, J.J. (2005) Co-activation of ERK, NF-kappaB, and GADD45betta in response to ionizing radiation. *J. Biol. Chem. 280,* 12593-12601.

[62] Kumar, P., Miller, A.I. and Polverini, P.J. (2004) p38 MAPK mediates γ-irradiation-induced endothelial cell apoptosis, and vascular endothelial growth factor protects endothelial cells through the phosphoinositide 3-kinase-Akt-Bcl-2 pathway. *J. Biol. Chem. 279,* 43352-43360.

[63] Choi, S.Y., Kim, M.J., Kang, C.M., Bae, S., Cho, C.K., Soh, J.W., Kim, J.H., Chung, H.Y., Lee, Y.S. and Lee, S.J. (2006) Activation of Bak and Bax through cabl-protein kinasedelta-p38 MAPK signaling in response to ionizing radiation in human non-small cell lung cancer cells. *J. Biol. Chem. 281,* 7049-7059.

[64] Pedraza-Alva, G., Koulnis, M., Charland, C., Thornton, T., Clements, J.L., Schlissel, M.S. and Rincon, M. (2006) Activation of p38 MAP kinase by DNA double-strand breaks in V(D)J recombination induces a G2/M cell cycle checkpoint. *EMBO J. 25,* 763-773.

[65] Creagh, E.M., Carmody, R.J. and Cotter, T.G. (2000) Heat shock protein 70 inhibits caspase-dependent and -independent apoptosis in Jurkat T cells. *Exp. Cell Biol. 257,* 58-66.

[66] Mosser, D.D., Caron, A.W., Bourget, L., Meriin, A.B., Sherman, M.Y. and Massie, B. (2000) The chaperone function of hsp70 is required for protection against stress-induced apoptosis. *Mol. Cell. Biol. 20,* 7146-7159.

[67] Steel, R., Doherty, J.P., Buzzard, K., Clemons, N., Hawkins, C.J. and Anderson, R.L. (2004) Hsp72 inhibits apoptosis upstream of the mitochondria and not through interactions with Apaf-1. *J. Biol. Chem. 279,* 490-499.

[68] Stankiewicz, A.R., Lachapelle, G., Foo, C.P., Radicioni, S.M. and Mosser, D.D. (2005) Hsp70 inhibits heat-induced apoptosis upstream of mitochondria by preventing Bax translocation. *J. Biol. Chem. 280,* 38729-38739.

[69] Beere, H.M., Wolf, B.B., Cain, K., Mosser, D.D., Mahboubi, A., Kuwana, T., Tailor, P., Morimoto, R.I., Cohen, G.M. and Green, D.R. (2000) Heat-shock protein 70 inhibits apoptosis by preventing recruitment of procaspase-9 to the Apaf-1 apoptosome. *Nat. Cell Biol. 2,* 469-475.

[70] Saleh, A., Srinivasula, S.M., Balkir, L., Robbins, P.D. and Alnemri, E.S. (2000) Negative regulation of the Apaf-1 apoptosome by Hsp70. *Nat. Cell Biol. 2,* 476-483.

[71] Li, C.Y., Lee, J.C., Ko, Y.G., Kim, J.I. and Sco, J.S. (2000) Heat shock protein 70 inhibits apoptosis down stream of cytochrome *c* release and upstream of caspase-3 activation. *J. Biol. Chem. 275,* 25663-25671.

[72] Hainaut, P. and Milner, J. (1992) Interaction of heat-shock protein 70 with p53 translated *in vitro*: evidence for interaction with dimeric p53 and for a role in the regulation of p53 conformation. *EMBO J. 11,* 3513-3520.

[73] Zylicz, M., King, F.W. and Wawrzynow, A. (2001) Hsp70 interactions with the p53 tumour suppressor protein. *EMBO J. 20,* 4634-4638.

[74] Hunt, C.R., Dix, D.J., Sharma, G.G., Pandita, P.K., Gupta, A., Funk, M. and Pandita, T.K. (2004) Genomic instability and enhanced radiosensitivity in Hsp70.1- and Hsp70.3-deficient mice. *Mol. Cell. Biol. 24,* 899-911.

[75] Ciavarra, R.P., Goldman, C., Wen, K.-K., Tedeschi, B. and Castora, F.J. (1994) Heat stress induces hsc70/nuclear topoisomerase I complex formation *in vivo*. Evidence for

hsc70-mediated, ATP-independent reactivation *in vitro*. *Proc. Natl. Acad. Sci. USA 91,* 1751-1755.

[76] Mao, Y. and Muller, M.T. (2003) Down modulation of topoisomerase I affects DNA repair efficiency. *DNA Repair (Amst.) 2,* 1115-1126.

[77] Mendez, F., Sandigursky, M., Franklin, W.A., Kenny, M.K., Kureekattil R., and Bases, R. (2000) Heat shock proteins associated with base excision repair enzymes in HeLa cells. *Radiat. Res. 153,* 186-195.

[78] Mendez, F., Kozin, E. and Bases, R. (2003) Heat shock protein 70 stimulation of the deoxyribonucleic acid base excision repair enzyme polymerase β. *Cell Stress Chaperones 8,* 153-161.

[79] Mendez, F., Sandigursky, M., Kureekattil, R.P., Kenny, M.K., Franklin, W.A. and Bases, R. (2003) Specific stimulation of human apurinic/apyrimidinic endonuclease by heat shock protein 70. *DNA Repair (Amst.) 2,* 259-271.

[80] Forsythe, H.L., Jarvis, J.L, Turner, J.W., Elmore, L.W. and Holt, S.E. (2001) Stable association of hsp90 and p23, but not hsp70, with active human telomerase. *J. Biol. Chem. 276,* 15571-15574.

[81] Sharma, GG., Gupta, A., Wang, H., Scherthan, H., Dhar, S., Gandhi, V., Iliakis, G., Shay, J.W., Young, C.S. and Pandita, T.K. (2003) hTERT associates with human telomeres and enhances genomic stability and DNA repair. *Oncogene 22,* 131-146.

[82] Pandita, T.K., Higashikubo, R. and Hunt, C.R. (2004) HSP70 and genomic stability. *Cell Cycle 3,* 591-592.

[83] Pandita, T.K. (2005) Role of HSPs and telomerase in radiotherapy. *Int J. Hyperthermia 21,* 689-694.

[84] Nadin, S.B., Vargas-Roig, L.M., Cuello-Carrion, F.D. and Ciocca, D.R. (2003) Deoxyribonucleic acid damage induced by doxorubicin in peripheral blood mononuclear cells: possible roles for the stress response and the deoxyribonucleic acid repair process. *Cell Stress Chaperones 8,* 361-372.

[85] Ehrnsperger, M., Graber, S., Gaestel, M. and Buchner, J. (1997) Binding of non-native protein to Hsp25 during heat shock creates a reservoir of folding intermediates for reactivation. *EMBO J. 16,* 221-229.

[86] Paul, C., Manero, F., Gonin, S., Kretz-Remy, C., Virot, S. and Arrigo, A.-P. (2002) Hsp27 as a negative regulator of cytochrome *c* release. *Mol. Cell. Biol. 22,* 816-834.

[87] Baek, S.-H., Min, J.-N., Park, E.-M., Han, M.-Y., Lee, Y.-S., Lee, Y.J. and Park, Y.-M. (2000) Role of small heat shock protein HSP25 in radioresistanse and Glutathione-redox cycle. *J. Cell. Physiol. 183,* 100-107.

[88] Park, S.H., Cho, H.N., Lee, S.J., Kim, T.H., Lee, Y., Park, Y.M., LeeY.J., Cho, C.K., Yoo, S.Y. and Lee, Y.S. (2000) Hsp25-induced radioresistance is associated with reduction of death by apoptosis: involvement of Bcl2 and the cell cycle. *Radiat. Res. 154,* 421-428.

[89] Cho, H.N., Lee, S.J., Park, S.H., Lee, Y.J., Cho, C.K. and Lee, Y.S. (2001) Overexpession of heat-shock protein 25 augments radiation-induced cell-cycle arrest in murine L929 cells. *Int. J. Radiat. Biol. 77,* 225-233.

[90] Cho, H.N., Lee, Y.J., Cho, C.K., Lee, S.J. and Lee, Y.-S. (2002) Downregulation of ERK2 is essential for the inhibition of radiation-induced cell death in HSP25 overexpressed L929 cells. *Cell Death Differ. 9,* 448-456.

[91] Yi, M.J., Park, S.H., Cho, H.N., Yong Chung, H., Kim, J.I., Cho, C.K., Lee, S.J. and Lee, Y.S. (2002) Heat-shock protein 25 (Hspb1) regulates manganese superoxide dismutase through activation of Nfkb (NF-kappaB). *Radiat. Res. 158,* 641-649.

[92] Fortin, A., Raybaud-Diogene, H., Tetu, B., Huot, J., Landry, J. and Deschenes, R. (2000) Overexpression of the 27 kDa heat shock protein is associated with thermoresitance and chemoresistance but not with radioresistance. *Int. J. Radiat. Oncol. Biol. Phys. 46,* 1259-1266.

[93] Park, K.M., Kramers, C., Vayssier-Taussat, M., Chen, A. and Bonventre, J.V. (2002) Prevention of kidney ischemia/reperfusion-induced functional injury, MAPK and MAPK kinase activation, and inflammation by remote transient ureteral obstruction. *J. Biol. Chem. 277,* 2040-2049.

[94] Kabakov, A.E., Budagova, K.R., Malyutina, Y.V., Latchman, D.S. and Csermely, P. (2004) Pharmacological attenuation of apoptosis in reoxygenated endothelial cells. *Cell Mol. Life Sci. 61,* 3076-3086.

[95] Rouse, J., Cohen, P., Trigon, S., Morange, M., Alonso-Llamazares, Zamanillo, D., Hunt, T. and Nebreda, A.R. (1994) A novel kinase cascade triggered by stress and heat shock that stimulates MAPKAP kinase-2 and phosphorylation of the small heat shock proteins. *Cell 78,* 1027-1037.

[96] Zhan, M. and Han, Z.C. (2004) Phosphatidylinositide 3-kinase/AKT in radiation response. *Histol. Histopathol. 19,* 915-923.

[97] Chinnaiyan, P., Allen, G.W. and Harari, P.M. (2006) Radiation and new molecular agents, Part II: targeting HDAC, HSP90, IGF-1R, PI3K, and Ras. *Semin. Radiat. Oncol. 16,* 59-64.

[98] Cataldi, A., Zauli, G., Di Pietro, R., Castorina, S. and Rana, R. (2001) Involvement of the pathway phosphatidylinositol-3-kinase/AKT-1 in the establishment of the survival response to ionizing radiation. *Cell Signal. 13,* 369-375.

[99] Soderlund, K., Perez-Tenorio, G. and Stal, O. (2006) Activation of the phosphatidylinositol 3-kinase/Akt pathway prevents radiation-induced apoptosis in breast cancer cells. *Int. J. Cancer 26,* 25-32.

[100] Cheng, J.C., Chou, C.H., Kuo, M.L. and Hsieh, C.Y. (2006) Radiation-enhanced hepatocellular carcinoma cell invasion with MMP-9 expression through PI3K/Akt/NF-kappaB signal transduction pathway. *Oncogene* (in press)

[101] Nakagomi, S., Suzuki, Y., Namikawa, K., Kiryu-Seo, S. and Kiyama, H. (2003) Expression of the activating transcription factor 3 prevents c-Jun N-terminal kinase-induced neuronal death by promoting heat shock protein 27 expression and Akt activation. *J. Neurosci. 23,* 87-96.

[102] Rane, M.J., Pan, Y., Singh, S., Powell, D.W., Wu, R., Cummins, T., Chen, Q., McLeish, K.R. and Klein, J.B. (2003) Heat shock protein 27 controls apoptosis by regulating Akt activation. *J. Biol. Chem. 278,* 27828-27835.

[103] Romashkova, J.A. and Makarov, S.S. (1999) NF-kappaB is a target of AKT in anti-apoptotic PDGF signaling. *Nature 401,* 86-90.

[104] Piva, R., Belardo, G. and Santoro, M.G. (2006) NF-kappaB: a stress-regulated switch for cell survival. *Antioxid. Redox Signal. 8,* 478-486.

[105] Parcellier, A., Schmitt, E., Gurbuxani, S., Seigneurin-Berny, D., Pance, A., Chantome, A., Plenchette, S., Khochbin, S., Solary, E. and Garrido, C. (2003) HSP27 is a ubiquitin- binding protein involved in I-κBα proteasomal degradation. *Mol. Cell. Biol. 23,* 5790- 5802.

[106] Bruey, J.M., Ducasse, C., Bonniaud, P., Ravagnan, L., Susin, S.A., Diaz-Latoud, C., Gurbuxani, S., Arrigo, A.P., Kroemer, G., Solary, E. and Garrido, C. (2000) Hsp27 negatively regulates cell death by interacting with cytochrome c. *Nat. Cell Biol. 2,* 645-652.

[107] Pandey, P., Farber, R., Nakazawa, A., Kumar, S., Bharti, A., Nalin, C., Weichselbaum, R., Kufe, D. and Kharbanda, S. (2000) Hsp27 functions as a negative regulator of cytochrome c-depeendent activation of procaspase-3. *Oncogene 19,* 1975-1981.

[108] Concannon, C.G., Orrenius, S. and Samali, A. (2001) Hsp27 inhibits cytochrome c-mediated caspase activation by sequestering both pro-caspase-3 and cytochrome c. *Gene Expr. 9,* 195-201.

[109] Belka, C. and Budach, W. (2002) Anti-apoptotic Bcl-2 proteins: structure, function and relevance for radiation biology. *Int. J. Radiat. Biol. 78,* 643-658.

[110] Le Bras, M., Clement, M.V., Pervaiz, S. and Brenner, C. (2005) Reactive oxygen species and the mitochondrial signaling pathway of cell death. *Histol. Histopathol. 20,* 205-219.

[111] Engel, R.H. and Evens, A.M. (2006) Oxidative stress and apoptosis: a new treatment paradigm in cancer. *Front. Biosci. 11,* 300-312.

[112] Park, Y.M., Han, M.Y., Blackburn, R.V. and Lee, Y.J. (1998) Overexpression of HSP25 reduces the level of TNF alpha-induced oxidative DNA damage biomarker, 8-hydroxyguanosine, in L929 cells. *J. Cell. Physiol. 174,* 27-34.

[113] Lee, Y.J., Cho, H.N., Jeoung, D.I., Soh, J.W., Cho, C.K., Bae, S., Chung, H.Y., Lee, S.J. and Lee, Y.S. (2004) HSP25 overexpression attenuates oxidative stress-induced apoptosis: roles of ERK1/2 signaling and manganese superoxide dismutase. *Free Radic. Biol. Med. 36,* 429-444.

[114] Neckers, L. and Ivy, S.P. (2003) Heat shock protein 90. *Curr. Opin. Oncol. 15,* 419-424.

[115] Whitesell, L. and Lindquist, S.L. (2005) HSP90 and the chaperoning of cancer. *Nat. Rev. Cancer 5,* 761-772.

[116] Isaacs, J.S., Xu, W. and Neckers, L. (2003) Heat shock protein 90 as a molecular target for cancer therapeutics. *Cancer Cell 3,* 213-217.

[117] Dai, C. and Whitesell, L. (2005) HSP90: a rising star on the horizon of anticancer targets. *Future Oncol. 1,* 529-540.

[118] Chiosis, G., Rodina, A. and Moulick, K. (2006) Emerging Hsp90 inhibitors: from discovery to clinic. *Anticancer Agents Med. Chem. 6,* 1-8.

[119] Soga, S., Shiotsu, Y., Akinaga, S. and Sharma, S.V. (2003) Development of radicicol analogues. *Curr. Cancer Drug Targets 3,* 359-369.

[120] Harashima, K., Akimoto, T., Nonaka, T., Tsuzuki, K., Mitsuhashi, N. and Nakano, T. (2005) Heat shock protein 90 (Hsp90) chaperone complex inhibitor, radicicol,

potentiated radiation-induced cell killing in a hormone-sensitive prostate cancer cell line through degradation of the androgen receptor. *Int. J. Radiat. Biol. 81,* 63-76.

[121] Russel, J.S., Burgan, W., Oswald, K.A., Camphausen, K. and Tofilon, P.J. (2003) Enhanced cell killing induced by the combination of radiation and the heat shock protein 90 inhibitor 17-allylamino-17-demethoxygeldanamycin: a multitarget approach to radiosensitization. *Clin. Cancer Res. 9,* 3749-3755.

[122] Ochel, H.J. and Gademann, G. (2006) In vitro combined modality treatment of prostate carcinoma cells with 17-(allylamino)-17-demethoxygeldanamycin and ionizing radiation. *Anticancer Res. 26,* 2085-2091.

[123] Kobayashi, S., Nantz, R., Kitamura, T., Higashikubo, R. and Horikoshi, N. (2005) Combined inhibition of extracellular signal-regulated kinases and HSP90 sensitizes human colon carcinoma cells to ionizing radiation. *Oncogene 24,* 3011-3019.

[124] Bull, E.E., Dote, H., Brady, K.J., Burgan, W.E., Carter, D.J., Cerra, M.A., Oswald, K.A., Hollingshead, M.G., Camphausen, K. and Tofilon, P.J. (2004) Enhanced tumor cell radiosensitivity and abrogation of G2 and S phase arrest by the Hsp90 inhibitor 17-(dimmethylaminoethylamino)-17-demethoxygeldanamycin. *Clin. Cancer Res. 10,* 8077-8084.

[125] Dote, H., Cerna, D., Burgan, W.E., Camphausen, K. and Tofilon, P.J. (2005) ErbB3 expression predicts tumor cell radiosensitization induced by Hsp90 inhibition. *Cancer Res. 65,* 6967-6975.

[126] Kamal, A., Thao, L., Sensintaffar, J., Zhang, L, Boehm, M.F., Fritz, L.C. and Burrows, F.J. (2003) A high-affinity conformation of Hsp90 confers tumour selectivity on Hsp90 inhibitors. *Nature 425,* 407-410.

[127] Edwards, E., Geng, L., Tan, J., Onishko, H., Donnely, E. and Hallahan, D.E. (2002) Phosphatidylinositol 3-kinase/Akt signaling in the response of vascular endothelium to ionizing radiation. *Cancer Res. 62,* 4671-4677.

[128] Zingg, D., Riesterer, O., Fabbro, D., Glanzmann, C., Bodis, S. and Pruschy, M. (2004) Differential activation of the phosphatidylinositol 3'-kinase/Akt survival pathway by ionizing radiation in tumor and primary endothelial cells. *Cancer Res. 64,* 5398-5406.

[129] Gu, Q., Wang, D., Wang, X., Peng, R., Liu, J., Jiang, T., Wang, Z., Wang, S. and Deng, H. (2004) Basic fibroblast growth factor inhibits radiation-induced apoptosis of HUVECs. I. The PI3K/AKT pathway and induction of phosphorylation of BAD. *Radiat. Res. 161,* 692-702.

[130] Lu, B., Shinohara, E.T., Edwards, E., Geng, L., Tan, J. and Hallahan, D.E. (2005) The use of tyrosine kinase inhibitors in modifying the response of tumor microvasculature to radiotherapy. *Technol. Cancer Res. Treatment 4,* 691-698.

[131] Kim, D.W., Huamani, J., Fu, A. and Hallahan, D.E. (2006) Molecular strategies targeting the host component of cancer to enhance tumor response to radiation therapy. *Int. J. Radiat. Oncol. Biol. Phys. 64,* 38-46.

[132] Muller, P., Ceskova, P. and Vojtesek, B. (2005) Hsp90 is essential for restoring cellular functions of temperature-sensitive p53 mutant protein but not for stabilization and activation of wild-type p53: implications for cancer therapy. *J. Biol. Chem. 280,* 6682-6691.

[133] Zou, J., Guo, Y., Guettouche, T., Smith, D.F. and Voellmy, R. (1998) Repression of heat shock transcription factor HSF1 activation by HSP90 (HSP90 complex) that forms a stress-sensitive complex with HSF1. *Cell 94,* 471-480.

[134] Bagatell, R., Paine-Murrieta, G.D., Taylor, C.W., Pulcini, E.J., Akinaga, S., Benjamin, I.J., and Whitesell, L. (2000) Induction of a heat shock factor 1-dependent stress response alters the cytotoxic activity of hsp90-binding agents. *Clin. Cancer Res. 6,* 3312-3318.

[135] Guo, F., Rocha, K., Bali, P., Pranpat, M., Fiskus, W., Boyapalle, S., Kumaraswamy, S., Balasis, M., Greedy, B., Armitage, E.S.M., Lawrence, N. and Bhalla, K. (2005) Abrogation of heat shock protein 70 induction as a strategy to increase antileukemia ativity of heat shock protein 90 inhibitor 17-allylamino-demethoxy geldanamycin. *Cancer Res. 65,* 10536-10544.

[136] Zaarur, N., Gabai, V.L., Porco, J.A., Calderwood, S. and Sherman, M.Y. (2006) Targeting heat shock response to sensitize cancer cells to proteasome and Hsp90 inhibitors. *Cancer Res. 66,* 1783-1791.

[137] Yokota, S., Kitahara, M. and Nagata, K. (2000) Benzylidene lactam compound, KNK437, a novel inhibitor of acquisition of thermotolerance and heat shock protein induction in human colon carcinoma cells. *Cancer Res. 60,* 2942-2948.

[138] Pervan, M., Pajonk, F., Sun, J.R., Withers, H.R. and McBride, W.H. (2001) Molecular pathways that modify tumor radiation response. *Am. J. Clin. Oncol. 24,* 481-485.

[139] Adams, J. (2004) The development of proteasome inhibitors as anticancer drugs. *Cancer Cell 5,* 417-421.

[140] Chauhan, D., Hideshima, T., Mitsiades, C. Richardson, P. and Anderson, K.C. (2005) Proteasome inhibitor therapy in multiple myeloma. *Mol. Cancer Ther. 4,* 686-692.

[141] Pajonk, F., Pajonk, K. and McBride, W.H. (2000) Apoptosis and radiosensitization of Hodgkin cells by proteasome inhibition. *Int. J. Radiat. Oncol. Biol. Phys. 47,* 1025-1032.

[142] Munshi, A., Kurland, J.F., Nishikawa, T., Chiao, P.J., Andreef, M. and Meyn, R.E. (2004) Inhibition of constitutively activated nuclear factor-kappaB radiosensitizes human melanoma cells. *Mol. Cancer Ther. 3,* 985-992.

[143] Grimes, K.R., Daosukho, C., Zhao, Y., Meigooni, A. and St Clair, W. (2005) Proteasome inhibition improves fractionated radiation treatment against non-small cell lung cancer: an antioxidant connection. *Int. J. Oncol. 27,* 1047-1052.

[144] Warren, G., Grimes, K., Xu, Y., Kudrimoti, M. and St Clair, W. (2006) Selectively enhanced radiation sensitivity in prostate cancer cells associated with proteasome inhibition. *Oncol. Rep. 15,* 1287-1291.

[145] Mimnaugh, E.G., Xu, W., Vos, M., Yuan, X., Isaacs, J.S., Bisht, K.S., Glus, D. and Neckers, L. (2004) Simultaneous inhibition of hsp90 and the proteasome promotes protein ubiquitination, causes endoplasmic reticulum-derived cytosolic vacuolization, and enhances antitumor activity. *Mol. Cancer Ther. 3,* 551-566.

[146] Russo, S.M., Tepper, J.E., Baldwin, A.S., Liu, R., Adams, J., Elliott, P. and Cusack, J.C. (2001) Enhancement of radiosensitivity by proteasome inhibition: implications for a role of NF-kappaB. *Int. J. Radiat. Oncol. Biol. Phys. 50,* 183-193.

[147] Edelman, M.J. (2005) The potential role of bortezomib in combination with hemotherapy and radiation in non-small cell lung cancer. *Clin. Lung Cancer 7 Suppl. 2,* S64-S66.

[148] Goel, A., Dispenzieri, A., Greipp, P.R., Witzig, T.E., Mesa, R.A. and Russell, S.J. (2005) PS-341-mediated selective targeting of multiple myeloma cells by synergistic increase in ionizing radiation-induced apoptosis. *Exp. Hematol. 33,* 784-795.

[149] Goel, A., Dispenzieri, A., Geyer, S.M., Greiner, S., Peng, K.W. and Russell, S.J. (2006) Synergistic activity of the proteasome inhibitor PS-341 with non-myeloablative 153-Sm-EDTMP skeletally targeted radiotherapy in an orthotopic model of multiple myeloma. *Blood 107,* 4063-4070.

[150] Hideshima, T., Podar, K., Chauhan, D., Ishitsuka, K., Mitsiades, C., Tai, Y.T., Hamasaki, M., Raje, N., Hideshima, H., Schreiner, G., Nguyen, A.N., Navas, T., Munshi, N.C., Richardson, P.G., Higgins, L.S. and Anderson, K.S. (2004) p38 MAPK inhibition enhances PS-341 (bortezomib)-induced cytotoxicity against multiple myeloma cells. *Oncogene 23,* 8766-8776.

[151] Navas, T.A., Nguyen, A.N., Hideshima, T., Reddy, M., Ma, J.Y., Haghnazari, E., Henson, M., Stebbins, E.G., Kerr, I., O'Young, G., Kapoun, A.M., Chakravarty, S., Mavunkel, B., Perumattam, J., Luedtke, G., Dugar, S., Medisherla, S., Protter, A.A., Schreiner, G.F., Anderson, K.S. and Higgins, L.S. (2006) Inhibition of p38alpha MAPK enhances proteasome inhibitor-induced apoptosis of myeloma cells by modulating Hsp27, Bcl-X(L), Mcl-1 and p53 levels in vitro and inhibits tumor growth in vivo. *Leukemia 20,* 1017-1027.

[152] Tytell, M. and Hooper, P.L. (2001) Heat shock proteins: new keys for the development of cytoprotective therapies. *Expert Opin. Ther. Targets 5,* 267-287.

[153] Coss, R.A. (2005) Inhibiting induction of heat shock proteins as a strategy to enhance cancer therapy. *Int. J. Hyperthermia 21,* 695-701.

[154] Ciocca, D.R. and Calderwood, S.K. (2005) Heat shock proteins in cancer: diagnostic, prognostic, and treatment implications. *Cell Stress Chaperones 10,* 86-103.

[155] Sakamoto, H., Mashima, T., Yamamoto, K. and Tsuruo, T. (2002) Modulation of heat-shock protein 27 (Hsp27) anti-apoptotic activity by methylglyoxal modification. *J. Biol. Chem. 277,* 45770-45775.

[156] Schmitt, E., Maingret, L., Puig, P.E., Rerole, A.L., Ghiringhelli, F., Hammann, A., Solary, E., Kroemer, G. and Garrido, C. (2006) Heat shock protein 70 neutralization exerts potent antitumor effects in animal models of colon cancer and melanoma. *Cancer Res. 66,* 4191-4197.

[157] Farid, N.R. (2001) P53 mutations in thyroid carcinoma: tidings from an old foe. *J. Endocrinol. Invest. 24,* 536-545.

[158] Little, J.B. (2003) Genomic instability and radiation. *J Radiol. Prot. 23,* 173-181.

[159] Smith, L.E., Nagar, S., Kim, G.J. and Morgan, W.F. (2003) Radiation-induced genomic instability: radiation quality and dose response. *Health Phys. 85,* 23-29.

[160] Sowa, M., Arthurs, B.J., Estes, B.J. and Morgan, W.F. (2006) Effects of ionizing radiation on cellular structures, induced instability and carcinogenesis. *EXS 96,* 293-301.

[161] Toulany, M., Kasten-Pisula, U., Brammer, I., Wang, S., Chen, J., Dittmann, K., Baumann, M., Dikomey, E. and Rodemann, H.P. (2006) Blockage of epidermal growth factor receptor-phosphatidylinositol 3-kinase-AKT signaling increases radiosensitivity of K-RAS mutated human tumor cells in vitro by affecting DNA repair. *Clin. Cancer Res. 12,* 4119-4126.

[162] Sierra-Rivera, E., Voorhees, G.J. and Freeman, M.L. (1993) Gamma irradiation increases hsp-70 in Chinese hamster ovary cells. *Radiat. Res. 135,* 40-45.

[163] Nogami, M., Huang, J.T., James, S.J., Lubinski, J.M., Nakamura, L.T. and Makinodan, T. (1993) Mice chronically exposed to low dose ionizing radiation possess splenocytes with elevated levels of HSP70 mRNA, HSC70 and HSP72 and with an increased capacity to proliferate. *Int. J. Radiat. Biol. 63,* 775-783.

[164] Nogami, M., Huang, J.T., Nakamura, L.T. and Makinodan, T. (1994) T cells are the cellular target of the proliferation-augmented effect of chronic low-dose ionizing radiation in mice. *Radiat. Res. 139,* 47-52.

[165] Agoff, S.N., Hou, J., Linzer, D.I. and Wu, B. (1993) Regulation of the human hsp70 promoter by p53. *Science 259,* 84-87.

[166] Tsutsumi-Ishii, Y., Tadokoro, K., Nanaoka, F., and Tsuchida, N. (1995) Response of heat shock element within the human HSP70 promoter to mutated p53 genes. *Cell Growth Differ. 6,* 1-8.

[167] Rutherford, S.L. and Lindquist, S. (1998) Hsp90 as a capacitor for morphological evolution. *Nature 396,* 336-342.

[168] Queitsch, C., Sangster, T.A. and Lindquist, S. (2003) Hsp90 as a capacitor of phenotypic variation. *Nature 417,* 618-624.

[169] Bertram, J.S. (2000) The molecular biology of cancer. *Mol. Aspects Med. 21,* 167-223.

[170] Calderwood, S.K., Khaleque, M.A., Sawyer, D.B. and Ciocca, D.R. (2006) Heat shock proteins in cancer: chaperones of tumorigenesis. *Trends Biochem. Sci. 31,* 164-172.

[171] Mosser, D.D. and Morimoto, R.I. (2004) Molecular chaperones and the stress of oncogenesis. *Oncogene 23,* 2907-2918.

[172] Garrido, C., Schmitt, E., Cande, C., Vahsen, N., Parcellier, A. and Kroemer, G. (2003) HSP27 and HSP70: Potentially oncogenic apoptosis inhibitors. *Cell Cycle 2,* 579-584.

[173] Ron, E. (2003) Cancer risks from medical radiation. *Health Phys. 85,* 47-59.

[174] Allan, J.M. and Travis, L.B. (2005) Mechanisms of therapy-related carcinogenesis. *Nat. Rev. Cancer 5,* 943-955.

[175] Fueger, G.F. (1986) Biological effects and risks of accidental radionuclide uptake. *Acta Med. Austriaca 13,* 86-93.

[176] Harrison, J.D. and Muirhead, C.R. (2003) Quantitative comparison of cancer induction in humans by internally deposited radionuclides and external radiation. *Int. J. Radiat. Biol. 79,* 1-13.

[177] Robbins, J. and Schneider, A.B. (2000) Thyroid cancer following exposure to radioactive iodine. *Rev. Endocr. Metab. Disord. 1,* 197-203.

[178] Hindie, E., Leenhardt, L., Vitaux, F., Colas-Linhart, N., Grosclaude, P. Galle, P., Aurengo, A. and Bok, B. (2002) Non-medical exposure to radioiodines and thyroid cancer. *Eur. J. Nucl. Med. Mol. Imaging 29 Suppl 2,* S497-S512.

[179] Ulrich, R.L. and Ponnaiya, B. (1998) Radiation-induced instability and its relation to radiation carcinogenesis. *Int. J. Radiat. Biol. 74,* 747-754.

[180] Kim, I.G. and Lee, Y.S. (2003) Radiation-induced tumorigenesis. *J. Biochem. Mol. Biol. 36,* 144-148.

[181] Khaleque, M.A., Bharti, A., Sawyer, D., Gong, J., Benjamin, I.J., Stevenson, M.A. and Calderwood, S.K. (2005) Induction of heat shock proteins by heregulin beta1 leads to protection from apoptosis and anchorage-independent growth. *Oncogene 24,* 6564-6573.

[182] Taira, T., Sawai, M., Ikeda, M., Tamai, K., Iguchi-Ariga, S.M. and Ariga, H. (1999) Cell cycle-dependent switch of up- and down-regulation of human hsp70 gene expression by interaction between c-Myc and CBF/NF-Y. *J. Biol. Chem. 274,* 24270-24279.

[183] Wu, G., Osada, M., Guo, Z., Fomenkov, A., Begum, S., Zhao, M., Upadhyay, S., Xing, M., Wu, F., Moon, C., Westra, W.H., Koch, W.M., Mantovani, R., Califano, J.A., Ratovitski, E., Sidransky, D. and Trink, B. (2005) DeltaNp63alpha up-regulates the Hsp70 gene in human cancer. *Cancer Res. 65,* 758-766.

[184] Teng, S.C., Chen, Y.Y., Su, Y.N., Chou, P.C., Chiang, Y.C., Tseng, S.F. and Wu, K.J. (2004) Direct activation of HSP90A transcription by c-Myc contributes to c-Myc-induced transformation. *J. Biol. Chem. 279,* 14649-14655.

[185] Farooqui-Kabir, S.R., Budhram-Mahadeo, V., Lewis, H., Latchman, D.S., Marber, M.S. and Heads, R.J. (2004) Regulation of Hsp27 expression and cell survival by the POU transcription factor Brn3a. *Cell Death Differ. 11,* 1242-1244.

[186] Lee, S.A., Ndisang, D., Patel, C., Dennis, J.H., Faulkes, D.J., D'Arrigo, C., Samady, L., Farooqui-Kabir, S., Heads, R.J., Latchman, D.S. and Budhram-Mahadeo, V.S. (2005) Expression of the Brn-3b transcription factor correlates with expression of HSP-27 in breast cancer biopsies and is required for maximal activation of the HSP-27 promoter. *Cancer Res. 65,* 3072-3080.

In: New Developments in Cell Apoptosis Research
Editor: Alan J. Corvin, pp. 159-178
ISBN: 1-60021-453-3

Chapter VI

Role of Mobile Genetic Element SINE Expression in Apoptosis of the Cell

Tatyana V. Nikitina and Lyudmila I. Tishchenko
Department of Biochemistry, Faculty of Biology and Soil Science,
St Petersburg State University Universitetskaya 7/9, 199034 St Petersburg, Russia
Tel. +7 (812) 328-21-82; e-mail: Tanik@tn3532.spb.edu

Abstract

There is increasing evidence that RNA products of the mobile genetic elements SINE (such as Alu sequences in primate cells or B1 and B2 elements in rodent cells) can play a considerable role in apoptosis of the cell. All these SINEs are represented in a great number in genomes, possess intragenic promoters and are transcribed by DNA-dependent RNA polymerase III. A series of experiments have shown that diverse stress influences cause different SINE expression activation in the cells. Depending on the severity of a damage caused by the stress, the cell can survive and recover or enter apoptosis. Accumulation of the SINE RNA products in the cell can serve as a pro-apoptotic factor through several mechanisms. (i) It was shown that stress influences not only cause SINE expression activation, but also induce expression of another class of mobile genetic elements – LINEs, coding for reverse transcriptase. Alu RNAs and some other SINE RNA products are the proper templates for reverse transcriptase and can actively replicate through retrotransposition and insert into the active chromatin sites. This in turn can result in disruption of normal gene expression pattern and, as a result, in apoptosis pathway triggering. (ii) In addition, Alu RNA was shown to be able to inhibit translation, which also can induce apoptosis. Alu RNA is highly homologous to 7SL RNA and can bind to SRP (signal recognition particle) proteins possessing affinity to ribosome. It is possible that such Alu-containing SRP bind to ribosomes independent of the presence of N-terminal export signal in nascent polypeptide, and inhibit total translation in the cell. On the other hand, Alu RNAs contain double strand sites and, upon the increase of their level in the cell, can regulate pro-apoptotic protein PKR (dsRNA-activated protein kinase). PKR activation also inhibits total translation. But the specific translation of newly synthesized mRNAs (these can be mRNAs of stress or/and pro-apoptotic proteins) in this case is activated. Considering the possible role of SINE RNA products in apoptosis, of particular interest is the problem of the activation of their transcription by RNA polymerase III during this process. This activation seems to be specific to SINEs, as the transcription of the other studied genes directed by RNA polymerase III is suppressed during apoptosis. The conceivable mechanisms of SINE transcription regulation in apoptosis are considered in this chapter on the basis of literature and our data.

Introduction

Sequencing of the entire human genome and the genomes of several other mammals testified that just a minor part of the genomic DNA contains protein-coding sequences (only about 2% in human) [1]. The rest of the DNA (more than 95%), except for regulatory sequences, consists of many types of repetitive elements and until recently has been considered as non-functional "junk" DNA. Among other sequenced genomes, the human genome is the most enriched with these repeats. The role of the repeats in different genomes is still unclear. But in the last few years great attention is paid to the elucidation of their functions, especially when the sequencing of the total genomes has shown little difference between protein-coding regions in lower and higher eukaryots (including human).

Of special interest are the repeats belonging to dispersed repeated sequences. They are not arranged in large blocks (in contrast to satellite DNA), but are spread throughout the entire genome. This type repeats, otherwise called medium reiterated frequency repeats (MER), are divided into two classes (or families): SINE – short interspersed elements, and LINE – long interspersed elements. The length of SINE is only about 90-400 bp, and the length of LINE can be as long as 7 thousand bp. The most studied SINEs in human and some other primate genomes are the so called Alu repeats, with the repetitive unit length of about 300 bp. There are about 1.1 million copies of Alus in the human genome, i.e. one Alu-sequence per about every 4 thousand bp of the genome; in total Alus comprise about 10% of the human genome DNA. SINEs are found in the genomes of some other mammals (for example, B1 and B2 elements in rodents, C-element in rabbit), and also in the genomes of several non-mammals – silk worm *Bombix mori* and eel [2; 3; 4; 5; 6]. B1, B2, Alus and other SINEs contain no open reading frames, are surrounded by the short direct repeats (10-20 bp long) in the genome. These short direct repeats are unique for every SINE. The common feature of all known SINEs is the presence of two sequences (A- and B-boxes), homological to the consensus sequence of the split internal (intragenic) prompter of the genes transcribed by RNA polymerase III – tRNA genes and 7SL RNA genes [2; 3; 6; 7]. It is believed that all known SINEs derived from the sequences of either tRNA genes, or 7SL RNA genes. Different SINEs were shown to be transcribed by RNA polymerase III, and then their RNA can serve as a template for the reverse transcriptase, encoded by the repeats from the LINE family. As a result of reverse transcription, cDNA is generated that can be integrated into the new locus of the genome, similar to what occurs in the case of retrovirus infection, owing to the second, endonuclease, activity of LINE-encoded reverse transcriptase [6].

It becomes clear that SINEs are not "junk" DNA, but have some important functions in the cell. So, it was shown that Alu repeats can serve as enhancers, silencers, insulators, alternative splicing sites, RNA editing sites, can be a part of mRNA and code for protein nuclear localization signal, or regulate transcription and translation of such mRNAs [8; 9; 10; 11; 12]. During mouse cell stress response B2 RNAs take part in RNA polymerase II-directed transcription regulation [13]. In this chapter we review data showing that SINE and their RNA products can play some substantial role in cell apoptosis. Concomitantly the close functional interplay of the mechanisms of the cell recovery after stress and the programmed cell death (apoptosis) is discussed. We also consider possible ways for SINE transcription regulation during stress and apoptosis.

SINE-RNAs in Stress and Apoptosis Interplay

For a long time cell stress and apoptosis were studied independently from each other. As this takes place, in cell stress studies the main attention was paid to the mechanisms of the cell recovery, and in apoptosis studies – to the mechanisms of cell death. As a result, the role of diverse heat shock proteins (such as the proteins of Hsp27, Hsp70 and Hsp90 subfamilies) in suppression of stress-induced pro-apoptotic pathways was elucidated, and, on the other hand, different mechanisms of pro-apoptotic pathways induction after severe cell injuries were explored [14]. But the data received allowed some researches to assume that there is a close functional interplay between such seemingly opposite processes as cell apoptotic death and cell recovery after stressful action. It was shown that the same stressful action on the cell cause practically concomitant induction of both pro- and anti-apoptotic pathways (reviewed in [14]). But it is not clear yet how this induction occurs, and what factors provide the interaction between pro- and anti-apoptotic pathways. We can speculate that among such factors are SINEs and their RNA products. Some data support this idea. For example, some SINE-RNAs (particularly, Alu-RNAs in primates, B1- and B2-RNAs in rodents, C-element RNA in rabbit, and Bm1-RNA in silk worm) were shown to behave as typical stress-induced factors upon the cell treatment with UV, heat, translation inhibitors, DNA damaging agents, γ-radiation, or upon virus infection: their synthesis is increasingly fast, and the levels of these RNAs in the cell rise transiently, on the period of the cell recovery [15; 16]. It is known that the damaging treatments listed above, depending on their severity and the type of the cell treated, can cause either stress followed by the recovery, or apoptosis. It can be suggested that after the cell damage caused by these factors SINEs can be activated and take part in cell growth and proliferation suppression, and also in the modulation of some mechanisms, leading, depending on the severity of the cell damage, to the cell recovery or to the cell death (apoptosis).

SINE Expression in Stressed Cells

It was demonstrated for the first time about 20 years ago that SINE-RNA level increased in the cell after stressor influences. For example, during SV40 virus infection in mouse cells and heat shock in Chinese hamster cells V79 and CHO and mouse cells NIH 3T3 the level of B2 RNA was shown to rise significantly [17; 18]. In human HeLa cells infected with herpes simplex virus (HSV) RNA polymerase III-directed transcription of Alu repeats was activated [19; 20]. Human immunodeficiency virus (HIV) protein Tat activates the transcription of co-transfected and endogenous Alu sequences in human cells HeLa and Jurkat [21]. It was also revealed that Alu transcription is activated in human cells HeLa and 293 upon their infection with adenovirus type 5 [22]. Heat shock and translation inhibitor cycloheximide cause transient elevation of B1 and B2 element expression in mouse cells, C-element expression in rabbit cells, and Alu expression in human HeLa cells. Importantly, this induction of Alu-RNA synthesis in HeLa is specific, as the levels of other RNA polymerase III-synthesised RNAs – 7SL, 7SK, 5S and 6U RNAs – do not change during heat shock [23]. The levels of B1 and B2 RNAs increase in mouse fibroblasts LA9 infected with parvovirus MVM (minute virus of mice) [24]. It is notable that the increase in SINE RNA product levels is characteristic not only for mammals, but also for evolutionary distant taxon – insects. It was demonstrated that different stressor treatments of silk worm *Bombix mori* cells BmN-4 – heat shock, virus infection and protein synthesis inhibition by cycloheximide – caused the increase in RNA products of Bm1-repeats (SINE which is innate to this species) [15]. These data, and the fact that SINEs were found in most studied eukaryots, made it possible to suggest a hypothesis that SINE transcription activation is an important mechanism of cell response to stressor treatments, which is typical of not only mammals, but also other taxonomic groups [15].

SINE Expression in Apoptotic Cells

The literature data on SINE transcript levels in apoptotic cells is rather limited. Although some investigation results allow suggesting that SINE expression is regulated specifically upon different cytotoxic treatments of the cell. For example, the data exist that severe heat shock (with cytotoxic effect) cause the increase in B2 RNA level, but not the level of heat shock proteins, in the cells of Chinese hamster. But in this case the type of cell death (apoptosis or necrosis) was not determined [25]. Besides, parvovirus MVM, activating B1 and B2 RNA synthesis in mouse cells, is able to induce apoptosis in these cells [24]. And prolonged HeLa cell incubation with DNA methylation inhibitor 5-azacytidine causes their death accompanied by Alu repeat overexpression [26]. In our study we used real-time RT-PCR method to show the elevation of Alu RNA level in total cellular RNA of human epidermoid carcinoma cells A431, induced to apoptosis by the high concentration (100 nM) of epidermoid growth factor (EGF) [27].

Noteworthy, UV and ionizing radiation cause programmed death of rat chloroleukaemia cells by induction of mobile L1 element retrotransposition [28]. In this work the authors did not measure the levels of SINE RNAs, but in some later research it was shown that (i) UV and γ-radiation activated Alu repeats expression in human cells; (ii) LINE expression activation and retrotransposition is followed by SINE retrotransposition activation [29]. The authors used a special system for Alu repeat retrotransposition detection: HeLa cells were co-transfected with expressing vector with human L1 element controlled by cytomegalovirus (CMV) promoter, and specially constructed plasmid with Alu repeat insertion, carrying neomycin resistance gene in backward orientation – *neo*^{Tet}. Alu repeat is controlled by its own internal promoter and 7SL RNA gene enhancer, and *neo* gene is controlled by its own promoter. Initially *neo*^{Tet} gene is inactive, as it contains *Tetrahymena* autosplicing intron interfering with transcription. The transcription of this gene is only possible after the complete cycle of the retrotransposition of the Alu repeat carrying *neo* gene: chimerical Alu-*neo*^{Tet} gene transcription directed by RNA polymerase III, Tet intron cut out as a result of autosplicing, reverse transcription on Alu-*neo*-template, and integration of Alu-*neo*-cDNA into HeLa cell genome (two last steps are directed by the L1 element-encoded protein). The HeLa cells with retrotransposed Alu elements were selected based on their resistance to neomycin [29]. Taking these data into account, we can suggest that in apoptotic rat chloroleukaemia cells studied earlier [28] activation of SINE (B1 and B2 elements) transcription and retrotransposition could take place. But this suggestion needs further investigation. The other authors revealed that human HeLa cell transfection with the plasmids carrying Alu repeat or 7SL RNA gene causes inhibition of DNA and protein synthesis in these cells and slows down their proliferation. The effect of these genes is specific, as the transfection of the cells with tRNAi^{Met}-carrying plasmid has no effect [30]. The cell treatment with DNA damaging agents causing genotoxic stress – cisplatin, etoposide and γ-radiation – induces B1 and B2 element expression in mouse cells FL5.12 and Alu repeats in human cells Jurkat. In addition, etoposide treatment of the mouse cells FL5.12 transfected with Alu repeat-carrying plasmid activates the synthesis of Alu RNAs. In all the cases the induction of SINE expression was specific, as under the conditions used the transcription of other RNA polymerase III transcribed genes – 5S rRNA and U6 snRNA genes – was not elevated [31]. In the same work it was demonstrated that in the cytoplasmic extract of the cells treated with DNA damaging agents the activity of reversed transcriptase (measured in RT-PCR *in vitro* system) was increased [31], that can favor to SINE retrotransposition. Panhandle PCR and inverse PCR methods were used to demonstrate the retrotransposition of human Alu elements into genomic DNA of mouse cells FL5.12 after their transfection with the Alu repeat-carrying genetic construction followed by topoisomerase II inhibition by etoposide [32]. Noteworthy, in the last two papers sited [31; 32] the authors used the cells transfected with the plasmids carrying the gene of anti-apoptotic protein Bcl-x_L to prevent apoptosis caused by genotoxic agents. This fact testifies indirectly that under the experimental conditions used some apoptotic pathways were activated in the cells, and it is possible that SINE RNAs are the part of these pathways.

Possible Role of Sines and Their RNA Products in Apoptosis

The data presented in the previous part of this chapter demonstrate that different damaging influences on the cell cause the increase in the levels of diverse SINE transcripts, which in turn leads to either cell recovery, or cell death (apoptosis). There are some hypotheses on the possible role of Alu RNAs, B1 and B2 RNAs in cell recovery after stress. But it can be assumed that SINE and their RNA products also play some role in the cell response to severe, cytotoxic damages, and possibly participate in apoptosis pathways realisation. But until now the role of SINE in apoptosis has not been investigated. The data received allowed one to propose some hypotheses about the functions of SINE and SINE transcripts in programmed cell death. So, during the apoptosis pathway realization the activation of SINE retrotransposition can cause genomic DNA structure damage; SINE RNAs can interact with ribosomes and protein kinase PKR, regulating mRNA translation; and finally, Alu repeat transcription activation can intensify the expression of the juxtaposed protein-coding genes. Importantly, the same mechanisms of SINE action can be used in both cell recovery and apoptosis, consistently with the suggestion about the close functional interrelation between these two processes (discussed later).

SINE Retrotransposition

The active retrotransposition of SINEs is possibly one of the mechanisms providing their effect as pro-apoptotic agents. The retrotransposition leads to the integration of SINEs into new genomic locations resulting in the normal gene expression violation, genotoxic stress and, in the end, activation of pro-apoptotic pathways. As far as SINEs have no protein-coding sequences, they use the enzymatic apparatus of other mobile elements – LINE – for retrotransposition. This mechanism of retrotransposition was demonstrated for Alu repeats in human cells HeLa in experiment *ex vivo*, when the cells were co-transfected with the plasmids carrying the sequences of Alu repeat and L1 (human LINE element) [29]. Similarly, when HeLa cells were co-transfected with the plasmids carrying mouse B1 or B2 repeats and L1, the retrotransposition of the mouse SINEs took place [33]. Noteworthy, the retrotransposition frequency in these experiments was practically the same for human Alu and mouse B2, different in sequences and origination, and was 80-100-fold lower for mouse B1, having the sequence relative to Alu repeats [33]. This fact can explain why in all the studied cases B2, but not Alu-like B1, plays the role of stress (and, possibly, pro-apoptotic) factor in rodent cells. So, the activation of LINE retrotransposition is followed by the activation of some SINE retrotransposition (such as Alus in primates, or B1 and B2 in rodents), that can use LINE enzymes for reverse transcription and genome integration. On the other hand, it was demonstrated that L1 retrotransposition activation in human cells can cause apoptosis. For example, human cancer cells HCT-116, SW480, and DLD-1, which have wild type p53 protein, transfected with the plasmid carrying human L1 element, underwent L1 retrotransposition activation resulting in p53-dependent apoptotic pathway induction. And p53 protein inactivation prevented apoptosis in spite of active L1 retrotransposition [34].

Later it was shown that active L1 retrotransposition in human carcinoma cells MCF-7 induced multiple DNA damages, mostly single strand nicks, which could not be totally repaired by the cell. As a result, apoptosis was induced in respond to DNA damage [35]. In the papers cited the authors did not test Alu repeat retrotransposition in the cells after LINE retrotransposition induction. But, as it was noticed earlier, there are some data on the concomitant induction of LINE and SINE retrotransposition. Evidently, SINEs can cause DNA damage accumulation and apoptosis induction through the same mechanism as LINEs. It should be noticed, that SINE retrotransposition seems to be more effective than that of LINEs: for instance, it was determined that two human individuals harbor in average 1283 Alu insertion polymorphisms, and only 180 L1 insertion polymorphisms [36]. This can be explained by the fact that human genome contains more retrotransposition-competent Alu repeats than L1 elements. It was counted that there are about 80-100 active L1 elements in the human genome [35]. The number of retrotransposition-competent Alu repeats is unknown. But it is clear that for the effective retrotransposition the Alu repeat must be transcriptionally active, which is characteristic to young Alu families, such as Ya5, Yb5, Yb8, and some others, appeared, in contrast to the older families, only about 3-19 million years ago [3], and accounting for 15% of 1.1 million Alu repeats in the human genome [6; 37]. On the other hand, the Alu retrotransposition frequency is known to be even more dependent on the presence of some specific sequences in Alu RNAs that on the quantity of these RNAs. And the most important is the presence of the poly(A)-sequence on the 3′-terminus (its length must be not shorter than 10-15 nucleotides, optimally – 40-50 nucleotides) [38; 39]. For example, the members of the most actively amplified Alu family Ya5 produce the RNAs which account for as little as <1% of all Alu RNAs in the cell [38]. It is possibly the specific features of the RNA sequence that favor retrotransposition of Alu Ya5. The human genome computer analysis revealed that only about 1000 Alu repeats (of 1.1 million) have poly(A)-sequence with the length of >40 nucleotides, optimal for L1-dependent retrotransposition. And only about 47% among them belong to the young families (i.e. potentially transcriptionally active) [38]. But even this quantity of the active Alu elements dramatically exceeds the number of the active L1 elements (see earlier). Besides, the Alu elements having poly(A)-sequence shorter than optimal but producing a lot of RNA molecules can play substantial role in genome structure perturbation. For instance, most Alu repeats from transcriptionally active young families Ya5 and Yb5 were demonstrated to contain 25-30 nucleotide long poly(A)-sequences [38; 39]. In *ex vivo* experiments on HeLa cells described earlier [33], the genetic constructions were used which carried Alu repeats with poly(A)-sequences providing retrotransposition frequency of $1–2\times10^{-3}$ [39]. It is lower that maximal frequency ($3–4\times10^{-3}$), characteristic of Alu repeats with >50 nucleotide long poly(A)-sequences [39], but high enough for Alu-dependent genomic DNA damage and apoptosis induction under high Alu RNA level conditions (e.g., upon cell damaging influences causing the activation of Alu repeat expression). It is known that for LINE retrotransposition two LINE-encoded proteins are needed (encoded in ORF1 and ORF2), but surprisingly only the protein encoded in ORF2 is needed for Alu retrotransposition (this protein has two enzymatic activities – reverse transcriptase and endonuclease) [29]. It was suggested that this feature of Alu repeats allows them to increase the level of retrotransposition significantly, as they can use protein products of both full length (having

ORF1 and ORF2, and actively retrotransposing) and short length (having only ORF2 and not possible to support their own retrotransposition) LINEs [40].

Alu insertion into new locations of the genome was revealed to be able to cause different consequences. In some cases these insertions do not produce catastrophic changes in genome functioning and are saved in the genome. It is believed that Alu insertion in a concrete locus is a accidental event. But there is at least one example when such an insertion caused the appearance of a new pro-apoptotic protein gene. Yi and co-authors found a cDNA copy of mRNA splice-variant for Bcl-rambo protein in the human lymph node cDNA library (Clontech, USA) [41]. Bcl-rambo protein is a member of Bcl-2 protein family, localized in mitochondria and having pro-apoptotic activity, provided by its specific C-terminal domain. The found splice-variant mRNA contains Alu-like insertion (98 nucleotide long) in protein-coding region, have no two exons, characteristic to 3′-terminus of the full length mRNA for Bcl-rambo protein, and code for a protein called by the authors Bcl-rambo beta. Hence, Bcl-rambo beta protein contains so called Alu-cassette (the amino acid sequence coded by Alu-like sequence in mRNA), and has no C-terminal domain providing pro-apoptotic functions in Bcl-rambo protein [41]. But surprisingly, the experiments on human cell HeLa and 293A transfection with the plasmids carrying Bcl-rambo beta protein-coding cDNA showed that this protein has even stronger pro-apoptotic activity than Bcl-rambo protein, and takes part in mitochondria-dependent apoptosis induction upon DNA damaging agent (etoposide and Taxol) treatment. It was suggested that Alu-cassette is responsible for pro-apoptotic features of Bcl-rambo beta protein, as the protein lost its pro-apoptotic activity after Alu-cassette removing [41]. It is well known that Alu repeat insertion into the protein-coding regions of the genes is a rather common event, called exonization [42; 43]. One could assume that Alu-cassettes can also provide pro-apoptotic features to some proteins other than Bcl-rambo beta, but this hypothesis needs further investigation for the identification of such proteins and elucidation of the mechanisms by which Alu-cassettes change their properties.

SINE-RNAs in mRNA Translation Regulation

It was shown that during both cell response to stressor influences and induction of apoptosis the inhibition of total protein synthesis takes place. On the other hand, in both cases the *de novo* synthesis of specific proteins (e.g., heat shock proteins or caspases) is needed for the realization of survival or death pathways. It was suggested that as the level of SINE RNAs increases after cell damaging influences, these RNAs can take part in both total translation inhibition and specific mRNA translation induction. This hypothesis is supported by the fact that all known SINEs derived from either the gene of 7SL RNA (the RNA component of SRP, signal recognition particle, participating in ribosome and nascent polypeptide transport to rough endoplasmic reticulum), or tRNA genes, i.e. genes of RNAs taking part in translation and having ability to interact with ribosomes and other protein synthesis participants.

Primate Alu repeats, derivatives of 7SL RNA genes, produce RNAs which have affinity to SRP-proteins: the interaction was shown with the protein complex SRP9/14 [44; 45]. Alu RNA contains the domain responsible in 7SL RNA for the interaction of SRP with the ribosome, and lacks the domain responsible for nascent protein localization signal recognition by 7SL RNA in SRP [44; 46]. It was recently determined that the complex of Alu RNA with SRP9/14 proteins inhibits the translation initiation *in vitro* on the mRNAs of different reporter proteins – cyclin, preprolactin, luciferase and PAI-2, but Alu RNA in the absence of SRP proteins activates the initiation of translation in the same system [47]. It can be assumed that the particle containing Alu RNA and SRP9/14 proteins binds ribosome and stops translation independent on the presence of the specific signal for endoplasmic reticulum localization in the nascent polypeptide. But the mechanism of the Alu RNA-dependent induction of translation is not clear yet.

In rodent cells B1 elements also derived from 7SL RNA gene and contain Alu-like domains, but because of the mononucleotide conversion G→T in location 24 their RNA products have lost the ability to bind SRP9/14 proteins [33]. But SRP protein binding is not the only possible mechanism of SINE RNA action on translation apparatus of the cell. As far as these RNAs have rather long poly(A)-sequences, they can bind poly(A)-binding protein (PABP), another important participant of the process of translation. In this case Alu and B1 RNAs could repress translation by competing with mRNAs for PABP binding [33; 39; 48].

Another mechanism for translation regulation in the stressed cell was allowed with the participation of Alu RNAs and protein kinase PKR, the protein regulated by double stranded RNAs [49]. But as it was noticed earlier in this chapter, the same mechanism can act in apoptosis induction, as far as PKR belongs to the pro-apoptotic proteins [50]. It was shown that Alu RNAs bind protein kinase PKR and inhibit its kinase activity *in vivo* in transfected human cells 293 and *in vitro* [49]. Later it was determined that the Alu RNA influence on the PKR is more complex and depends on the concentration of RNA: when the ratio is 0.02-2.0 pmol of dimeric Alu RNA to 2 pmol of PKR – the kinase activity is activated, but the ratio >100 pmol of dimeric Alu RNA to 2 pmol of PKR causes the inhibition of the kinase activity [50]. Besides, this action on PKR activity is not specific to Alu RNAs, as other RNAs with double stranded structure act on this kinase similarly [50; 51]. But evidently, in case when the level of Alu RNAs (in human cells) or B1 and B2 RNAs (in rodent cells) is elevated, it is these RNAs that are to be the main regulators of the PKR kinase.

It was revealed lately that upon overexpression of wild type PKR in mouse cells NIH3T3, this protein kinase takes part in the pathway of "delayed" cell death, concomitantly activating two "controversial" programs – firstly, cell survival program through NF-κB pathway, and later –apoptosis program through the phosphorylation of translation initiation factor eIF-2α. Noteworthy, the first (survival) program does not depend on PKR kinase activity, and, hence, needs no double stranded RNAs, but the second program (apoptotic) requires PKR activation by double stranded RNAs [52]. It can be assumed that in the damaged cell gradually accumulating SINE RNAs can act as time limiting factors for the PKR-activated pathways during the stress response development. Just after the damaging influence, when SINE RNA concentration is rather low yet, PKR activates cell survival program through NF-κB pathway independent on its kinase activity, and later, when the level of SINE RNAs becomes high enough for PKR kinase activation, this kinase induces apoptotic pathway, but in the end, when the level of SINE RNAs is too high, PKR kinase activity is inhibited, as the realization of the activated pathways needs no PKR participation any more. As far as it is known that SINE RNA accumulation in the damaged cell is temporal [23], their inhibiting action on PKR is also time limited.

SINE RNAs accumulating in the damaged cell can not only influence total protein synthesis, but also specifically and temporary activate translation of the newly synthesized mRNAs using some yet unknown mechanism independent on PKR. For example, after the co-transfection of human cells HeLa and 293 with the plasmids carrying Alu repeat and reporter luciferase gene, there was temporal and selective increase in luciferase synthesis, while the total translation and the level of luciferase mRNA stayed unchanged. Similar results were shown for some other reporter genes – β-galactosidase gene and adenovirus E1A protein gene [12]. It is evident that some mechanism allows the cell to switch temporally to the synthesis of the proteins needed for the recovery after the damage (e.g., heat shock proteins), or for apoptosis induction (e.g., caspases). This mechanism seems not to be specific to any concrete mRNAs [12], and what proteins are to be synthesized depends on what mRNA synthesis has been induced in the cell. The specific activation of newly synthesized mRNA translation was demonstrated in mouse cells either – the translation was activated there by B1 and B2 RNAs. Surprisingly, in human cells transfected with the plasmids carrying mouse B1 or B2 repeats, it was B1 RNA (Alu-like), but not B2 RNA (tRNA-like), that activated specific mRNA translation. Possibly, in human cells some factors are absent needed for tRNA-like SINE RNA action in specific translation activation [12].

RNA Polymerase II-directed Transcription Inhibition by RNA Products of Mouse B2 Elements

It was determined that in heat shocked mouse cells NIH3T3 the level of B2 RNA increased, and that B2 RNA inhibited RNA polymerase II-directed transcription *in vitro* and *in vivo* [53]. As we can see, B2 RNA accumulation takes place in case of different cytotoxic influences on the cell [24; 25; 31]. It is possible that in all these cases the mechanism of RNA polymerase II-directed transcription inhibition by B2 RNA could act. It was revealed that B2 RNA with high affinity and specificity bound RNA polymerase II, associated into a pre-initiation transcription complex on the promoter *in vitro*, and this binding blocked the mRNA synthesis on the initiation stage. The mRNA synthesis *in vitro* resumed after the B2 RNA cleavage by RNase I [13]. These experimental data allowed the authors to suggest that B2 RNA binds RNA polymerase II, associated with pre-initiation complex, which has been transiently inhibited earlier by some other factors in response to cell damaging conditions. So, B2 RNA action fastens pre-initiation complexes in inactive state until the removal of B2 RNA [13].

Active Transcription of Alu Repeats Can Cause the Activation of Juxtaposed Protein-coding Genes

There are some cases known when the Alu repeat transcription activation causes the activation of the juxtaposed gene transcription. It was shown, for example, for the human proliferation associated antigen p120 gene [54]. As we mentioned yet, after cell damaging influences a lot of silent in normal conditions Alu repeats become transcriptionally active [16]. In case these Alu repeats are situated near some pro-apoptotic protein genes, it is possible that Alu transcription activation can make these protein-coding genes active. The influence of Alu transcription efficiency on the juxtaposed gene transcription can be caused, for example, by the chromatin structure re-organization in the flanking regions of the Alu repeats. So, it was shown that Alu repeats can position nucleosomes in 5′-flanking regions (see later in this chapter), which can influence the binding of protein regulators to the sites inside these flanking regions, and, hence, modulate the transcription of the genes near the Alu repeat [55]. But it should be taken into account that the relative position of Alu repeats and pro-apoptotic protein genes in the human genome has not been studied yet.

Mechanisms of SINE-element Expression Regulation in Apoptosis

It has been revealed that in the conditions of cell growth and proliferation inhibition caused by the lack of nutrients and growth factors the repression of most RNA polymerase III-transcribed genes – tRNA and 5S RNA genes – takes place (reviewed in [56]). In this case the cell represses protein synthesis and enters the state of the rest, which is usually thought to be the cause of the decrease in translation-incorporated tRNA and 5S rRNA gene expression. It was demonstrated that SINEs in the quiescent (resting) cells are also almost totally represses [2]. It could be suggested that in the case of the cell stress other situation takes place: damages are accumulating in the cell which need to be repaired. In this case the arrest of cell cycle also occurs, but the cell does not become quiescent, and develops a new program for the damage repair and, if the recovery is impossible, enters apoptosis. During this process global transcription and translation are repressed, but specific expression (transcription and translation) of some genes (e.g. heat shock protein genes in cell recovery, or pro-apoptotic protein genes in case of apoptosis) is activated. Evidently, the cell stops growth and proliferation, and there is no need to form new ribosomes and other translation proteins and RNAs. That is why the expression of tRNA genes and most other RNA polymerase III-transcribed genes is repressed during the stress, which is characteristic of both lower (yeast) and higher eukaryots. But on the other hand, as it was described in the previous parts of this chapter, the expression of many SINEs (Alu repeats in human, B1 and B2 repeats in rodent) is activated in the stressed cells (see earlier). But we still do not know what factors provide such the difference in the transcription levels for the diverse RNA polymerase III-transcribed genes.

As it was mentioned earlier, SINEs are transcribed by RNA polymerase III. They obtain internal promoters similar to tRNA gene promoters and comprising conservative sites – A- and B-boxes. There are two basal transcription factors needed for pre-initiation complex formation on such promoters – TFIIIC, binding A- and B-boxes, and TFIIIB, interacting to TFIIIC and recruiting RNA polymerase III to the pre-initiation complex [7; 57; 58]. But in spite of the similarity between tRNA gene and SINE promoters, the data obtained show that their expression regulation in response to stressor influences differs substantially. It could be assumed that this difference is provided by the action of some specific transcription factors, different from basal transcription factors TFIIIB and TFIIIC, in SINE transcription. These factors could be activated or synthesized in the cell in response to stressor treatments, and participate in pre-initiation complex formation on SINE promoters resulting in effective SINE transcription in case when most RNA polymerase III-transcribed genes become silent. Unfortunately now we do not know a lot of the transcription factors for RNA polymerase III, other than its basal transcription factors, and most among known factors (e.g. pRb, p53, c-Myc) are non-specific to RNA polymerase III, and also participate in RNA polymerase I and/or II-directed transcription.

There are some data on RNA polymerase III-transcribed gene repression (usually tRNA and 5S RNA genes are studied) in growth suppression conditions. So, in yeast, which has no SINE-like elements in their genome, the general repression of RNA polymerase III-driven transcription is provided by the protein Maf1. This protein is activated in yeast cells grown in nutrient depletion conditions, after the cell treatment with rapamycin (modeling nutrient deployment), or in response to secretion suppression [59]. It was revealed that protein Mafl is dephosphorylated, transported from cytoplasm to nucleus, associates with RNA polymerase III and represses tRNA gene transcription [60; 61]. The other results showed that Mafl prevent *de novo* interaction of TFIIIB with DNA template. The authors suggested that in nutrient deployment conditions Mafl is activated owing to TOR-pathway inhibition, but in case of DNA damages (genotoxic stress) the activation is thought to be provided through the signal pathway, including protein kinase CKII [59]. The genome database searching revealed several possible homologues for the protein Mafl in higher eukaryots, including human [62]. But the role of this protein in higher eukaryots is still unknown. Possibly, it has lost its function of the global RNA polymerase III-driven transcription repressor. For instance, it is known that in mammal cells RNA polymerase III-transcribed genes are repressed during some special cell cycle stages by other protein factors – tumor suppressor proteins pRb и p53 [56; 63].

As it was demonstrated, overexpression of p53 gene in human cells H1299, with the level not sufficient for apoptosis induction, caused the repression of tRNA and 5S rRNA gene transcription. These genes have no p53-binding sites, and the suppressor protein acts by specific interaction to TBP (TATA-box binding protein), a subunit of TFIIIB, and this interaction prevents TFIIIB binding to the gene promoter [64]. It is also possible that p53 can act RNA polymerase III transcription apparatus indirectly. So, in cell cycle arrested human fibroblasts TR9-7 p53 was reported to cause degradation of one of TFIIIB subunits – Brf1, resulting in RNA polymerase III-directed transcription repression [65]. Noteworthy, p53 influences differently on diverse promoters: the weaker the promoter, the stronger the p53 repression. The most susceptible are weak Alu repeat promoters, as they have no enhancer elements in 5'-flanking regions [63]. It is known that in higher eukaryots protein p53 takes part in cell cycle arrest and apoptosis induction in stressor conditions [66]. It can be suggested that it is p53 that represses tRNA and 5S RNA genes in stressed and apoptotic cells. But some other mechanisms are also possible. For instance, it was shown that hypoxic stress caused no p53 level increase in rat cardiomyocytes, and RNA polymerase III-driven transcription repression was achieved owing to decrease in RNA polymerase III recruitment to pre-initiation complex, as basal transcription factor TFIIIB had been bound by repressor protein pRb [67].

Tumor suppressor protein pRb (retinoblastoma protein) binds to TFIIIB during G_0-(quiescence stage) and G_1-stages of the cell cycle, and prevents TFIIIB interaction with gene promoters, resulting in RNA polymerase III-driven transcription repression. It was also demonstrated that other pocket proteins – p107 and p130 – act on TFIIIB in similar way [68]. The repression of transcription by pRb can take place even after RNA polymerase III recruitment into pre-initiation complex [69]. Interestingly, pRb uses different mechanisms of transcription repression for the RNA polymerase III-transcribed genes with different promoters: pRb binds external promoter of human U6 snRNA gene, but does not interact directly with internal promoter of adenovirus VAI gene, acting indirectly through protein-protein interactions with RNA polymerase III transcription apparatus [69].

Taking into account the data reported, what mechanisms can participate in specific SINE expression activation during stress and apoptosis, when the other genes with similar promoters are getting repressed? One of the possible mechanisms is the action of some specific protein factors. This mechanism is studied rather well for the case of SINE expression induction during virus infection. For instance, in nuclear extracts from HeLa cells infected by adenovirus, the RNA polymerase III activity was elevated on exogenous templates of virus VA RNA gene and cloned human tRNA and 5S RNA genes. In the same study the increase in *in vivo* transcription of virus RNA gene was demonstrated in HeLa cells co-transfected with the plasmids carrying genes of VA RNA and adenovirus protein E1A, which suggests the important role of this virus protein in RNA polymerase III transcription activation [70]. But in these conditions the level of transcription factor TFIIIC was shown to stay unchanged, and the increase was in the level of its activated – phosphorylated – form that associates promoter faster and form more stable complex with it [71]. In the other research it was demonstrated that human cell HeLa and Jurkat transfection with the plasmids carrying human immunodeficiency virus (HIV) protein Tat gene activates the transcription of co-transfected and endogenous Alu repeats in these cells. In this case the activation of TFIIIC was also determined [21]. HeLa cell infection by herpes simplex virus (HSV) or transfection with expression plasmid carrying immediate-early virus protein ICP27 gene caused the elevation of endogenous Alu repeat transcription by RNA polymerase III, but not tRNA and 5S RNA genes. In this case the protein ICP27 was shown to activate factor TFIIIC not by phosphorylation, as protein E1A, but possibly by the activation of this factor's subunit synthesis [19; 20].

In several studies devoted to the influence of virus infection on RNA polymerase III-directed transcription it was pointed out that virus proteins activated transcription on exogenous (transfected) templates, but not on endogenous tRNA and 5S RNA genes [19; 20; 72]. That is why the suggestion was drawn that the main factor for virus and other exogenous RNA gene transcription activation is the accessibility of these genes to transcription factor binding. Most probably, the decisive feature of these templates is that they, in contrast to most endogenous genes transcribed by RNA polymerase III, are not incorporated into repressive chromatin structure in stressor conditions [72]. It was also suggested that Alu repeats in human cells, contrary to tRNA and 5S RNA genes, become more accessible for transcription factor binding during virus infection, which account for the expression activation of these SINEs [20]. The main factors influencing the accessibility of SINEs to transcription factor binding are known to be DNA hyper- or hypo-methylation and SINE-DNA organization into more compact (inactive) or more open (active) chromatin structure.

The chromatin structure influence on SINE transcription was studied on Alu repeats. It was demonstrated that *in vitro* transcription on Alu-templates was totally repressed in the presence of core histone octamers $(H2A/H2B/H3/H4)_2$ [73]. In the same study it was determined that Alu repeats contain nucleosome positioning signals and are possible to position nucleosomes in *in vitro* chromatin reconstitution experiments [73]. Later in *in vivo* experiments the nucleosome formation was demonstrated on a specific sequence including 5′-terminus of the Alu repeat's left monomer (position from -80 to +69), as far as nucleosome positioning in 5′-flanking region of Alu repeats [55]. The nucleosome positioned on the 5′-terminus of the Alu's left monomer involves the A-box of the promoter and prevents the binding of TFIIIC with the juxtaposed B-box of the promoter, leading to Alu repeat silencing [16; 55]. In heat shocked human cells HeLa the increase in Alu RNA level correlates with the increase in the accessibility of Alu chromatin to the restriction endonucleases, which have restriction sites inside the region of Alu repeat occupied by the nucleosome in repressed chromatin. This accessibility increase, and also Alu RNA level increase, is temporal – in 8 hours after the stressor influence the accessibility of Alu chromatin starts decreasing and reaches the minimal level in 24 hours [16]. Interestingly, in this case the kinetics of LINE1 chromatin accessibility to restriction endonucleases is similar to that of Alu chromatin [16]. But as we have discussed earlier in this chapter, Alu repeats use the enzymes encoded in LINEs for the retrotransposition. Based on the data received, Kim and others suggested a model for SINE expression activation during heat shock induced stress [16]: as SINEs have weaker promoters than other RNA polymerase III-transcribed genes and usually have no activating cis-sequences in 5′-flanking regions, their transcription is repressed rather efficiently by the "closed" chromatin structure. But in the conditions of stress the chromatin of the part of Alu repeats transforms into activated "open" structure. In spite of the fact that only minor part (0.1%) of Alu repeats are getting accessible to restriction endonucleases in heat stressed cells, owing to the multiple copies of these repeats in the genome, even this number of transcriptionally active Alu genes (about 2,200 per the cell) is enough for the elevation of Alu RNA level in the cell [16; 74].

The factors influencing the formation of open or closed chromatin structure on a concrete Alu repeat were investigated. It seems that one of the most important factors favoring the closed Alu chromatin formation is cytosine methylation in dinucleotides CpG, of which Alu repeats are very rich (Alu repeats contain 1/3 of all potential sites for DNA methylation in the human genome) [1; Z75]. Cytosine methylation was shown to repress Alu transcription in cell-free system based on nuclear extract from HeLa cells [75]. Total repression of *in vitro* transcription on Alu templates took place even in the presence of only two of four core histones, forming tetramers $(H3/H4)_2$ (i.e. in the absence of core histones H2A и H2B) [73]. Besides, the treatment of HeLa cells with DNA methyltransferase inhibitor 5-azacytidine causes the Alu-DNA hypomethylation and the Alu RNA level increase *in vivo* [26]. The other fact supporting the important role of DNA hypo- and hypermethylation in Alu repeat expression regulation is the data on far more active transcription of Alu repeats in human cells K562, where these repeats are constitutively hypomethylated, in comparison to the other human cells (HeLa and 293), where Alu repeats are hypermethylated. Moreover, owing to the increased level of Alu RNAs in K562 cells, only a minor additional elevation of Alu RNA level takes place in response to the stress caused by heat shock, adenovirus infection, and cycloheximide treatment [74]. Noteworthy, although in stressor conditions both non-methylated and hypermethylated exogenous (transfected) Alu template transcription is activated *in vivo* (in K562 and HeLa cells), the level of transcription is much lower for hypermethylated templates, in comparison to non-methylated [74]. It can be concluded that in stressed cells K562 and HeLa the Alu repeat transcription is activated by the mechanism that can not totally abolish Alu repeat repression caused by their DNA methylation [74]. Possibly, the Alu-DNA methylation patterns, different for diverse cells, are formed independently of the stressor conditions, but it is hypomethylated in a given cell Alu repeats that are the main targets for the binding of the factors activating their transcription during cell recovery after stressor influences. For instance, it was demonstrated that in different type cells (HeLa, K562 and 293) the stressor treatment caused the activation of different sets of Alu-repeats [76].

Summarizing, it can be assumed that in case of cell stress and apoptosis the repression of 5S RNA and tRNA gene transcription owing to factor TFIIIB repression by the proteins pRb и p53 [56; 67] is accompanied by the SINE transcription induction through the activation or *de novo* synthesis of factor TFIIIC. TFIIIC is the first factor that binds SINE promoters, followed by the recruitment of other transcription factors and RNA polymerase III. In mammals three subunits of TFIIIC have histone acetyltransferase activity; hence, this factor can influence the formation of more active "open" SINE chromatin structure, making it more accessible for other protein binding [77]. It should be also taken into account that some other factors can be important for SINE transcription regulation. For example, different regulatory post-translational modifications of RNA polymerase III and its basal transcription factors (discussed in [58]). But the precise mechanisms of SINE transcription regulation in response to stressor and damaging influences are still to be investigated.

Conclusion

Owing to the numerous data obtained during the last several years and concerning the structure and functions of SINEs, it became clear that these repetitive elements are not “junk” DNA. SINEs take a substantial part not only in such fundamental cellular processes as transcription, translation, splicing, but also in cell response to different stressor and damaging influences that can cause either the cell’s recovery or its death (apoptosis). It seems to be SINE expression that participates in the choice of the cell fate in unfavorable conditions. All these considerations show that it is of great importance to study the mechanisms of SINE transcription regulation, and also the action of SINE RNA products on the pathways and processes designating the fate of the cell.

Acknowledgments

The authors were supported by the Grant from the President of RF for the young PhD (MK-1434.2005.4) and the Grant of St Petersburg in the field of scientific and scientific-technical activity (2006).

References

[1] International Human Genome Sequencing Consortium. (2004). Finishing the euchromatic sequence of the human genome. *Nature*, 431, 931-945.

[2] Schmid, C.W. (1998). Does SINE evolution preclude Alu function? *Nucleic Acids Res,* 26, 4541-4550.

[3] Mighell, A.J., Markham, A.F., Robinson, P.A. (1997). Alu sequences. FEBS Lett, 417, 1-5.

[4] Prak, E.T., Kazazian, H.H. (2000). Mobile elements and the human genome. *Nat Rev Genet,* 1, 134-144.

[5] Deininger, P.L., Batzer, M.A. (2002). Mammalian retroelements. *Genome Res,* 12, 1455-1465.

[6] Grover, D., Kannan, K., Brahmachari, S.K., Mukerji, M. (2005). ALU-ring elements in the primate genomes. *Genetica*, 124, 273-289.

[7] Nikitina, T.V., Tishchenko, L.I. (2005b). RNA polymerase III transcription apparatus: structure and transcription regulation. *Mol Biol (Mosk),* 39, 179-192.

[8] Kim, D.D.Y., Kim, T.T.Y., Walsh, T., Kobayashi, Y., Matise, T.C., Buyske, S., Gabriel, A. (2004). Widespread RNA editing of embedded Alu elements in the human transcriptome. *Genome Research*, 14, 1719-1725.

[9] Buzdin, A.A. (2004). Retroelements and formation of chimeric retrogenes. *Cell Mol Life Sci,* 61, 2046-059.

[10] Donze, D., Kamakaka, R.T. (2001). RNA polymerase III and RNA polymerase II promoter complexes are heterochromatin barriers in Saccharomyces cerevisiae. *EMBO J,* 20, 520-531.

[11] Huang, W., Flint, S.J. (2003). Unusual properties of adenovirus E2E transcription by RNA polymerase III. *J Virol,* 77, 4015-4024.

[12] Rubin, C.M., Kimura, R.H., Schmid, C.W. (2002). Selective stimulation of translational expression by Alu RNA. *Nucleic Acids Res,* 30, 3253-3261.

[13] Espinoza, C.A., Allen, T.A., Hieb, A.R., Kugel, J.F., Goodrich, J.A. (2004). B2 RNA binds directly to RNA polymerase II to repress transcript synthesis. *Nature Struct Mol Biol,* 11, 822-829.

[14] Beere, H. M. (2005). Death versus survival: functional interaction between the apoptotic and stress-inducible heat shcock protein pathways. *J Clinical Invest,* 115, 2633-2639.

[15] Kimura, R.H., Choudary, P.V., Schmid, C.W. (1999). Silk worm Bm1 SINE RNA increases following cellular insults. *Nucleic Acids Res*, 27, 3380-3387.

[16] Kim, C., Rubin, C.M., Schmid, C.W. (2001). Genome-wide chromatin remodeling modulates the Alu heat shock response. *Gene*, 276, 127-133.

[17] Singh, K., Carey, M., Saragosti, S., and Botchan, M. (1985). Expression of enhanced levels of small RNA polymerase III transcripts encoded by the B2 repeats in simian virus 40-transformed mouse cells. *Nature*, 314, 553-556.

[18] Fornace, A. J. and Mitchell, J. B. (1986). Induction of B2 RNA polymerase III transcription by heat shock: enrichment for heat shock induced sequences in rodent cells by hybridization subtraction. *Nucleic Acids Res,* 14, 5793-5811.

[19] Jang, K.L., Latchman, D.S. (1989). HSV infection induces increased transcription of Alu repeated sequences by RNA polymerase III. *FEBS Letters*, 258, 255-258.

[20] Jang, K.L., Latchman, D.S. (1992). The herpes simplex virus immediate-early protein ICP27 stimulates the transcription of cellular Alu repeated sequences by increasing the activity of transcription fator TFIIIC. *Biochem J*, 284, 667-673.

[21] Jang, K.L., Collins, M.K., Latchman, D.S. (1992). The human immunodeficiency virus tat protein increases transcription of human Alu repeated sequences by increasing the activity of the cellular transcription factor TFIIIC. *J Acqir Immune Defic Syndr,* 5, 1142-1147.

[22] Panning, B., Smiley, J. R. (1993). Activation of RNA polymerase III transcription of human Alu repetitive elements by adenovirus type 5: requirement for the E1b 58-kilodalton protein and products of E4 open reading frames 3 and 6. *Mol Cel Biol,* 13, 3231-3244.

[23] Liu, W.-M., Chu, W.-M., Choudary, P. V., and Schmid, C. W. (1995). Cell stress and translational inhibitors transiently increase the abundance of mammalian SINE transcripts. *Nucleic Acids Res,* 23, 1768-1765.

[24] Williams, W. P., Tamburic, L., and Astell, C. R. (2004). Increased levels of B1 and B2 SINE transcripts in mouse fibroblast cells due to minute virus of mice infection. *Virology*, 327, 233-241.

[25] Fornace, A.J., Alamo, I., Hollander, M.C., Lamoreaux, E. (1989). Induction of heat shock protein transcripts and B2 transcripts by various stresses in Chinese hamster cells. *Exp Cell Res,* 182, 61-74.

[26] Liu, W.-M., Maraia, R. J., Rudin, C. M., and Schmid, C. W. (1994). Alu transcripts: cytoplasmic localization and regulation by DNA methylation. *Nucleic Acids Res,* 22, 1087-1095.

[27] Nikitina, T.V., Nazarova, N.Iu., Tishchenko, L.I., Tuohimaa, P., Sedova, V.M. (2003). Use of the real-time RT-PCR method for investigation of small stable RNA expression level in human epidermoid carcinoma cells A431. *Tsitologiia*, 45, 392-402.

[28] Servomaa, K., Rytomaa, T. (1990). UV light and ionizing radiation cause programmed death of rat chloroleukaemia cells by inducing retrotranspositions of a mobile DNA element (L1Rn). *Int J Radiat Biol,* 57, 331-343.

[29] Dewannieux, M., Esnault, C., Heidmann, T. (2003). LINE-mediated retrotransposition of marked Alu sequences. *Nature Genetics,* 35, 41-48.

[30] Sakamoto, K., Fordis, C. M., Corsico, C. D., Howard, T. H., and Howard, B. H. (1991). Modulation of HeLa cell growth by transfected 7SL RNA and Alu gene sequences. *J Biol Chem,* 266, 3031-3038.

[31] Rudin, C. M., Thompson, C. B. (2001). Transcriptional activation of short interspersed elements by DNA-damaging agents. *Genes Chromosomes Cancer,* 30, 64-71.

[32] Hagan, C. R., Sheffield, R., and Rudin, C. M. (2003). Human Alu element retrotransposition induced by genotoxic stress. *Nature Genetics,* 35, 219-220.

[33] Dewannieux, M., Heidmann, T. (2005a). L1-mediated retrotransposition of murine B1 and B2 SINEs recapitulated in cultured cells. *J Mol Biol,* 349, 241-247.

[34] Haoudi, A., Semmes, O.J., Mason, J.M., Cannon, R.E. (2004). Retrotransposition-competent human LINE-1 induces apoptosis in cancer cells with intact p53. *J Biomed Biotech,* 4, 185-194.

[35] Belgnaoui, S.M., Gosden, R.G., Semmes, O.J., Haoudi, A. (2006). Human LINE-1 retrotransposition induces DNA damage and apoptosis in cancer cells. *Cancer Cell Int,* 6, 13.

[36] Bennett, E.A., Coleman, L.E., Tsui, C., Pittard, W.S., Devine, S.E. (2004). Natural genetic variation caused by transposable elements in humans. *Genetics,* 168, 933-951.

[37] Batzer, M.A., Deininger, P.L., Hellmann-Blumberg, U., Jurka, J., Labuda, D., Rubin, C.M., Schmid, C.W., Zietkiewicz, E., Zuckerkandl, E. (1996). Standardized nomenclature for Alu repeats. *J Mol Evol,* 42, 3-6.

[38] Roy-Engel, A.M., Salem, A.-H., Oyeniran, O.O., Deininger, L., Hedges, D.J., Kilroy, G.E., Batzer, M.A., Deininger, P.L. (2002). Active Alu element "A-tails": size does matter. *Genome Res,* 12, 1333-1344.

[39] Dewannieux, M., Heidmann, T. (2005b). Role of poly(A) tail length in Alu retrotransposition. *Genomics*, 86, 378-381.

[40] Schmid, C.W. (2003). Alu: a parasite's parasite? *Nat Genet,* 35, 15-16.

[41] Yi, P., Zhang, W., Zhai, Z., Miao, L., Wang, Y., Wu, M. (2003). Bcl-rambo beta, a special splicing variant with an insertion of an Alu-like cassette, promotes etoposide- and Taxol-induced cell death. *FEBS Lett,* 61-68.

[42] Nekrutenko, A., Li, W.-H. (2001). Transposable elements are found in a large number of human protein-coding genes. *TRENDS Genet,* 17, 619-621.

[43] Krull, M., Brosius, J., Schmitz, J. (2005). Alu-SINE exonization: en route to protein coding function. *Mol Biol Evol,* 22, 1702-1711.

[44] Siegel, V., Walter, P. (1996). Removal of the Alu structural domain from signal recognition particle leaves its protein translocation activity intact. *Nature*, 320, 81-84.

[45] Sarrowa, J., Chang, D.-Y., Maraia, R.J. (1997). The decline in human Alu retroposition was accompanied by an asymmetric decrease in SRP9/14 binding to dimeric Alu RNA and increased expression of small cytoplasmic Alu RNA. *Mol Cel Biol,* 17, 1144-1151.

[46] Bovia, F., Strub, K. (1996). The signal recognition particle and related small cytoplasmic ribonucleoprotein particles. *J Cell Sci,* 109, 2601-2608.

[47] Hasler, J., Strub, K. (2006). Alu RNP and Alu RNA regulate translation initiation in vitro. *Nucleic Acids Res,* 34, 2374-2385.

[48] Kondrashov, A.V., Kiefmann, M., Ebnet, K., Khanam, T., Muddashetty, R.S., Brosius, J. (2005). Inhibitory effect of naked neural BC1 RNA or BC200 RNA on eukaryotic in vitro translation systems is reversed by poly(A)-binding protein (PABP). *J Mol Biol,* 353, 88-103.

[49] Chu, W.-M., Ballard, R., Carpick, B.W., Williams, B.R.G., Schmid, C.W. (1998). Potential Alu function: regulation of the activity of double-stranded RNA-activated kinase PKR. *Mol Cel Biol,* 18, 58-68.

[50] Williams, B.R.G. (1999). PKR; a sentinel kinase for cellular stress. *Oncogene*, 18, 6112-6120.

[51] Robertson, H.D., Mathews, M.B. (1996). The regulation of protein kinase PKR by RNA. *Biochimie*, 78, 909-914.

[52] Donze, O., Deng, J., Curran, J., Sladek, R., Picard, D., Sonenberg, N. (2004). The protein kinase PKR: a molecular clock that sequentially activates survival and death programs. *EMBO J,* 23, 1-8.

[53] Allen, T.A., Von Kaenel, S., Goodrich, J.A., Kugel, J.F. (2004). The SINE-encoded mouse B2 RNA represses mRNA transcription in response to heat shock. *Nature Struct Mol Biol,* 11, 816-821.

[54] Lowe, D., Henning, D., Reddy, R. (1999). An RNA polymerse III transcription unit located in the upstream control regions of the human proliferating-cell nucleolar protein p120 gene is transcribed in vitro and in vivo. *Eur J Biochem,* 266, 112-122.

[55] Englander, E.W., Howard, B.H. (1995). Nucleosome positioning by human Alu elements in chromatin. *J Biol Chem,* 270, 10091-10096.

[56] White, R.J. (2005). RNA polymerases I and III, growth control and cancer. *Mol Cel Biol,* 6, 69-78.

[57] Schramm, L., Hernandez, N. (2002). Recruitment of RNA polymerase III to its target promoters. *Genes Dev,* 16, 2593-2620.

[58] Nikitina, T.V., Tishchenko, L.I. (2005a). Computational search for potential post-translational modification sites in human RNA polymerase III subunits. *Mol Biol (Mosk),* 39, 437-444.

[59] Upadhya, R., Lee, J., Willis, I.M. (2002). Mafl is an essential mediator of diverse signals that repress RNA polymerase III transcription. *Mol Cell,* 10, 1489-1494.

[60] Oficjalska-Pham, D., Harismendy, O., Smagowicz, W.J., Gonzalez de Peredo, A., Boguta, M., Sentenac, A., Lefebvre, O. (2006). General repression of RNA polymerase III transcription is triggered by protein phosphatase type 2A-mediated dephosphorylation of Mafl. *Mol Cell,* 22, 623-632.

[61] Roberts, D.N., Wilson, B., Huff, J.T., Stewart, A.J. (2006). Dephosphorylation and genome-wide association of Mafl with pol III-transcribed genes during repression. *Mol Cell,* 22, 633-644.

[62] Pluta, K., Lefebvre, O., Martin, N.C., Smagowicz, W.J., Stanford, D.R., Ellis, S.R., Hopper, A.K., Sentenac, A., Boguta M. (2001). Maflp, a negative effector of RNA polymerase III in Saccharomyces cerevisiae. *Mol Cell Biol,* 21, 5031-5040.

[63] White, R.J. (2004). RNA polymerase III transcription and cancer. Oncogene, 23, 3208-3216.

[64] Crighton, D., Woiwode, A., Zhang, C., Mandavia, N., Morton, J.P., Warnock, L.J., Milner, J., White, R.J., Johnson, D.L. (2003). p53 represses RNA polymerase III transcription by targeting TBP and inhibiting promoter occupancy by TFIIIB. *EMBO J,* 22, 2810-2820.

[65] Eichhorn, K., Jackson, S.P. (2001). A role for TAF3B2 in the repression of human RNA polymerase III transcription in nonproliferating cells. *J Biol Chem,* 276, 21158-21165.

[66] Vousden, K.H., Lu, X. (2002). Live or let die: the cell's response to p53. *Nat Rev Cancer,* 2, 594-604.

[67] Ernens, I., Goodfellow, S.J., Innes, F., Kenneth, N.S., Derblay, L.E., White, R.J., Scott, P.H. (2006). Hypoxic stress suppresses RNA polymerase III recruitment and tRNA gene transcription in cardiomyocytes. *Nucleic Acids Res,* 34, 286-294.

[68] Scott, P.H., Cairns, C.A., Sutcliffe, J.E., Alzuherri, H.M., McLees, A., Winter, A.G., White, R.J. (2001). Regulation of RNA polymerase III transcription during cell cycle entry. *J Biol Chem,* 276, 1005-1014.

[69] Hirsch, H.A., Jawdekar, G.W., Lee, K.-A., Gu, L., Henry, R.W. (2004). Distinct mechanisms fir repression of RNA polymerase III transcription by the retinoblastoma tumor suppressor protein. *Mol Cell Biol,* 24, 5989-5999.

[70] Hoeffler, W.K., Roeder, R.G. (1985). Enhancement of RNA polymerase III transcription by the E1A gene product of adenovirus. *Cell*, 41, 955-963.

[71] Hoeffler, W.K., Kovelman, R., Roeder, R.G. (1988). Activation of transcription factor IIIC by the adenovirus E1A protein. *Cell,* 53, 907-920.

[72] Gaynor, R.B., Feldman, L.T., Berk, A.J. (1985). Transcription of class III genes activated by viral immediate early proteins. *Science*, 230, 447-450.

[73] Englander, E.W., Wolffe, A.P., Howard, B.H. (1993). Nucleosome interactions with a human Alu element. Transcriptional repression and effects of template methylation. *J Biol Chem,* 268, 19565-19573.

[74] Li, T.-H., Kim, C., Rubin, C.M., Schmid, C.W. (2000). K562 cells implicate increased chromatin accessibility in Alu transcriptional activation. *Nucleic Acids Res,* 28, 3031-3039.

[75] Liu, W.-M., Schmid, C. W. (1993). Proposed roles for DNA methylation in Alu transcriptional repression and mutational inactivation. *Nucleic Acids Res*, 21, 1351-1359.

[76] Li, T.-H., Schmid, C.W. (2001). Differential stress induction of individual Alu loci: implication for transcription and retrotransposition. *Gene*, 276, 135-141.

[77] Kundu, T.K., Wang, Z., Roeder, R.G. (1999). Human TFIIIC relieves chromatin-mediated repression of RNA polymerase III transcription and contains an intrinsic histone acetyltransferase activity. *Mol Cell Biol,* 19, 1605-1615.

In: New Developments in Cell Apoptosis Research
Editor: Alan J. Corvin, pp. 179-192
ISBN: 1-60021-453-3

Chapter VII

The Role of Endonucleases in Chromosomal DNA Fragmentation Associated with Apoptosis and Necrosis

Yoshihiro Higuchi*
Department of Molecular Pharmacology, Kanazawa University Graduate School of Medical Science, Kanazawa 920-8640, Japan

Abstract

Chromosomal DNA dysfunction plays a role in mammalian cell death. Oxidative stress induces chromatin dysfunction such as single strand and double strand DNA fragmentation leading to cell death through apoptosis or necrosis. More than 1 Mb giant DNA fragments, 200-800 kb or 50-300 kb high molecular weight (HMW) DNA fragments and internucleosomal low molecular weight (LMW) DNA fragments are produced not only by oxidative stresses such as ROS, ionizing irradiation and ultraviolet radiation, but also by some chemotherapeutic drugs that induce apoptosis or necrosis in several types of mammalian cells. The mechanismof DNA fragmentation during cell death is poorly understood, although several endonucleases have been implicated in the process. Some nucleases involved in chromosomal DNA fragmentation in apoptosis and necrosis are classified.

Keywords: Apoptosis, Endonucleases, Giant DNA fragmentation, High molecular weight DNA fragments, Low molecular weight DNA fragments, Necrosis, Oxidative stress.

* Correspondence to: Yoshihiro Higuchi, Department of Molecular Pharmacology, Kanazawa University School of Medical Science, 13-1 Takara-machi, Kanazawa, Ishikawa 920-8640, Japan. Phone; +81-76-265-2186, Fax; +81-76-234-4227; E-mail; higuchiy@med.kanazawa-u.ac.jp

1. Introduction

Cellular genomes are continually subjected to endogenous and environmentally induced structural alterations. Our environment contains a multitude of substances that are carcinogenic and which, in many cases, are thought to act via direct damage to DNA. Such damage can manifest itself as gross chromosomal abnormalities inducing cell death. Although the classification of cell death has proven difficult, two distinct patterns of cell death have been identified based on the morphology of dying cells, and on the DNA fragmentation or damage. In multi cellular organisms, programmed cell death is an essential mechanism in morphogenesis, development, differentiation, and homeostasis, which occurs primarily through an evolutionarily conserved process termed apoptosis. Typical apoptosis is distinguished from necrosis as the former involves the activation of specific pathways that result in characteristic morphological features including DNA fragmentation, chromatin condensation, cytoplasmic and nuclear pycnosis, and the formation of apoptotic bodies [1-3].

Mammalian cell death is induced through chromosomal DNA damage by ionizing radiation, ultra violet (UV) radiation, anticancer drugs and various triggers of apoptosis. Reactive oxygen species (ROS) such as hydrogen peroxide (H_2O_2), hydroxyl radicals ($^{\cdot}OH$) and superoxide anions (O_2^-) have been shown to damage chromosomal DNA [4] and other cellular components, resulting in DNA degradation, protein denaturation, and lipid peroxidation. However, the mechanisms behind these cellular effects are rather complex, and are not yet fully understood. Our purpose is to review the chromosomal DNA fragments such as giant DNA fragments, high molecular weight (HMW) DNA fragments, and internucleosomal low molecular weight (LMW) DNA fragments and to reflect upon their significance in cell death by apoptosis and necrosis. We refer to the radical-induced DNA lesion, as well as to the enzymatic digestion of chromosomal DNA during cell death.

2. Chromatin Structure and Pattern of Chromosomal DNA Fragmentation

A mammalian cell nucleus contains almost 50 cm of DNA requiring more than a 50,000-fold reduction in length to fit in the nucleus and nuclear matrix. The nuclear matrix is an important structural component in a variety of nuclear functions and nuclear morphology, including DNA organization, DNA-replication, RNA synthesis and nuclear regulation. DNA loop domains of chromatin are attached to the nuclear matrix at their base and this organization is maintained throughout both interphase and metaphase. These loops are 50-150 kb long and are equivalent in size to the replicon [5]. The haploid human genome contains 3000 megabase (Mb) pairs of DNA with a mean chromosomal size of 130 Mb.

Chromosomal DNA fragmentation is caused by two types of DNA breaks that are classified into single strand and double strand DNA breaks. Single-strand cleavage of DNA has been suggested to occur during apoptosis. At the level of the nuclear scaffold, single strand DNA breaks were detected in HL-60 cells treated with camptothecin, a topoisomerase I inhibitor and inducer of apoptosis, but these were rapidly repaired after drug removal [6].

Internucleosomal DNA cleavage occurred after the repair of these single-strand cuts, suggesting that single-strand breaks at higher levels of DNA organization may not play an active role during apoptosis but can perhaps act as signals to induce the process. Clearly, the role that single stranded DNA breaks play during apoptosis requires additional studies. Double strand DNA breaks are generally thought to have a greater biological consequence than single strand DNA breaks because they can lead directly to chromosomal aberrations, and more frequently to the loss of genetic information [7]. Double-strand DNA breaks are 20 times less frequent than single-strand DNA breaks and are more difficult to measure at the physiological level. The determination of double strand DNA break by gel electrophoresis has been described in a previous report [8].

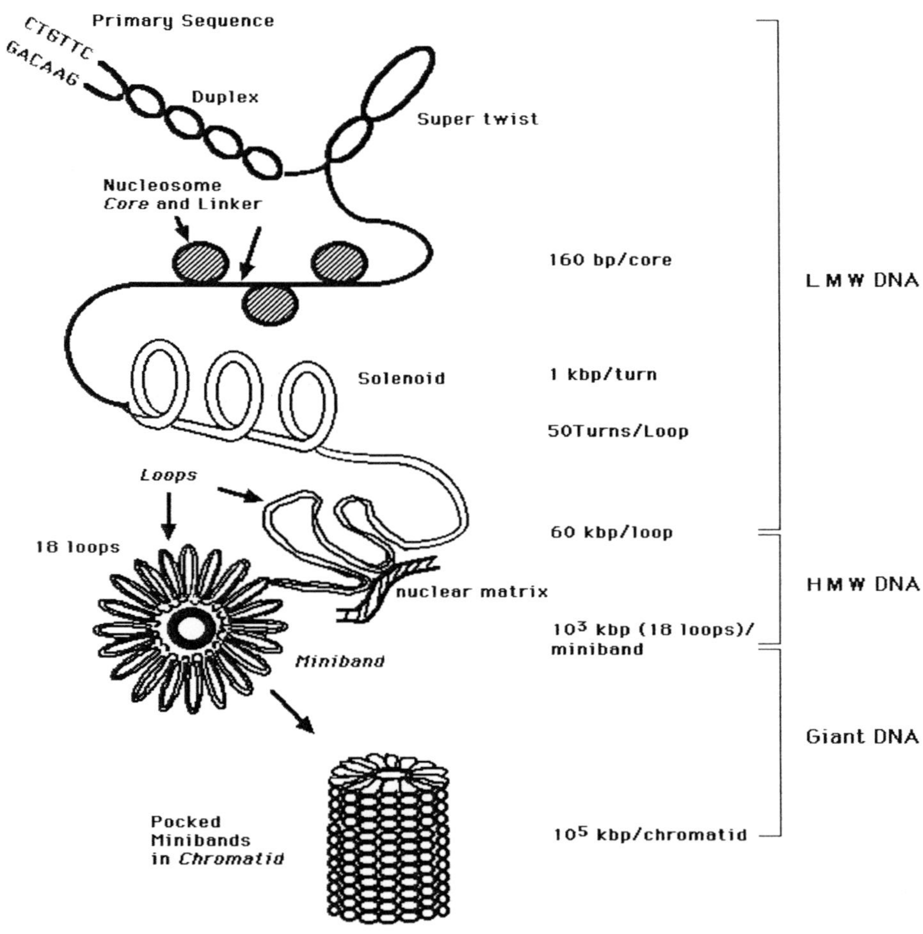

Figure 1. The structure of chromosomes.

Chromosomal DNA fragments of more than 1 Mb in size are double strand DNA breaks and are classified as giant DNA fragments. DNA degradation accompanied by DNA fragmentation producing 1-2 Mb and 200-800 kb DNA fragments were observed during cell death in cells treated with various agents that can produce ROS [4, 9], or under GSH depletion [10]. Chromosomal DNA fragments which are 200-800 kb and 50-300 kb in size are called high molecular weight (HMW) DNA fragments. The 1-2 Mb giant DNA and 200-800 kb HMW fragments that may represent features of high-order chromatin structure such as minibands and loops of DNA [11], lead to apoptosis ascertained by internucleosomal DNA fragmentation [12].

3. Reactive Oxygen Species (ROS)-mediated Chromosomal DNA Fragmentation

One of the most serious cellular lesions induced by ionizing radiation is the random DNA double-strand break. This lesion jeopardizes a chromosome's physical integrity, which is essential for its correct segregation during mitosis and meiosis, as well as its informational redundancy, which is critical for maintaining accurate encoding of cellular components. We have hardly any imformation about the *in vivo* mechanism of ROS production by ionizing radiation on chromatin DNA in the nuclei of cells. Ionizing radiation such as X-rays and γ-rays are, in general, thought to produce hydroxyl radicals from water molecules in or around the target sites in the DNA, and these in turn attack DNA and break it down [13]. DNA damage caused by radiation-produced ROS *in vivo* or in cultured cell systems is classified into DNA cleavages such as single strand breaks, and double strand breaks and nucleotide base oxidative modifications [14]. In addition, the involvement of ROS in the induction of apoptosis has been suggested in several cell lines [12, 15].

Some groups have reported the size distribution of radiation-induced DNA fragments in mammalian cells such as Chinese hamster ovary cells [16] and L-1210 mouse leukemia cells [17]. In L-1210 cells irradiated at 1-50 Gy, double-strand DNA break fragments, calculated from marker chromosomes to be in the range of 0.1 to 12.6 Mb, have been demonstrated [17]. In another X-ray irradiation study on T-24 human bladder carcinoma cells, similar DNA fragments to the DNA fragments found in the X-ray irradiated L-1210 cells were observed in the 1-2 Mb but not the 0.1-1 Mb range [4]. Giant DNA and HMW DNA fragments ranging from 100 kb to 10 Mb are also distinctly produced by ultraviolet irradiation [76]. The formation of 1-2 Mb giant and 200-800 kb HMW DNA fragments prior to internucleosomal DNA fragmentation are caused by H_2O_2 [4, 9].

Some antibiotics that possess quinone moieties as part of their chemical structure are widely used as a drug to treat various human cancers. Most of these drugs produce ROS at various cellular sites *in vivo* [18-20]. Neocarzinostatin and bleomycin, both of which are radiomimetic chemicals and anticancer drugs generating ROS *in vivo*, produce not only 1-2 Mb and 200-800 kb DNA fragments but also apoptotic internucleosomal DNA fragments during cell death (Fig. 2). Cellular DNA cleavage into HMW DNA fragments during apoptosis is highly reminiscent of topoisomerase II-mediated HMW DNA fragmentation in cells [21, 22]. In fact, the pattern of HMW DNA fragmentation by topoisomerase II poison, and that produced in apoptotic cells induced by other stimuli, is found to be similar [23]. However, the relationship between topoisomerase II and the excision of chromosomal loops during apoptotic cell death is still unclear. Recently, several new mechanisms, including DNA structural modifications, enzyme modifications, oxidative stress and acidic pH environment have also been shown to activate topoisomerase II mediated DNA cleavage

[24]. Topoisomerase II inhibitors such as VM-26 and etoposide also produced 800 kb-1Mb giant DNA fragments together with 200-600 kb and less than 100 kb DNA fragments in thymocytes or Du145 human prostatic carcinoma cells [25, 26]. Furthermore, although DNA damage caused by exogenously added ROS appears to activate apoptosis, it is important to elucidate the role of ROS in apoptosis. However, not only the mode of action of ROS but also the role of such chromosomal giant DNA degradation remains to be elucidated.

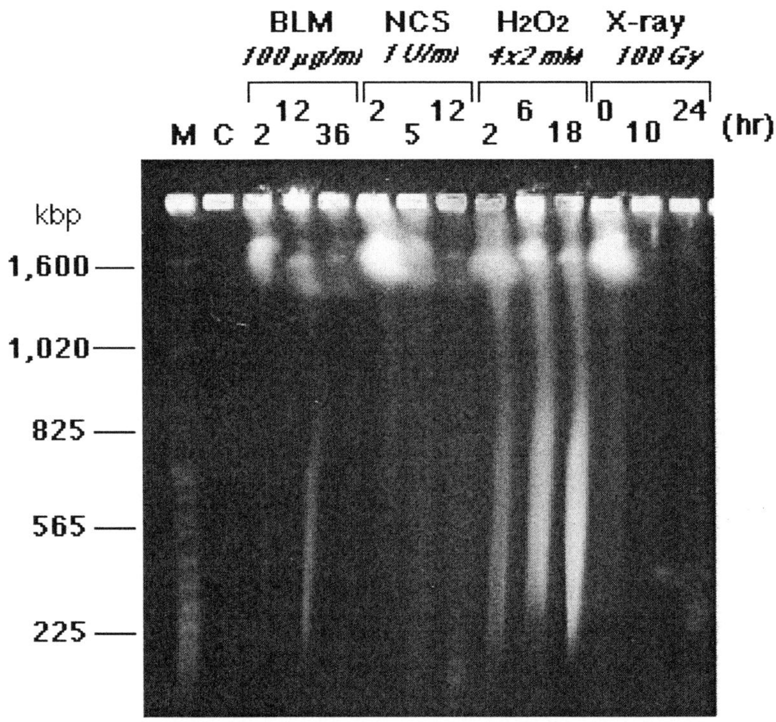

Figure 2. Giant DNA fragmentation in T24 human bladder carcinoma cells.

4. DNA Fragmentation in Apoptosis or Necrosis

Apoptosis and necrosis are two distinct forms of cell death that have profoundly different implications for the surrounding tissue. Apoptosis is characterized by chromatin condensation, activation of some proteases and fragmentation of DNA at internucleosomal linker sites giving rise to discrete bands of multiples of 180-200 base pairs [27]. These enzymatic events encompass a vast array of chromosomal degradation states in the cell with the ultimate common consequence being cell death [28]. In contrast, necrosis is a passive process, an uncontrolled event and typified by cell and organelle swelling with spillage of the intracellular contents into the extra cellular milieu. Necrosis results from the loss of homeostasis and the cell contents, which is dispersed, may then have adverse effects on neighboring tissues [29].

Nuclear apoptosis corresponds to the most striking morphological modification associated with the execution phase of apoptosis [30]. It consists of an early perinuclear chromatin condensation that leads to a massive nuclear breakdown. In some cells, it is accompanied by karyorhexis, the segmentation of the nucleus into dense nuclear parts further emitted in apoptotic bodies. These morphological alterations are accompanied by early DNA cleavage that sequentially generates high-molecular-weight (HMW) DNA fragments of 1 Mb and 50 kb [31]. Additional 700 kb and 300 kb fragments can also be introduced in response to certain stimuli, but their appearance seems correlated to a genotoxic effect rather than to a typically apoptotic event [32]. One of the hallmarks of apoptosis is the digestion of genomic DNA by an endonuclease, generating a ladder of small fragments of double-stranded DNA. Single-strand nicks were found to be very frequent in the internucleosomal regions, but also to occur in the core particle-associated DNA. DNA fragmentation induced during apoptosis is not due to a double-strand cutting enzyme as previously postulated, but rather is the result of single-strand breaks. This ensures the dissociation of the DNA molecule at sites where cuts are found within close proximity [33]. There is a two-step process of DNA fragmentation in apoptosis: DNA is first cleaved into large fragments of 50-300 kb that are subsequently cleaved into smaller oligonucleosomes in some, but not all cells [34-36]. Significantly, only the first stage is considered essential for cell death since some cells, for example human MCF7 breast carcinoma cells and human NT2 neuronal cells, do not show this behavior but still display normal nuclear morphological apoptotic changes. Some inducers of apoptosis such as etoposide and glucocorticoids have provided formation of 50-300 kb HMW DNA fragments prior to internucleosomal DNA fragmentation [37] in apoptotic MCF-7 cells induced by etoposide [34], and in mouse L-929 cells induced by tumor necrosis factor (TNF-α) [21]. These DNA fragment formations have been observed in HeLa nuclei treated with apoptosis-inducing factor (AIF) [38]. Apoptosis has also been widely observed in some cells treated with anticancer drugs [39], and other cell death processes induced by some biological events such as depletion of nutrients [40]. However, not much has been reported about the involvement of not only 1-2 Mb giant DNA fragments and its significance or role in apoptosis.

In some cases of apoptosis, ROS may be involved not only as inducers of DNA damage but also as specific second messengers in the signal transduction pathway, whereas in other situations they may be a side effect of either the experimental system or changes in the cellular redox status as a result of ROS-independent apoptosis signaling pathways [41]. Therefore, it is still unclear whether endogenous ROS are really involved in DNA degradation leading to apoptosis. Not only 1-2 Mb giant DNA and 100-800 kb HMW DNA fragments but also internucleosomal DNA fragments were observed in C6 cells under glutamate or BSO-induced GSH depletion [10, 12]. This DNA fragmentation as a process was associated with intracellular accumulated ROS under glutamate-induced GSH-depletion [42]. However, the relationship between GSH-depletion and active oxygen-induced DNA damage remains to be clarified. Polyunsaturated fatty acids with a homoconjugated *cis-cis*-pentadienyl system, such as linoleic acid and arachidonic acid, are substrates for lipoxygenases. Such polyunsaturated fatty acids enhance not only lipid peroxidation followed by numerous forms of base damage and the formation of 8-hydroxy-2'-deoxyguanosine (8-OH-dG) [43], but also the formation of giant DNA fragments in chromosomal DNA under both glutamate- and BSO-induced GSH depletion [44]. Arachidonic acid is metabolized into substances controlling cell survival and moreover is oxidized to its hydroperoxides not only by lipoxygenases or cyclooxygenases but also by a chemical reaction under aerobic conditions [45]. Arachidonic acid converts apoptosis to necrosis representing the disappearance of internucleosomal DNA fragments under BSO- fragments and reduction of internucleosomal DNA fragments [46, 47].

5. Characteristics of Nucleases Involved in DNA Fragmentation

The different mediators that can be responsible for DNA fragmentation by this are summarized in Table 1.

Table 1. Nucleases and Nuclease-like Factors Involved in Chromosomal DNA Fragmentation

Mediator (Nucleases)	Type of DNA fragmentation	3'and 5' termini	References
Topoisomerase IIα	HMW	3'OH and 5'P	65
AIF (apoptosis inducing factor)	HMW	Undetermined	59、67
DFF-40	HMW and LMW	3'OH and 5'P	71
CAD (caspase activated DNase)	HMW and LMW	3'OH and 5'P	54
DNAS1L3	HMW and LMW	3'OH and 5'P	55
P25	HMW and LMW	3'OH and 5'P	60
97 kDa endonuclease	HMW and LMW	Undetermined	72
DNase I	LMW	3'OH and 5'P	73
DNase γ	LMW	3'OH and 5'P	74
Endonuclease G	LMW	3'OH and 5'P	52
NUC18	LMW	3'OH and 5'P	61
DNase II	LMW	3'P and 5'OH	75, 77
L-DNase II	LMW	3'P and 5'OH	58, 63
Cathepsin B	LMW	Undetermined	57
Acinus	Chromatin condensation	Undetermined	76

HMW and LMW indicate high molecular weight DNA fragments and low, molecular weight DNA fragments, respectively.

In apoptosis, internucleosomal DNA degradation in which some endonucleases are involved has been observed [48, 49] and several studies on the enzyme activation process are in progress [50, 51]. Cells may also detach parts of their cytoplasm, which sometimes includes highly condensed fragments of the karyorrhectic nucleus. The dying cells also activate catabolic enzymes that ensure digestion of critical cellular components from the inside. Such catabolic hydrolases include a class of specific protein-cleaving enzymes (caspases), as well as DNA-digesting enzymes (DNases), both of which participate directly or indirectly in nuclear pyknosis [56]. The single strand-specific nuclease, DNase I, is thought to be specific for some type of DNA structure. Recently, a new type of endonuclease involved in apoptosis has been reported. This nuclease is endonuclease G, a mitochondrion-specific nuclease that translocates to the nucleus during apoptosis. Endonuclease G cleaves chromatin DNA into nucleosomal fragments independently of caspases [52]. Sahara *et al.* [53] have suggested that caspase-3 cleaves Acinus, which is the precursor of a chromatin-condensation factor. Another chromatin condensation factor, caspase-activated DNase (CAD) [54], for example, uses its DNase activity to cleave chromatin at the boundaries between nucleosomes. In this way, CAD generates stretches of DNA of about 200 base pairs long or multiples thereof. The nuclear localization of DNAS1L3 is a unique feature of this enzyme that distinguishes it from other suggested apoptotic nucleases [55]. Our data indicated that this nuclease cleaves DNA into both oligonucleosomal and high molecular weight fragments [55]. We now provide additional evidence that DNAS1L3 is regulated by poly(ADP-ribose) polymerase (PARP) and that it may be responsible for apoptotic DNA degradation. A few proteins responsible for caspase-independent chromatin condensation have, in fact, been identified. Apoptosis-inducing factor, AIF, is a flavoprotein that is normally confined to the space between the outer and inner mitochondrial membranes [38]. AIF translocates from the mitochondrion to the nucleus, where it causes partial chromatin condensation in the periphery of the nucleus. AIF causes degradation of DNA into fragments greater than around 50 kb in length. Another chromatin condensation factor, which translocates from the cytoplasm to the nucleus, is L-DNase II. When added to isolated nuclei, L-DNase II causes marked chromatin condensation and cleaves the chromatin into nucleosome-sized fragments. Yet other proteins that might contribute to chromatin condensation and internucleosomal DNA fragmentation are endonuclease-γ [56] and cathepsin B [57]. These proteins could be activated on their release from the lysosomes of apoptotic cells. During the last decade, intensive research has attempted to describe the activation of the mediators of nuclear apoptosis. The candidate apoptotic nucleases have their own properties and differ in their structure, cation dependence, pH optima, and mechanisms of activation [58]. Some mediators have an optimal activity in the presence of Mg^{2+} ions (CAD), both Mg^{2+} and Ca^{2+} (DNase I, DNase γ, NUC-18), whereas other factors are cation-independent (such as DNase II-like enzymes or apoptosis-inducing factor, AIF). They also differ in their pH of optimal activity, since some factors, like DNase I or CAD, require a neutral pH to be fully activated, whereas others, like DNase II and L-DNase II, only cleave DNA in an acidic environment. This dichotomy is associated with the generation of 3'OH/5'P and 3'P/5'OH termini, respectively (Table 1). Finally, certain factors require caspase activation (CAD) [54], whereas others are caspase-independent (AIF) [59] or require serine protease activation (L-DNase II) [58]. The significance of the multiplicity of candidate endonucleases that cleave genomic DNA in HMW and LMW fragments is attributed to different requirements. Although some factors, like AIF or DNase I, are ubiquitous, other factors, such as CAD or L-DNase II, are cell-type specific. For example, DNA fragmentation of apoptotic rat thymocytes can be mediated by the activation of NUC-18 and CAD [60-62], and other mediators, such as L-DNase II and an unknown Ca^{2+}-, Mg^{2+}-dependent endonuclease, seem to be involved [58, 63]. The requirement of different synergic mediators in the same apoptotic cell seems necessary to induce the proper and complete development of DNA fragmentation and condensation [64-66]. CAD has been involved in the execution of nuclear apoptosis, in association with factors that initiate the fragmentation into

HMW fragments, such as AIF [67], and topoisomerase IIα [65], or in synergy with enzymes that generate oligonucleosomal fragmentation, such as DNase γ-like endonucleases [68].

The terminal deoxynucleotidyl transferase-mediated dUTP-biotin nick end labeling (TUNEL) assay has been used to label the 3' ends of nicked or fragmented DNA in apoptosis using [14-biotin]-dCTP and the terminal deoxynucleotide transferase enzyme. The labeled DNA is detected with horseradish peroxidase-conjugated streptavidin and diaminobenzidine. DNase I action can present a positive result in the TUNEL assay, and therefore, the TUNEL assay may be positive in both apoptosis and necrosis. DNase II could also mediate DNA cleavage independently of any phagocytosis, particularly during spontaneous apoptosis of aging human neutrophils [69], and lovastatine-treated HL60 cells [70]. In addition, L-DNase II, is implicated during the apoptosis of staurosporine-treated L1210 cells and in longterm-cultured HeLa cells [58, 63]. These mediators introduce 3'P and 5'OH termini in cleaved DNA fragments (Table 1) that cannot be detected by the TUNEL assay. Etoposide- and staurosporine-treated apoptotic DT40 cells, which activate an unknown endonuclease that introduces HMW fragments with 3'P ends, were undetectable by the TUNEL assay [66]. Consequently, the absence of positivity in the TUNEL assay does not signify the absence of apoptosis- associated DNA breaks, but could reflect the requirement of DNase II-like endonucleases. It is worth noting that nuclear apoptosis in one cell can be mediated by the activation of different enzymes leading to a fraction of DNA fragments with 3'OH/5'P ends and another with 3'P/5'OH termini. It should be emphasized that apoptotic cells could be detected by the TUNEL assay if sufficient 3'OH ends were introduced in DNA fragments. Since the TUNEL assay can efficiently detect nuclear apoptosis in human neutrophil cells, known to be mainly driven by acidic nucleases [69], endonucleases generating 3'OH and 5'P also could be activated in these cells. Because their relevance does not depend on the precise nature of 3' and 5' ends, they could probably be appropriate for detecting nuclear apoptosis mediated by DNase II-like mediators and could be adequate substitutes for the TUNEL technique.

Acknowledgments

This work was supported in part by a Grant-in-Aid for Scientific Research (C10680580 and C15590268) from the Ministry of Education, Culture, Sports, Science and Technology of Japan and the Hokkoku Foundation for Cancer Research in Japan

References

[1] Wyllie AH. Glucocorticoid-induced thymocyte apoptosis is associated with endogenous endonuclease activation. *Nature* 1980; 284: 555-556.

[2] Yuan J, Shaham S, Ledoux S, Ellis HM, Horvitz HR. The C. elegans cell death gene ced-3 encodes a protein similar to mammalian interleukin-1 beta-converting enzyme. *Cell* 1993;75: 641-652.

[3] Vaux DL, Korsmeyer SJ. Cell death in development. *Cell* 1999; 96: 245-254.

[4] Higuchi Y, Matsukawa S. Appearance of 1-2 Mb giant DNA fragments as an early common response leading to cell death induced by various substances which cause oxidative stress. *Free Radic Biol Med* 1997; 23: 90-99.

[5] Pardoll DM, Vogelstein B, Coffey DS. A fixed site of DNA replication in eucaryotic cells. *Cell* 1980; 19: 527-536.

[6] Yoshida A, Ueda T, Wano Y, Nakamura T. DNA damage and cell killing by camptothecin and its derivative in human leukemia HL-60 cells. *Jpn J Cancer Res* 1993; 84: 566-573.

[7] Bryant PE. Enzymatic restriction of mammalian cell DNA using Pvu II and Bam HI; evidence for the double strand break origin of chromosomal aberrations. *Int J Radiation Biol* 1984; 46: 57-65.

[8] Lobrich M, Ikepeme S, Kiefer J. DNA double-strand break measurement in mammalian cells by pulsed-field gel electrophoresis: an approach using restriction enzymes and gene probing. *Int J Radiat Biol* 1994; 65: 623-630.

[9] Matsukawa S, Higuchi Y. The nature of giant DNA molecules produced from nuclear chromosome DNA by active oxygen producing agents. In Oxidative damage and repair. *New York: Pergamon Press,* 1991. 197-201.

[10] Higuchi Y, Matsukawa S. Active oxygen-mediated chromosomal 1-2 Mb giant DNA fragmentation into internucleosomal DNA fragmentation in apoptosis of glioma cells induced by glutamate. *Free Radic Biol Med* 1998; 24: 418-426.

[11] Cohen GM, Sun X, Fearnhead H, MacFarlane M, Brown DG, Snowden RT, Dinsdale D. Formation of large molecular weight fragments of DNA is a key committed step of apoptosis in thymocytes. *J Immunology* 1994; 153: 507-516.

[12] Higuchi Y, Matsukawa S. Glutathione depletion induces chromosomal giant DNA and high molecular weight DNA fragmentation associated with apoptosis through lipid peroxidation and protein kinase C activation in C6 glioma cells. *Arch Biochem Biophys* 1999; 363: 33-42.

[13] Ward JF. DNA damage produced by ionizing radiation in mammalian cells. In Cohn WE and Moldave K, editors. Prog Nucleic Acid Res and Mol Biol New York: Academic Press, 1988. p95-125.

[14] Halliwell B, Aruoma OI. DNA damage by oxygen-derived species—Its mechanism and measurement in mammalian systems. *FEBS Lett* 1991; 281: 9-19.

[15] Clutton S. The importance of oxidative stress in apoptosis. *Brit Med Bull* 1997; 53: 662-668.

[16] Wlodek DBJ, Olive JPL. Comparison between pulsed-field and constant gel electrophoresis for measurement of DNA double-strand breaks in irradiated Chinese hamster ovary cells. *Int J Radiat Biol* 1991; 60: 779-790.

[17] Erixon K, Cedervall B. Linear induction of DNA double-strand breakage with X-ray dose, as determined from DNA fragment size distribution. *Radiation Res* 1995; 142: 153-162.

[18] Doroshow JH. Role of hydrogen peroxide and hydroxyl radical formation in the killing of Ehrlich tumor cells by anticancer quinones. *Proc Natl Acad Sci USA* 1986; 83: 4514-4518.

[19] Teicher BA, Herman TS, Holder SA. Combined modality therapy with bleomycin, hyperthermia and radiation. *Cancer Res* 1988; 48: 6291-6297.

[20] Chin D, Goldberg IH. Generation of superoxide free radical by neocarzinostatin and its possible role in DNA damage. *Biochemistry* 1986; 25: 1009-1015.

[21] Lagarkova MA, Iarovaia OV, Razin SV. Large-scale fragmentation of mammalian DNA in the course of apoptosis proceeds via excision of chromosomal DNA loops and their oligomers. *J Biol Chem* 1995; 270: 20239-41.

[22] Wang JC. DNA topoisomerases. *Annu Rev Biochem* 1996; 65: 635-692.

[23] Bellamy CO. p53 and apoptosis. *Brit Med Bull* 1997; 53: 522-538.

[24] Li TK, Chen AY, Yu C, Mao Y, Wang H, Liu LF. Activation of topoisomerase II-mediated excision of chromosomal DNA loops during oxidative stress. *Genes Development* 1999; 13: 1553-1560.

[25] Rusnak JM, Calmels TPG, Hoyt DG, Kondo Y, Yalowich JC. Genesis of discrete higher order DNA fragments in apoptotic human prostatic carcinoma cells. *J Pharm Exp Ther* 1996; 49: 244-252.

[26] Filipski J, Leblanc J, Youdale T, Sikorska M, Walker PR. Periodicity of DNA folding in higher order chromatin structures. *EMBO J* 1990; 9:1319-1327.

[27] Carson DA, Ribeiro JM. Apoptosis and disease. *Lancet* 1993; 341: 1251-1254.

[28] Bortner CD, Oldenburg NE, Cidlowski JA. The role of DNA fragmentation in apoptosis. *Trends Cell Biol* 1995; 5: 21-26.

[29] Schwartz LM, Smith SW, Jones ME, Osborne BA. Do All Programmed Cell Deaths Occur Via Apoptosis? *Proc Natl Acad Sci USA* 1993; 90: 980-984.

[30] Arends MJ, Morris RG, Willie AH. Apoptosis. The role of the endonuclease. *Am J Pathol* 1990; 136: 593-608.

[31] Bacso Z, Everson RB, Eliason JF. The DNA of V-binding apoptotic cells is highly fragmented. *Cancer Res* 2000; 60: 4623-4628.

[32] Ploski JE, Aplan PD. Characterization of DNA fragmentation events caused by genotoxic and non-genotoxic agents. *Mutat Res* 2001; 473: 169-180.

[33] Peitsch MC, Muller C, Tschopp J. DNA fragmentation during apoptosis is caused by frequent single-strand cuts. *Nucleic Acids Res* 1993; 21: 4206-9.

[34] Oberhammer F, Wilson JW, Dive C, Morris ID, Hickman JA, Wakeling E, Walker PR, Sikorska M. Apoptotic death in epithelial cells: Cleavage of DNA to 300 and/or 50 kb fragments prior to or in the absence of internucleosomal fragmentation. *EMBO J* 1993; 12: 3679-3684.

[35] Brown DG, Sun XM, Cohen GM. Dexamethasone-induced apoptosis involves cleavage of DNA to large fragments prior to internucleosomal fragmentation. *J Biol Chem* 1992; 268: 3037-3039.

[36] Huang P, Robertson LE, Wright S, Plunkett W. High molecular weight DNA fragmentation: A critical event in nucleotide analogue-induced apoptosis in leukaemia cells. *Clin Cancer Res* 1995; 1: 1005-1013.

[37] Zhivotovsky B, Orrenius S. Involvement of Ca^{2+} in the formation of high molecular weight DNA fragments in thymocyte apoptosis. *Biochem Biophys Res Commun 1994;* 202: 120-127.

[38] Susin SA, Lorenzo HK, Zamzami N, Marzo I, Snow BE, Brothers GM, Mangion J, Jacotot E, Costantini P, Loeffler M, Larochette N, Goodlett DR, Aebersold R, Siderovski DP, Penhinger JM, Kroemer G. Molecular characterization of mitochondrial apoptosis-inducing factor. *Nature* 1999; 397: 441-446.

[39] Kaufmann SH. Induction of endonuclolytic DNA cleavage in human acute myelogeneous leukemia cells by etoposide, camptothecin and other cytotoxic anticancer drugs. *Cancer Res* 1989; 49: 5870-5875.

[40] Batistatou A, Greene LA. Aurintricarboxylic acid rescues PC12 cells and sympathetic neurons from cell death caused by nerve growth factor deprivation: Correlation with suppression of endonuclease activity. *J Cell Biol* 1991; 115: 461-471.

[41] Jacobson MD. Reactive oxygen species and programmed cell death. *Trends Biochem Sci* 1996; 21: 83-86.

[42] Walker PR, Weaver VM, Lach B, LeBlanc J, Sikorska M. Endonuclease activity associated with high molecular weight and internucleosomal DNA fragmentation in apoptosis. *Exp Cell Res* 1994; 213: 100-106

[43] Higuchi Y. Polyunsaturated fatty acids promote 8-hydroxy-2'-deoxyguanosine formation through lipid peroxidation under the glutamate-induced GSH depletion in rat glioma cells. *Arch Biochem Biophys* 2001; 392: 65-70.

[44] Higuchi Y, Yoshimoto T. Promoting effect of polyunsaturated fatty acids on chromosomal giant DNA fragmentation associated with cell death induced by glutathione depletion. *Free Radical Res* 2004; 38: 649-658.

[45] Spiteller G. Enzymic lipid peroxidation-A consequence of cell injury. *Free Radic Biol Med* 1996; 21: 1003-1009.

[46] Higuchi Y, Yoshimoto T. Archidonic acid converts the glutathione depletion-induced apoptosis to necrosis by promoting lipid peroxidation in glioma cells. *Arch Biochem Biophys* 2002; 400: 133-140.

[47] Higuchi Y. Glutathione depletion-induced chromosomal DNA fragmentation associated with apoptosis and necrosis. *J Cell Mol Med* 2004; 8: 455-464.

[48] Arends MJ, Murris RG,Wyllie AH. Apoptosis: The role of the endonucleases. *Am J Pathol* 1990; 136: 593-608.

[49] Lecoeur H. Nuclear apoptosis detection by flow cytometry: Influence of endogenous endonucleases. *Exp Cell Res* 2002; 277:1-14.

[50] Walker PR, Leblanc J, Smith B, Pandey S, Sikorska M. Detection of DNA fragmentation and endonuclease in apoptosis. *Methods* 1999; 17: 329-338.

[51] Zamzami N, Kroemer G. Condensed matter in cell death. Nature 1999; 401: 127-128.

[52] Li LY, Wang X. Endonuclease G is an apoptotic DNase when released from mitochondria. *Nature* 2001; 412: 95-99.

[53] Sahara S, Aoto M, Eguchi Y, Imamoto N, Yoneda Y, Tsujimoto Y. Acinus is a caspase-3-activated protein required for apoptotic chromatin condensation. *Nature* 1999; 401: 168-173.

[54] Enari M, Sakahira H, Yokohama H, Okawa K, Iwamatsu A, Nagata S. Caspase-activated DNase that degrades DNA during apoptosis and its inhibitor ICAD. *Nature 1998;* 391: 43-50.

[55] Yakovlev AG, Wang G, Stoica BA, Simbulan-Rosenthal CM, Yoshihara K, Smulson ME. Role of DNAS1L3 in Ca^{2+} - and Mg^{2+} -dependent cleavage of DNA into oligonucleosomal and high molecular mass fragments. *Nucleic Acids Res 1999;* 27:1999-2005.

[56] Shiokawa D, Ohyama H, Yamada T, Tanuma S. Purification and properties of DNase gamma from apoptotic rat thymocytes. *Biochem J 1997*; 326: 675-81.

[57] Vancompernolle K, Van Herreweghe F, Pynaert G, Van de Craen M, De Vos K, Totty N, Sterling A, Fiers W, Vandenabeele P, Grooten J. Atractyloside-induced release of cathepsin B, a protease with caspase-processing activity. *FEBS Letters 1998*; 438: 150-158.

[58] Counis MF, Torriglia A. DNases and apoptosis. Biochem Cell Biol 2000; 78: 405-414.

[59] Susin SA, Zamzami N, Castedo M, Daugas E, Wang HG, Geley S, Fassy F, Reed JC, Kroemer G. The central executioner of apoptosis: Multiple connexions between protease activation and mitochondria in Fas/Apo-1/CD95- and ceramide-induced apoptosis. *J Exp Med 1997;* 186: 25-37.

[60] Hughes FM, Evans-Storm RB, Cidlowski J.A. Evidence that non-caspase proteases are required for chromatin degradation during apoptosis. *Cell Death Differ 1998;* 5: 1017-1027.

[61] Gaido ML, Cidlowski JA. Identification, purification and characterization of a calcium dependent endonuclease (NUC18) from apoptotic rat thymocytes. *J Biol Chem 1991;* 266: 18580-18585.

[62] Chen D, Stetler RA, Cao G, Pei W, O'Horo C, Yin XM, Chen J. Characterization of the rat DNA fragmentation factor 35/inhibitor of caspase-activated DNase (short form). *J Biol Chem 2000;* 275: 38508-38517.

[63] Torriglia A, Negri C, Chaudun E, Prosperi E, Courtois,Y, Counis MF, Scovassi AI. Differential involvement of DNases in HeLa cell apoptosis induced by etoposide and long term culture. *Cell Death Differ 1999*; 6: 234-244.

[64] McManus JP, Rasquinha I, Tuor U, Preston E. Detection of higher-order 50- and 10-Kb DNA fragments before apoptotic internucleosomal cleavage after transient cerebral ischemia. *J Cereb Blood Flow Metab 1997;* 17: 376-387.

[65] Durrieu F, Samejima K, Fortune JM, Kandels-Lewis S, Osheroff N, Earnshaw WC. DNA topoisomerase IIα interacts with CAD nuclease and is involved in chromatin condensation during apoptosis. *Curr Biol 2001;* 10: 923-926.

[66] Samejima K, Shigenobu T, Earnshaw WC. CAD/DFF40 nuclease is dispensable for high molecular weight DNA cleavage and stage I chromatin condensation in apoptosis. *J Biol Chem 2001;* 276: 45427.

[67] Susin SA, Daugas E, Ravagnan L, Samejima K, Zamzami N, Loeffler M, Costantini P, Ferri KF, Irinopoulou T, Prevost MC, Brothers G, Mak TW, Penninger J, Earnshaw WC, Kroemer G. Two distinct pathways leading to nuclear apoptosis. *J Exp Med* 2000; 192: 571-579.

[68] Zhao QL, Kondo T, Noda A, Fujiwara Y. Mitochondrial and intracellular free calcium regulation of radiation- induced apoptosis in human leukemic cells. *Int J Radiat Biol 1999;* 75: 493-504.

[69] Gottlieb RA, Heidi AG, Engler RL, Babior BM. The acid deoxyribonuclease of neutrophils: A possible participant in apoptosis-associated genome destruction. *Blood 1995;* 86: 2414-2418.

[70] Wang IK, Lin-Shiau SY, Lin JK. Induction of apoptosis by lovastatin through activation of caspase-3 and DNase II in leukaemia HL-60 cells. *Pharmacol Toxicol 2000;* 86: 83-91.

[71] Liu X, Zou H, Slaughter C, Wang X. DFF, a heterodimeric protein that functions downstream of caspase-3 to trigger DNA fragmentation during apoptosis. *Cell 1997;* 89: 175-184.

[72] Pandley S, Walker PR, Sikorska M. Identifi-cation of a novel 97 kDa endonuclease capable of internucleosomal DNA cleavage. *Biochemistry 1997;* 36: 711-720.

[73] Peitsch MC, Polzar B, Stephan H, Crompton T, Mac-Donald HR, Mannherz HG, Tschopp J. Characterization of the endogenous deoxyribonuclease involved in nuclear DNA degradation during apoptosis (programmed cell death). *EMBO J 1993;* 12: 371-377.

[74] Liu QY, Pandey S, Singh RK, Lin W, Ribecco M, Borowy-Borowski H, Smith B, LeBlanc J, Walker PR, Sikorska M. DNase Y: A rat DNase I gene coding for a constitutively expressed chromatin-bound endonuclease. *Biochemistry 1998;* 37: 10134-10143.

[75] Barry M, Eastman A. Identification of deoxyribonuclease II as an endonuclease involved in apoptosis. *Arch Biochem Biophys 1993;* 300: 440-450.

[76] Higuchi Y, Mizukami Y, Yoshimoto T. Ultraviolet irradiation-induced chromosomal giant DNA fragmentation leads to apoptosis providing internucleosomal DNA fragments in glioma cells. *Ann NY Acad Sci 2003;* 1010: 326-330.

[77] Kawane K, Fukuyama H, Kondoh G, Takeda J, Ohsawa Y, Uchiyama Y, Nagata S. Requirement of DNase II for definitive erythropoiesis in the mouse fetal liver. *Science 2001*; 292:1546-1549.

In: New Developments in Cell Apoptosis Research
Editor: Alan J. Corvin, pp. 193-212
ISBN: 1-60021-453-3

Chapter VIII

Overgrowth-induced Apoptosis by *dco*[3], a *Drosophila* Mutant Allele for Casein Kinase I δ/ε

***Takashi Adachi-Yamada*[1, 2, 3, *], *Toshiyuki Harumoto*[2], *Atsushi Okabe*[2], *Masahiko Takemura*[2], *Kousuke Maeda*[2], *Makoto Umemori*[2], *Ayumu Tagami*[2] and *Masami Shimoda*[4]**

Graduate School of Science and Technology[1],
Graduate School of Human Science[2],
SORST, Japan Science and Technology Agency[3],
Kobe University, Kobe 657-8501, Japan
National Institute of Agrobiological Sciences[4], Tsukuba 305-8634, Japan

Abstract

Growth, differentiation and apoptosis are mutually affected during organogenesis. Elucidating these interactions should be important for understanding the developmental regulation of morphogenesis, homeostasis and organ size control, all of which are commonly found in multicellular organisms. We report here that the overgrowth phenotype by *dco*[3], an allele of *discs overgrown* gene which is the *Drosophila* ortholog of mammalian Casein kinase I δ/ε, causes apoptosis during development of the wing imaginal disc. This apoptosis strongly depends on c-Jun N-terminal Kinase (JNK), a stress-activated protein kinase. When apoptosis was blocked by inactivating JNK, the overgrowth phenotype became prominently malignant, indicating that the role of this apoptosis is for elimination of the overgrown tissue. To find the initial event during this apoptosis, we generated a genetic mosaic in the developing wing imaginal disc in which partial cell populations show the overgrown phenotype by *dco*[3]. Interestingly, JNK

[*] Correspondence: yamadach@kobe-u.ac.jp

activation was induced at the either side of the boundary between the *dco*^{3} mutant and normal cell populations. Furthermore, activation of JNK outside of the *dco*^{3} cell population (that is normal cell population) contributed to the efficient removal of the *dco*^{3} cell population. Although this type of nonautonomous apoptosis has widely been known to be a result of abrupt change in morphogen signal gradient by such as Decapentaplegic (Dpp, a TGF-β superfamily member) and Wingless (Wg, a Wnt family member), the nonautonomous apoptosis by *dco*^{3} was not accompanied by aberration of these morphogen signal levels. In particular, unlike the well-established relationship between Dco and several components of Wg signal pathway, *dco*^{3} did not significantly affect Wg signaling in the wing imaginal disc. These results demonstrated that the nonautonomous apoptosis that begins at the boundary between abnormal and normal cell populations is not only found in the case of inappropriate morphogen signaling but also associated with deregulated tissue growth such as tumorigenesis to restore and assure normal morphogenesis.

Introduction

Exclusion of inappropriate or mutant cells from the developing field is an important step for proper development of various tissues. For these purposes, cell proliferation, cell differentiation and apoptosis should be mutually affected. For example, in the genetically amenable organism *Drosophila*, various morphogenetic signalings have well been known to regulate tissue growth as well as cell type specification [Burke and Basler, 1996; Johnston and Sanders, 2003; Rogulja and Irvine, 2005]. Also, abnormally frequent removal of the cells from the epithelium induces the compensatory cell proliferation in a proactive manner for maintenance of proper cell number [Huh et al., 2004; Perez-Garijo et al., 2004; Perez-Garijo et al., 2005; Ryoo et al., 2004].

Cell competition which was discovered in the imaginal disc development in *Drosophila*, is also one of such interplaying mechanisms and may be applied to various kinds of animal tissue development [Morata and Ripoll, 1975]. Through this mechanism, a cell population with an abnormally slow rate of cell proliferation disappears during development. It has been considered that this phenomenon is achieved by reduced reception of extracelluar factor (such as Dpp) which promotes both cell survival and cell proliferation, in the abnormal cell population [Moreno et al., 2002a].

In contrast to cell competition to exclude slowly proliferating cells, it should be difficult to remove fast proliferating cell population, such as cancer, from the normally proliferating cell field. Because, to overcome the abnormally active cell proliferation signal, an immensely strong apoptotic signal is required for complete removal of the abnormal cells. However, many kinds of tumors in mammals have been known to gain a resistance to apoptotic signals, through accumulation of mutations in apoptosis-inducible genes (e.g. p53) or hyperactivation of apoptosis repressible-genes (e.g. Bcl-2) [Johnstone et al., 2002]. On the other hand, it has also been known that cells are harboring multiple barriers to prevent tumor growth, such as telomeric function for longevity restriction, DNA damage checkpoint to inhibit accumulation of mutations, maintenance of cell polarity and adhesion in the epithelial sheet, and NK cell-mediated immune response to cancer cells [Bissell and Radisky, 2001; Gomez-Lazaro et al., 2004; Greenberg, 2005; Orange and Ballas, 2006].

Some of these may also be common to the *Drosophila* system. For example, it has been known that the oncogenic form of Ras (Ras^{V12}) provides a survival ability to cells by inactivating Hid, a propoptotic gene product [Bergmann et al., 1998]. Also, Warts, a conserved tumor-suppressor gene product is known to possess an ability to suppress apoptosis [Harvey et al., 2003; Wu et al., 2003]. We have thus focused on *dco*^{3}, an allele of *dco* which is the *Drosophila* ortholog of mammalian Casein kinase I δ/ε, which has been known to induce a tumor-like hyperplasia [Zilian et al., 1999]. We have found that this overgrowth is an exceptional case without gaining apoptosis resistance and may be useful for studying about proliferation-inducible cell death. The *dco*^{3}-induced overgrown cell population led to the nonautonomous activation of JNK along the boundary between tumor and normal cells, which contributed to removal of the tumor cell population, suggesting a common mechanism to exclude abnormal cells from the normal cell field.

Results

Overgrowth by *dco*^{3} Causes Apoptosis with Activation of JNK and Caspase-3

The gene for *Drosophila* Casein kinase I δ/ε, *dco* is expressed ubiquitously in the imaginal discs (Fig. 1A). Although most of its loss-of-function alleles show a reduced proliferation phenotype, a particular *dco* allele *dco*^{3} conversely displays an overgrown disc and deregulated growth arrest phenotypes when combined with various authentic loss-of-function *dco* alleles [Zilian et al., 1999]. The molecular lesion in *dco*^{3} has been mapped to two amino acid residues, one is Arg-4 replaced with Cys, the other is Glu-74 replaced with Lys. In spite of its particular phenotype, its overgrowth phenotype was rescued by wild type *dco*$^+$ gene as is the cases for other *dco*$^-$ alleles, suggesting that *dco*^{3} is also a loss-of-function allele [Zilian et al., 1999].

To see whether apoptosis is involved in this overgrowth phenotype, we examined the activation of JNK and Caspase-3, a cysteine protease executing apoptosis, in an allelic combination of *dco*^{2}/*dco*^{3} in which the wing disc shows an overgrowth phenotype during third instar larva (Fig. 1B-B''). Both JNK and Caspase-3 activation were observed widely in the wing blade region of the wing disc. Although patterns of JNK and Caspase-3 activation do not always coincide in many cases [Adachi-Yamada and O'Connor, 2004], this case exceptionally shows well matching of the location of JNK-active cells with that of Caspase-3-active cells. This suggests that activation of JNK and Caspase-3 occur at the same time in these cells.

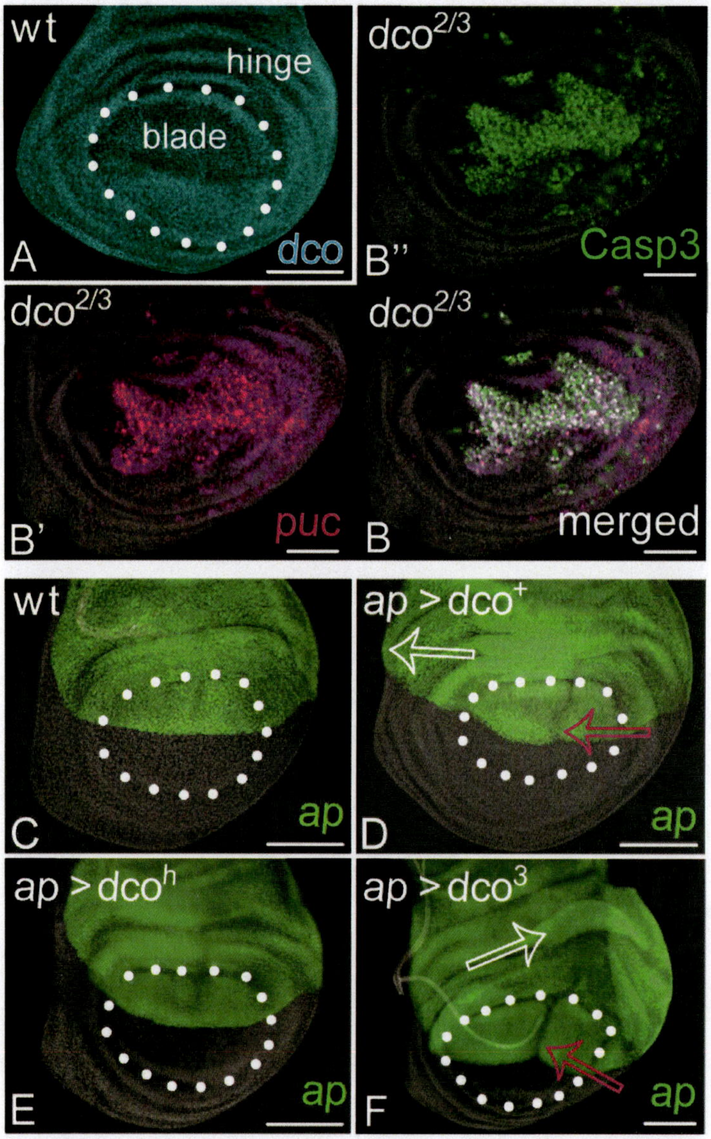

Figure 1. Characterization of *dco*^{3}. (A) Nuclear localization of β-GAL indicates ubiquitous expression of *dco-lacZ* in the wing disc of wild type (*dco*^{P538}/+). (B-B'') Overgrown wing disc by *dco*^{2}/*dco*^{3} allelic combination shows a blade-specific apoptosis with simultaneous activation of JNK (*puc-lacZ*, magenta) and Caspase-3 (green). Normal wind discs have been known to show no or only a few apoptotic cells. (C) A wild type wing disc. The dorsal compartment is visualized by *UAS-GFP* expression driven by *apterous-GAL4*. (D) A wing disc where *dco*$^{+}$ is overexpressed in the dorsal compartment. The size of the dorsal compartment is weakly enlarged. The expanded edge of the overgrown area is indicated by white arrow. An epithelial folding associated with the overgrowth is indicated by magenta arrow. (E) A wing disc where dominant-negative *dco* (*dco*^{h}) is overexpressed in the dorsal compartment. The size of the dorsal compartment weakly shrinks. (F) A wing where *dco*^{3} is overexpressed in the dorsal compartment. The dorsal compartment shows a prominent overgrowth. Anterior is to the left, and dorsal is up. Images of wild type discs are magnified 1.5 times larger than those in the mutant discs exhibiting overgrowth throughout this paper. Scale bars represent 100 μm. Dotted line indicates the boundary between the hinge and blade regions.

Overexpression of *dco*^{3} Shows an Overgrowth Phenotype

To gain insight into the overgrowth phenotype by *dco*^{3}, we generated the *UAS-dco*^{3} transgenic flies. When wild type *dco* (*UAS-dco*$^+$) is overexpressed in the dorsal compartment in the wing disc, it shows a mild overgrowth (Fig. 1D). In contrast, when a dominant negative version of *dco* (*UAS-dco*^{h}, [Xu et al., 2005]), was similarly overexpressed, the dorsal compartment showed a moderate shrinkage (Fig. 1E). Thus we tested overexpression of *UAS-dco*^{3} (Fig. 1F). As a result, the dorsal compartment showed an overgrowth of which degree is greater than that by *UAS-dco*$^+$ overexpression. This indicates that the overgrowth by *dco*^{3} is due to high level activity of *dco*, although its mutant phenotype was complemented by *dco*$^+$ as mentioned above. This complementation may probably be explained by the autoregulatory function of Dco to repress its own activity [Longenecker et al., 1998; Rivers et al., 1998](see Discussion).

Dpp, Wg and Hh Signals in *dco*^{2}/*dco*^{3} Overgrown Disc

Various morphogens have been known to regulate cell proliferation as well as provide positional information [Burke and Basler, 1996; Duman-Scheel et al., 2002; Johnston and Sanders, 2003; Rogulja and Irvine, 2005]. Thus we examined changes in signal levels of the three morphogens Dpp, Wg and Hedgehog (Hh), all of which are acting in the wing disc.

Dpp is a long-range morphogen that emanates from a band just anterior to the anteroposterior (AP) compartment boundary. It diffuses to outer cell population and gives rise to the signaling in these cells. As a result, one of the Dpp target genes *optomotor-blind* (*omb*) is expressed in response to Dpp signaling in the prospective wing region except for its lateral subregion. In the normal wing disc, expression of *omb* is strictly repressed by expression of *brinker* (*brk*) thereby creating a complementary expression pattern of both gene (Fig. 2A'''). In lateral wing, there is a zone where faded expressions of both *omb* and *brk* are overlapping (Fig. 2A'''', [Campbell and Tomlinson, 1999]).

The zone of *omb* expression in *dco*^{2}/*dco*^{3} is slightly narrower than that in wild type (Fig. 2B'). However this reduction is not completely explained by elevation of *brk* expression, which is found at the same time. Because an unconventional gap between the *omb*- and *brk*-expressing regions can be seen in this mutant. Accordingly, the regulation of *omb* expression in *dco*^{2}/*dco*^{3} is partially free from the restriction by *brk*, suggesting that the *omb* expression is affected by other factor(s) in this case. With respect to the regulation of cell proliferation, slight reduction of *omb* expression is not thought to be the cause of overgrowth based on the reduced growth phenotype brought by *omb* mutation [del Alamo Rodriguez et al., 2004].

Wg is another long-range morphogen that emanates from the region at the dorsoventral (DV) compartment boundary. It diffuses to the dorsal and ventral cell populations and acts as morphogen. Wg is also expressed in the hinge region as two concentric rings and defines cell fate in this region [Rodriguez Dd Ddel et al., 2002]. Thus one of the Wg target genes *Dfrizzled-3* (*fz3*) is expressed around the DV boundary and hinge region (Fig. 2C''). As a particular feature in Wg signaling to be focused, it has been known that Dco phosphorylates various signaling molecules in this pathway, such as Dishevelled (Dsh) and Armadillo (Arm, a β-catenine homolog), to alter the signaling intensity [Cong et al., 2004; Yanagawa et al., 2002].

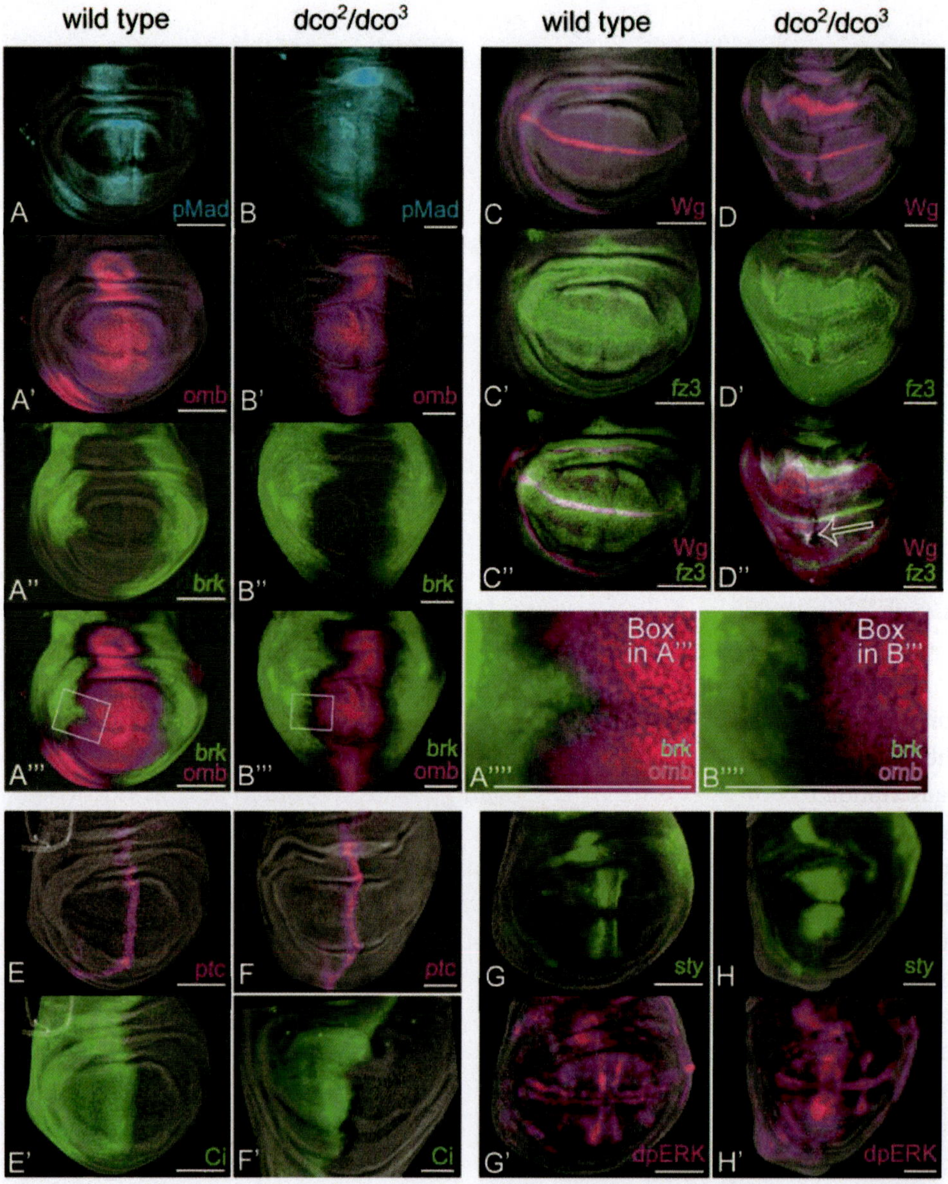

Figure 2. Activities in various morphogenetic signal pathways in global *dco²/dco³* mutant. (A-A'''') Wild type expression of p-Mad (cyan), *omb-lacZ* (red), and *brk-GAL4* (green), all of which reflect the Dpp signal activity. The box in A''' is magnified in A'''' to show a complementary expression of *omb* and *brk*. (B-B'''') *dco²/dco³*. The box in B''' is magnified in B'''' to show a gap between *omb* and *brk*-expressing areas (C-C'') Wild type expression of Wg (green) and *fz3-lacZ* (magenta), an indicator of Wg signal. (D-D'') *dco²/dco³*. Ectopic Wg expression associated with compensatory proliferation is indicated by arrow. (E and E') Wild type expression of *ptc-lacZ* (magenta) and accumulation of Ci (green). Both represent the Hh signal activity. (F and F') *dco²/dco³*. No significant change is observed. (G and G') Wild type expression of *sty-GAL4* (*sty*, green) and dual phosphorylation of ERK (dpERK, magenta). Both represent the Ras-ERK signal activity. (H and H') *dco²/dco³*. Signal intensity around the central wing region is elevated.

However, despite this well documented knowledge, expression of *fz3-lacZ* in *dco*^{2}/*dco*^{3} mutant are basically the same as that in wild type. Exceptionally, we can see a scattered elevation of *fz3-lacZ* expression which is also merged with ectopic expression of Wg (Fig. 2D-D''). Based on its sporadic pattern, it is considered that this elevation of Wg signaling is involved in compensatory proliferation accompanied by apoptosis [Huh et al., 2004; Perez-Garijo et al., 2004; Ryoo et al., 2004], and does not affect the global growth of the *dco*^{2}/*dco*^{3} wing disc.

We also examined Hh, a short-range morphogen inducing Dpp expression. The intensity of Hh signal can be monitored by *patched* (*ptc*)-*lacZ* expression and Cubitus interruptus (Ci) accumulation (Fig. 2E and F). Expression of *ptc* is known to be induced by higher level of the Hh signal whereas accumulation of Ci responds to more moderate level of the Hh signal. Unlike the cases for previous reports in which Dco modifies Hh signaling [Apionishev et al., 2005; Price and Kalderon, 2002], both the *ptc-lacZ* expression Ci accumulation did not change in *dco*^{2}/*dco*^{3}. This indicated that *dco*^{2}/*dco*^{3} does not affect Hh signling.

Together with these, *dco*^{2}/*dco*^{3} mutation slightly influences Dpp signaling but does not affect others. Thus, it is not presumed that these signals cause the overgrowth phenotype in *dco*^{2}/*dco*^{3}.

Ras-ERK Signal is Upregulated by *dco*^{2}/*dco*^{3}

Ras-ERK pathway provides a cell proliferation signal commonly in all of the imaginal epithelial tissues [Diaz-Benjumea and Hafen, 1994]. We thus examined whether Ras signal is altered in the *dco*^{2}/*dco*^{3} wing disc by monitoring expression of *sprouty* (*sty*) gene of which transcription has been known to be induced by Ras signal [Casci et al., 1999; Reich et al., 1999]. It has also been known that activation of Ras is followed by accumulation of dual phosphorylation form of Extracellular signal-Regulated Kinase (dpERK), mediated by the classical MAP kinase cascade. As a result, both the *sty* expression and dpERK accumulation in *dco*^{2}/*dco*^{3} were elevated around the central region of the wing disc (Fig. 2G and H). Therefore, Ras-ERK activity may help to promote the overgrowth by *dco*^{2}/*dco*^{3} at least in part. However, all of the aspects in *dco*^{2}/*dco*^{3} phenotypes can not be explained by Ras-ERK activity. For example, overgrowth by *dco*^{2}/*dco*^{3} is found throughout the prospective wing region whereas elevation of *sty* and dpERK levels was observed only in the central wing disc region. In addition, activation of Ras-ERK activity has been known to provide anti-apoptotic activity [Bergmann et al., 1998].

Apoptosis Eliminates *dco*^{2}/*dco*^{3} Cell Population from Wild Type Cell Background

Cells receiving growth signals or in the process of tumorigenesis often lose competence to choose apoptotic fate [Bergmann et al., 1998; Johnstone et al., 2002]. However, the cells in the *dco*^{2}/*dco*^{3} overgrown disc manifest a prominent apoptosis as shown above. This led us to examine whether the sporadically occurring *dco*^{2}/*dco*^{3} cell population is removed when it is located in wild type cell background. We thus made a genetic mosaic in which most of the dco^2/dco^3 cells are rescued by dco^+ gene.

At 3 days (72 hours) after mosaic induction, most of the dco^2/dco^3 cell population disappeared from the wing blade region (Fig. 3A) where severe apoptosis was observed in the case of simple mutant dco^2/dco^3 (Fig. 1B-B''). Thus the apoptosis can overcome tumorigenesis in this situation. At the same time, around the marginal zone of the wing blade, we can see some cell population in which JNK activation has just been initiated. In the vicinity of these cells, JNK is activated in cells on either side of the boundary between mutant and rescued cells (Fig 3A' and A''). This type of nonautonomous activation of JNK can be observed when various morphogenetic signals are altered [Adachi-Yamada et al., 2005; Adachi-Yamada and O'Connor, 2002; Adachi-Yamada and O'Connor, 2004].

However, in minor cases when removal of the mutant cell populations is delayed, some large mutant cell populations show an autonomous activation of JNK and Caspase-3 (Fig. 3A'''). Therefore, it is interpreted that there are two ways in activation of apoptotic pathways for removal of dco^2/dco^3 cells. One is the nonautonomous activation found at early stage when the cell population size is still small. The other is the autonomous activation found at later stage when the cell population size becomes larger.

Dpp and Wg Signalings are Stable Around the dco^2/dco^3 Cell Population in Mosaic Disc

Nonautonomous activation of JNK around the dco^2/dco^3 cell population led us to predict that alteration of some morphgen signalings is involved in the process of cell elimination. We thus monitored Dpp and Wg signaling by detecting *omb-lacZ* and *fz3-lacZ* expression as mentioned above (Fig. 4). In these experiments, we found that the dco^2/dco^3 cell population frequently remains even at 4 days after mosaic induction. This is probably caused by restoring the gene dosage of *puckered* (*puc*) which encodes a repressor of JNK. In above mentioned cases to see JNK activation, we always use *puc-lacZ*, an enhancer-trap mutant allele of *puc*, to monitor JNK activity. However, in the present case to see morphogen signalings, we needed to remove *puc-lacZ* to see *omb-lacZ* and *fz3-lacZ* reporters expession.

In the cases with relatively large mutant cell populations, they occasionally showed a moderate reduction in both *omb-lacZ* and *fz3-lacZ* expression (Fig. 4A and D). However this is rather minor cases. In most examples with relatively small mutant cell populations, in which JNK and Caspase-3 were presumably activated, they did not show significant changes in both *omb-lacZ* and *fz3-lacZ* expression (Fig. 4B and E). We also checked the cell populations homozygous for dco^2, an authentic loss-of-function allele of *dco*, with a reduced cell proliferation phenotype, but did not also detect alteration in both *omb-lacZ* and *fz3-lacZ* expression (Fig. 4C and F). Consequently, morphogen signals by Dpp and Wg do not seem to play major roles in nonautonomous activation of JNK around the dco^2/dco^3 cell population.

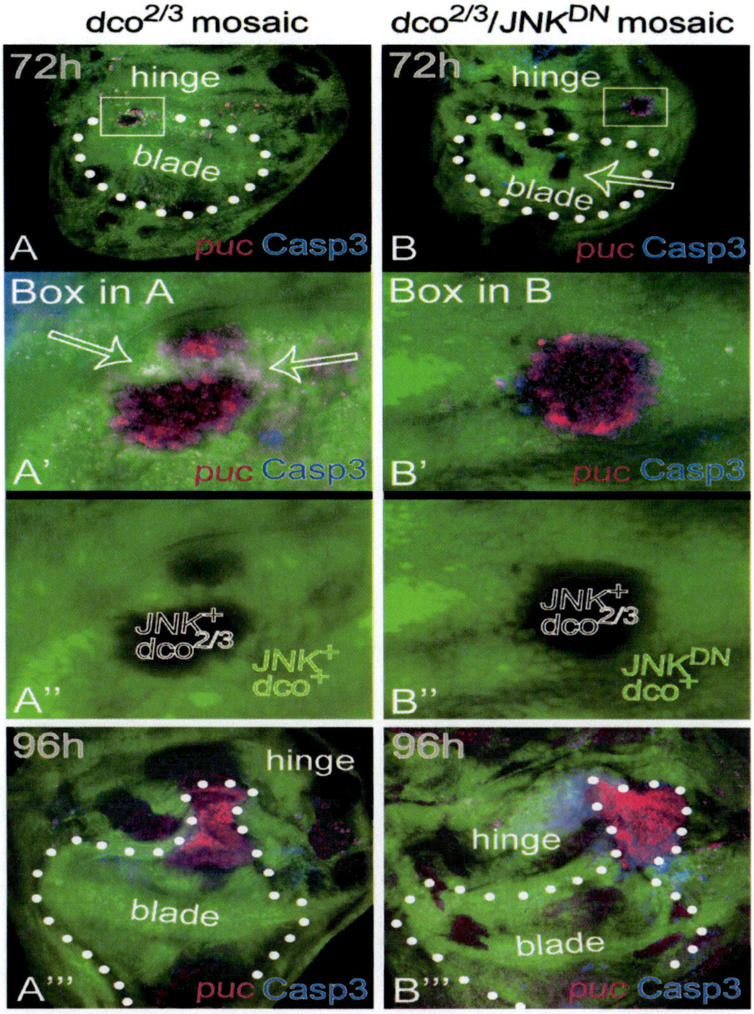

Figure 3. Apoptosis to eliminate *dco*^{2}/*dco*^{3} cell population in mosaic disc. (A-A''') Large areas of the *dco*^{2}/*dco*^{3} wing disc were rescued by mosaic expression of *UAS-dco*$^+$ (green). Some clones without flip-out remained as mutant cell population. Expression of *puc-lacZ* (magenta) and activation of Caspase-3 (blue) were monitored. (A) 72 hours after mosaic induction. Most of the mutant cell populations have already disappeared from the wing blade region. Dotted line indicates the boundary between the hinge and blade regions. (A' and A'') High magnification image of a representative cell population in A (boxed), where JNK activation is initiated around the boundary. White arrows indicate nonautonomous activation of JNK outside of the mutant cell population. (A''') 96 hours after mosaic induction. When a large mutant cell population remained in the wing blade region, JNK and Caspase-3 are autonomously activated later in development. (B-B''') Elimination of *dco*^{2}/*dco*^{3} cell population in mosaic disc requires JNK activity in the wild type background cells. Large areas of the *dco*^{2}/*dco*^{3} wing disc were rescued by mosaic expression of *UAS-dco*$^+$ (green). In addition, *UAS-JNK*^{DN} is simultaneously induced to inactivate JNK in the wild type background cells. As a result, some clones without flip-out remained as mutant cell population with normal JNK activity. The rest of cells, that were rescued by *dco*$^+$ and shown in green, lost JNK activity. (B) 72 hours after mosaic induction. Some mutant cell populations can still be found in the wing blade region (arrow). (B' and B'') High magnification image of a representative cell population in B (boxed), where JNK activation is initiated just inside the boundary. (B''') A large mutant cell population that remained in the wing blade region autonomously activate JNK and Caspase-3 later (96 hours after mosaic induction).

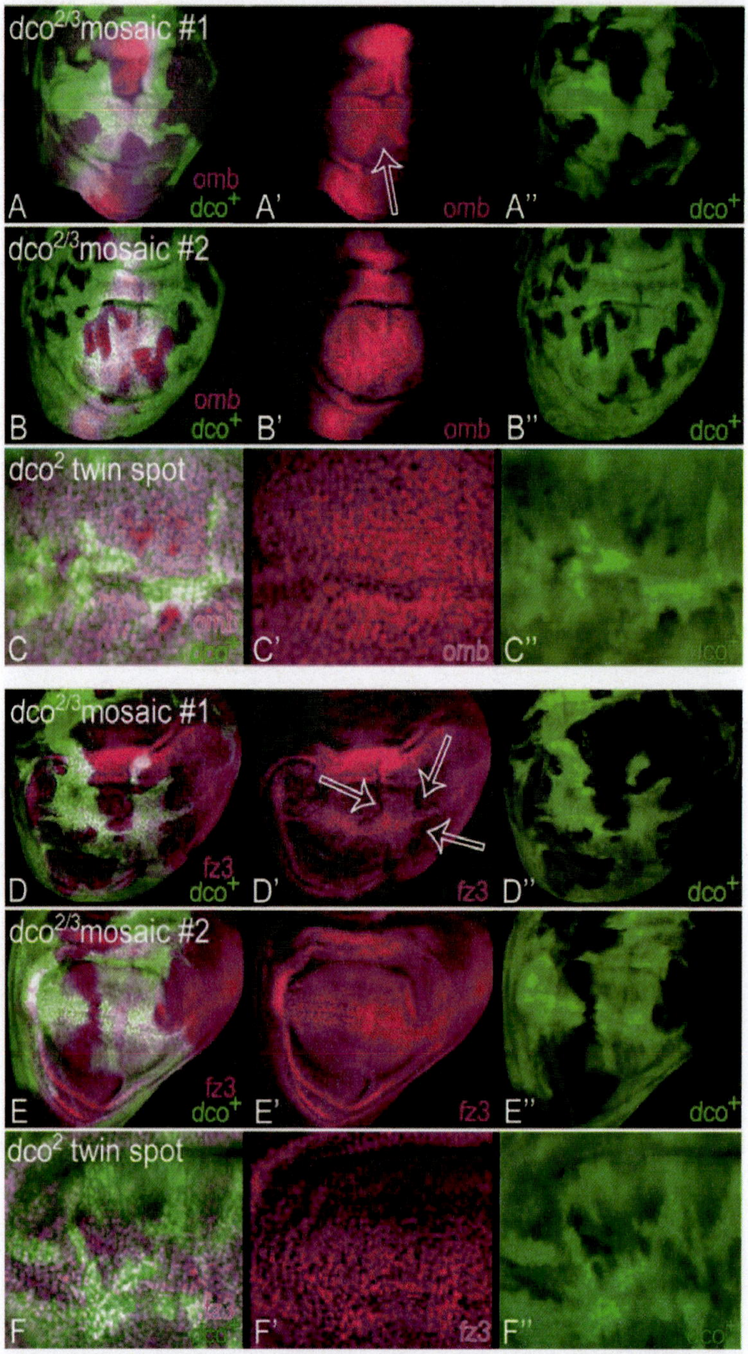

Figure 4. Dpp and Wg signaling in *dco* mutant in mosaic disc. 72-96 hours after mosaic induction. (A-C) Expression of *omb-lacZ*, an indicator of Dpp signaling. (D-F) Expression of *fz3-lacZ*, an indicator of Wg signaling. (A and D) Rare examples showing moderately weakened signal levels in relatively large mutant cell populations (arrows). (B and E) Common examples showing faint or no changes in signal levels in relatively small mutant cell populations. (C and F) Conventional twin spot mosaic using an authentic loss-of-function mutant dco^2. Magnification is three times larger than the others. Cells homozygous for can be recognized by absence of GFP (green). No significant changes are observed in Dpp and Wg signaling.

JNK Activity in Adjacent Wild Type Cells Contributes to Efficient Removal of the *dco*^{2}/*dco*^{3} Cell Population

We focused on the activation of JNK in wild type cells located just outside of the *dco*^{2}/*dco*^{3} cell population. There may be two opposing possibilities in role of this nonautonomous JNK activation. One is an active role in which the nonautonomous JNK activation contributes to removal of the inside mutant cells. The other is a passive role in which nonautonomous JNK activation is just a concomitant phenotype by contact between different types of cells and no positive role in removal of the inside mutant cells. To test these possibilities, JNK activity in wild type cells background was inactivated by expression of dominant-negative JNK (*UAS-JNK*^{DN}) together with *UAS-dco*$^{+}$ gene as the rescue construct (Fig. 3B-B''').

Interestingly, some *dco*^{2}/*dco*^{3} cell populations failed to disappear from the wing blade region even at 4 days after mosaic induction (Fig. 3B). This indicates that JNK activity in wild type cells located just outside of the mutant cell population contributes to efficient removal of the mutant cells inside. However, autonomous activation of JNK in later stage, another way for JNK activation as mentioned above, can be observed even in this genotype (Fig. 3B'''). Thus the mutant cells surrounded by JNK-inactive wild type cells may disappear at later stage, and the nonautonomous activation of JNK works to accelerate disappearance of the mutant cells or elevate efficiency of successful removal.

Epithelial Integrity and Cytoskeltal Organization of *dco*^{2}/*dco*^{3} Cell Population

It has been known that, in the columnar epithelium of the imaginal disc, cells undergoing apoptosis by reduced Dpp signaling shift basally after generating a cyst-like rearrangement [Gibson and Perrimon, 2005; Shen and Dahmann, 2005]. At this time, changes in cytoskeltal organization such as reduction of apical microtubule level are observed. Consequently, the shape of cells is affected and their apical surfaces are facing each other in the cyst. We therefore examined the epithelial integrity of the *dco*^{2}/*dco*^{3} cell population, which is often sorted out from the rescued cell background, together with detection of F-actin and Tubulin accumulation. Reconstitution of vertical sections from multiple X-Y confocal images revealed that the positions of the *dco*^{2}/*dco*^{3} cell population often shift basally with typical traits as the cyst-like structure. However, unlike the case in basal extrusion of cell population with decreased Dpp signaling, the cells rather showed upregulated Tubulin accumulation (Fig. 5A'). Therefore the molecular mechanism in cytoskeltal organization to form cyst-like structure is not considered to be identical to that in the case of the decreased Dpp signaling.

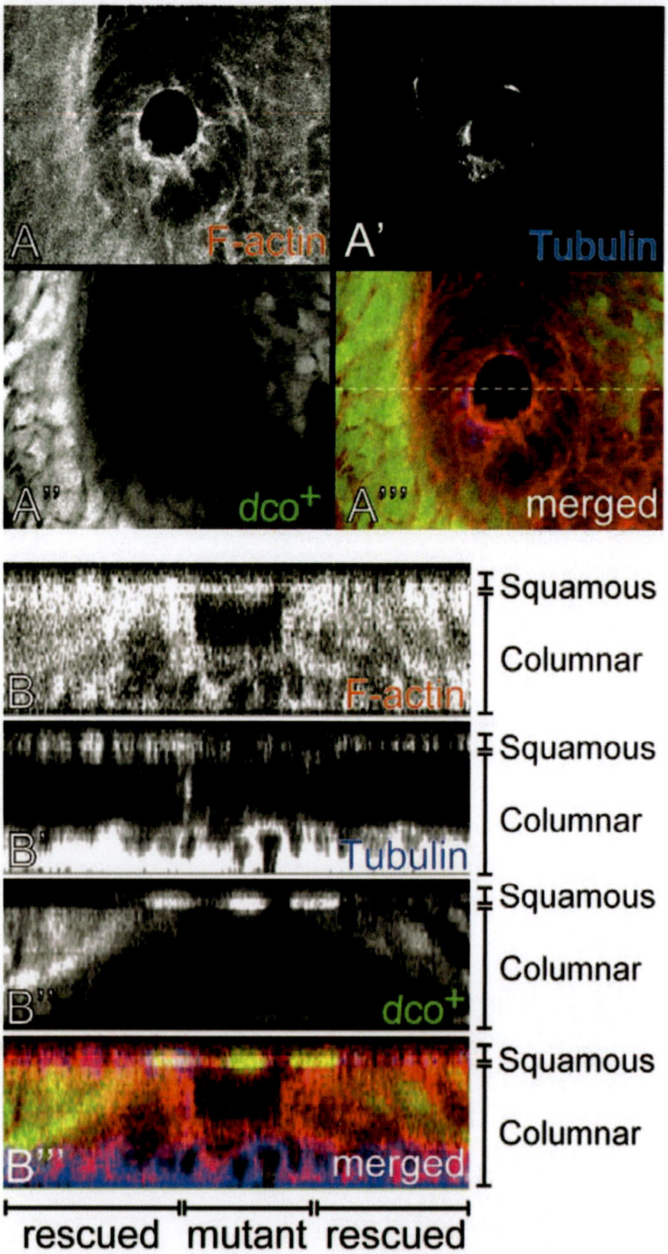

Figure 5. Laser confocal analysis of cytoskeltal organization in dco^2/dco^3 cell population. Mosaic discs were obtained as in Figure 3 and 4. Mutant cell populations are recognized by the absence of GFP (GFP) which labels rescued background cells. Accumulations of F-actin (red) and α-Tubulin (blue) were simultaneously detected. (A-A''') A representative example of the basally extruded dco^2/dco^3 cell population with a cyst-like structure. High accumulation of F-actin and moderate accumulation of Tubulin can be observed at the apical cell portions which are rearranged as a circle. (B-B''') An X-Z optical cross section of the cyst in A'''. The X-Y position of the X-Z section is indicated by broken line in A'''. The apical surface of the columnar epithelium is to the up. Vertical positions of the squamous cell layer (peripodial membrane) and the columnar cell layer (disc epithelium) are indicated on the right. Accumulation of F-actin traces the outlines of all the cells. The cyst cells that are basally extruded are surrounded by the rescued cells (dco^+) that are normally positioned.

Apoptosis to Prevent Malignant Overgrowth of *dco*^{2}/*dco*^{3} Cell Population

Although autonomous and nonautonomous apoptosis are reproducibly associated with occurrence of *dco*^{2}/*dco*^{3} cell population, the significance of these apoptosis have not been clarified. To seek the role of these apoptosis, we blocked JNK within the dorsal half of the wing disc of *dco*^{2}/*dco*^{3}, and compared phenotypes between the dorsal and ventral compartment. Consequently, the overgrowth in the dorsal compartment became more severe than that in the ventral compartment, indicating that the apoptosis ameliorates malignant overgrowth by *dco*^{2}/*dco*^{3} (Fig. 6A-C). Interestingly, cell proliferation rate in each compartment is undistinguishable (Fig. 6D-D''). Thus the cell proliferation is independent of JNK activity that extensively promotes apoptosis. This finding is consistent with many cases in human tumorigenesis, in which apoptosis prevents tumorgrowth [Johnstone et al., 2002].

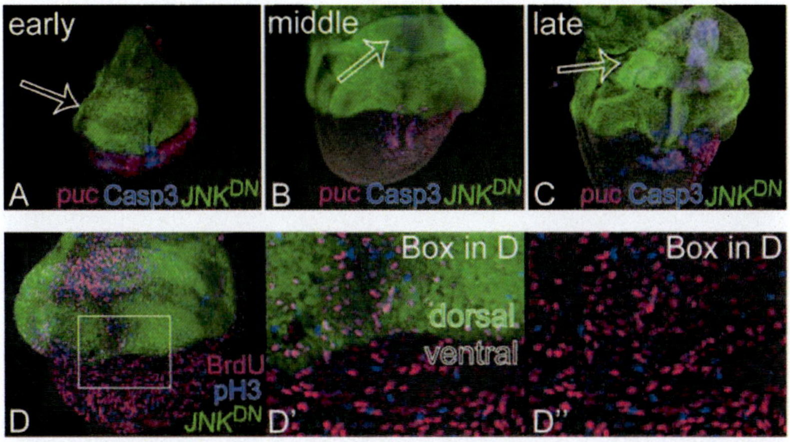

Figure 6. Blockade of apoptosis induces a malignant overgrowth without frequent occurrence of cell proliferation. (A-C) When JNK activity in the dorsal compartment (green) is inactivated by expression of dominant-negative JNK (JNK^{DN}), the dorsal cell population displays an overgrowth more severely with age (arrows). Activation of JNK (*puc-lacZ*, magenta) and Caspase-3 (blue) are predominant in the ventral compartment. (A) Early 3rd instar larva. (B) Middle 3rd instar larva. (C) Late 3rd instar larva. (D) BrdU incorporation (magenta) and phospho-histone H3 (green) in the same genotype are shown. (D' and D") High magnification images of the boxed region in D. On the basis of comparison between the dorsal and ventral compartment, inactivation of JNK does not significantly affect cell proliferation rate. High level of BrdU incorporation shown in the dorsal hinge in D is due to a superimpose of three layers of the folded epithelia, which was resulted from overgrowth.

Effect of *dco*^{2}/*dco*^{3} Mutation on Maintenance of Organ Identity

Homeotic transformation is also known to induce nonautonomous activation of JNK as revealed by ectopic leg/antenna formation through *spineless* (*ss*) or *Distal-less* (*Dll*) overexpression in the wing disc [Adachi-Yamada et al., 2005]. We thus checked the stability of organ identity in *dco*^{2}/*dco*^{3} cells. As shown in the simple global mutant *dco*^{2}/*dco*^{3}, expression of *sd*, of which expression can be used as a marker for the wing identity, was not decreased (data not shown). Similarly, expression of *Dll*, high level induction of which is known to elicit transformation to the leg [Gorfinkiel et al., 1997], was not increased (data not shown). Also in the case of the mosaic disc with dco^2/dco^3 cell population, both the *sd* and Dll expression levels were essentially the same between the mutant cells and the rescued cells (data not shown). Thus the organ identity is stable during nonautonomous JNK activation caused by occurrence of dco^2/dco^3 cell population.

Discussion

Dual Function of Dco in Regulation of Cell Proliferation

The dco^3 allele shows an unconventional phenotype to enhance cell proliferation unlike most of the loss-of-function *dco* alleles, which manifest reduced cell proliferation phenotypes. A simple expectation to explain this controversy is that the dco^3 allele is a rare gain-of-function allele which should show an opposite phenotype. However a previous report about complementation by dco^+ gene [Zilian et al., 1999] suggested that this overgrowth is also one of the loss-of-function phenotypes. By contrast, our present experiment shows that overexpression of the dco^3 mutant gene leads to a prominent overgrowth which is opposite to the phenotype by overexpression of a dominant-negative allele dco^h. Thus we predict that Dco protein has an autoregulatory function to repress itself by forming a dimer as reported [Longenecker et al., 1998; Rivers et al., 1998]. Dco^3 may only lose the function of this autoregulation. Probably Dco^3-Dco^+ heterodimer maybe normally active whereas Dco^3-Dco^3 homodimer is hyperactive. Accordingly, increasing the ratio of Dco^3-Dco^3 homodimer to Dco^+-Dco^+ homodimer and Dco^3-Dco^+ heterodimer results in hyperactivation of this gene in total.

Wg- and Hh- Independent Activation of Cell Proliferation

Dco and its mammalian homolog Casein kinase I δ/ε have been known to phosphorylate various signaling molecules in Wg and Hh pathways such as Dsh, Arm, and Ci [Cong et al., 2004; Yanagawa et al., 2002]. As a result, it stimulates ubiquitination-dependent proteolysis of such signaling molecules thereby modifying the signal intensies caused by Wg and Hh. However, Dco is not absolutely required for both pathways and is thought to just modify their intensities. Because the null mutant phenotype of *dco* is not identical to that in mutant for Wg and Hh signaling factors. The mutant for *dco* shows a severe defect in epithelial growth in strong alleles as well as in circadian rhythm defect in mild alleles whereas the mutants for Wg and Hh signaling factors extensively show defects in pattern formation. Thus it suggests that Dco has functions to phosphorylate numerous proteins affecting various phenomena and there should be additional pathways which Dco modifies for regulation of epithelial cell proliferation. Consistently, both Wg and Hh signaling pathways are not influenced by dco^2/dco^3 mutation as shown (Fig. 2 and 4). It is an interesting issue to explore novel factors under the control of Dco in cell proliferation regulation. Our genetic screen to find such factors suggests that the Warts-Hippo-Salvador pathway has a strong relation with *dco* (data not shown).

Significance of Nonautonomous Activation of JNK

Intriguingly, JNK activity in the wild type cells juxtaposed to the *dco*^{2}/*dco*^{3} mutant cell population contributes to removal of the mutant cells. Although it is difficult to explain how an intracellular signaling molecule JNK elicits such a nonautonomous effect, we currently infer two reasons for the mechanism. One is that activation of JNK in cells around the boundary destroys the epithelial integrity of the mutant cell population, which occurs prior to cell death in many cases in the wing cell apoptosis [Gibson and Perrimon, 2005; Hipfner and Cohen, 2003; Shen and Dahmann, 2005]. The other is that JNK might induce activation of a diffusible apoptotic signal. Consistently our preliminary result indicates that local activation of JNK leads to nonautonomous induction of expression of Egr, a *Drosophila* homolog of the mammalian diffusible apoptosis factor TNF-α [Igaki et al., 2002; Moreno et al., 2002b]. In either cases, cell shape change, which has well been documented as a result of JNK activation [Glise et al., 1995; Homsy et al., 2006], may be involved in a part of the whole process of the cell removal.

Generality of Nonautonomous Activation of Apoptosis in Removal of Abnormal Cells

Nonautonomous activation of apoptosis to remove abnormal cell population in the columnar epithelium of the imaginal disc was first observed in the case when decreased or increased level of signaling by the morphogen Dpp occurs to provide inappropriate positional information [Adachi-Yamada and O'Connor, 2002; Milan et al., 2002]. However such nonautonomous type of apoptosis was subsequently observed in a different case when the organ identity was transformed from wing to leg/antenna by forced expression of a master gene *spineless* [Adachi-Yamada et al., 2005]. At this time, the morphogen signaling level was not significantly affected. This suggested that the nonautonomous activation of apoptotic signal along the boundary between mutant and wild type cells is not specifically linked to the alteration of morphogen signaling. Furthermore, we now show an example displaying nonautonomously activated apoptosis of the overgrown cell population by *dco*^{2}/*dco*^{3}, in which both the morphogen signaling and organ identity are not affected. Thus it has unequivocally been established that the nonautonomous activation of apoptosis that begins from the boundary between mutant and wild type cells is one of the common ways to induce cell removal of various abnormal cells for maintenance or restoration of normal morphology of tissues.

Two mechanisms have been proposed to explain how the apoptotic signal is nonautonomously induced around the mutant/wild type boundary. The first hypothesis is focusing on cell-cell interaction via membrane protein(s) such as cell adhesion molecule [Adachi-Yamada et al., 2005; Adachi-Yamada and O'Connor, 2002; Milan et al., 2002]. It was based on the observation that the mutant cells usually showed an assembly with a roundly shaped outline suggesting their mutual affinity. In fact, some LRR transmembrane proteins that can regulate cell adhesion property have been reported to play a role in regulation of nonautonomous apoptosis signals. The second hypothesis dealt with a physical trigger of apoptosis by mechanical stress associated with destroyed epithelial integrity of the mutant cell population [Gibson and Perrimon, 2005]. In the present case of dco^2/dco^3 cells, we do not address which effect mainly contributes to the nonautonomous activation of JNK. However, some morphological traits about the round shape assembly and clear sorting out of the dco^2/dco^3 cell population allowed us to presume that both mechanisms are involved in this case of cell removal.

Experimental Procedures

Generation of *UAS-dco*^{3} Transgenic Fly

Genomic DNA was purified from 20 individuals of the third instar larvae of dco^{P538}/dco^3 transheterozygote using DNeasy Tissue Kit (Qiagen). dco^{P538} is a *dco* allele caused by insertion of the transposon P-element into the 3rd intron of the *dco* gene [Zilian et al., 1999]. An ORF-containing segment, which is within the single 4th exon in this case, was amplified by LA (long and accurate)-PCR technique (TaKaRa, JAPAN). The position of the forward primer was designed to be interrupted by the P-element insertion site in dco^{P538} to avoid amplification from the dco^{P538} allele. Thus the dco^3 ORF was specifically amplified as a 2.1 kb DNA and the 1.7 kb fragment corresponding to its 3' portion was cloned within the UAS-containing plasmid pUAST [Brand and Perrimon, 1993]. Germline transformation was performed by standard method (Bestgene, Inc.).

Generation of dco^2/dco^3 Cell Population in Rescued Cell Background

It has been known that the dco^3-induced overgrowth of the wing disc arises during the third instar larval stages [Zilian et al., 1999]. Thus we planned to rescue the dco^2/dco^3 mutant by mosaically inducing the dco^+ gene before starting the tumorigenic growth. Accordingly, at the second instar larval stage (72 hours after egg laying), heat shock-inducible flip-out reaction was induced by exposure to a temperature of 37 degrees (of Celsius) for approximately 1 hour. The wing discs were dissected at 72 or 96 hours after mosaic induction. The genotype of this mosaic fly was *hs-FLP / + ; act-FRT-stop-FRT-GAL4* (AyGAL4, [Ito et al., 1997]), *UAS-GFP / UAS-dco*$^+$; dco^2 / dco^3. The third chromosome with dco^2 was sometimes recombined with *puc-lacZ* in accordance with the intended use, such as to monitor JNK activity.

Detection of Cell Proliferation

Cell proliferation was detected by incorporation of 5-Bromo-2'-deoxyuridine (BrdU) or observation of phosphorylated Histone-H3. Incorporation of BrdU was performed as described [Rogulja and Irvine, 2005] using BrdU (Sigma), mouse anti-BrdU antibody (Becton Dickinson), and RNase-free DNase I (Promega).

Detection of JNK Activation and Apoptosis

Caspase-3 plays a central role in many types of apoptosis while JNK affects a limited group of apoptosis systems [Kanda and Miura, 2004]. In the *Drosophila* wing disc, activation of JNK usually links to the activation of Caspase-3 [Adachi-Yamada and O'Connor, 2002]. Puc is a protein phosphatase specifically inactivating JNK, and its transcription is induced by a JNK signal thereby making a negative-regulatory loop [Martin-Blanco et al., 1998]. The *puc-lacZ* expression was monitored by immunofluorescence using mouse anti-β-Galactosidase antibody (Promega) and Cy3-labelled anti-mouse Ig (Jackson ImmunoResearch). The Anti-cleaved Caspase-3 antibody (Cell Signaling Technology) specifically binds to the cleaved and activated forms of mammalian Caspase-3 and its *Drosophila* ortholog Drice.

Detection of Cytoskeltal Proteins

F-actin detection by phalloidin and microtubule detection by anti-α-Tubulin antibody were carried out as described [Gibson and Perrimon, 2005; Shen and Dahmann, 2005].

Microscopy

Most of microscopic observations were performed by Leica fluorescence microscope DMRA with the deconvolution system Q550FW. In the case of laser confocal analyses, Nikon Digital Eclipse C1 system was used.

Acknowledgments

We are grateful to B. A. Hay (*UAS-DIAP1*), Y. Hiromi (*AyGAL4*), Y. H. Inoue (balancers), J. Jiang (*ptc* alleles), G. Marques (*ap-GAL4*), E. Martín-Blanco (*puc* alleles), A. Martinez-Arias (*puc* alleles), S. Noselli (*hep*^{r75}), G. Struhl (*hs-FLP*), K. Takahashi (*puc* alleles), D. Yamamoto (*puc* alleles), GETDB in National Institute of Genetics Japan (*sty-GAL4*, *brk-GAL4*), the Bloomington Stock Center (others) for the fly strains; Bestgene Inc. for generating transgenic fly; R. A. Holmgren for anti-Ci antibody; P. ten Dijke for anti-phospho-Mad antibody; and Developmental Study Hybridoma Bank in the University of

Iowa for anti-Wg antibody. This work is supported by grants from Japan Science and Technology Agency and Ministry of Education, Science, Sports and Culture in Japan.

References

Adachi-Yamada, T., Harumoto, T., Sakurai, K., Ueda, R., Saigo, K., O'Connor, M. B. and Nakato, H. (2005). Wing-to-Leg homeosis by spineless causes apoptosis regulated by Fish-lips, a novel leucine-rich repeat transmembrane protein. *Mol Cell Biol*, *25*, 3140-3150.

Adachi-Yamada, T. and O'Connor, M. B. (2002). Morphogenetic apoptosis: a mechanism for correcting discontinuities in morphogen gradients. *Dev Biol*, *251*, 74-90.

Adachi-Yamada, T. and O'Connor, M. B. (2004). Mechanisms for removal of developmentally abnormal cells: cell competition and morphogenetic apoptosis. *J Biochem (Tokyo)*, *136*, 13-17.

Apionishev, S., Katanayeva, N. M., Marks, S. A., Kalderon, D. and Tomlinson, A. (2005). Drosophila Smoothened phosphorylation sites essential for Hedgehog signal transduction. *Nat Cell Biol*, *7*, 86-92.

Bergmann, A., Agapite, J., McCall, K. and Steller, H. (1998). The Drosophila gene hid is a direct molecular target of Ras-dependent survival signaling. *Cell*, *95*, 331-341.

Bissell, M. J. and Radisky, D. (2001). Putting tumours in context. *Nat Rev Cancer*, *1*, 46-54.

Brand, A. H. and Perrimon, N. (1993). Targeted gene expression as a means of altering cell fates and generating dominant phenotypes. *Development*, *118*, 401-415.

Burke, R. and Basler, K. (1996). Dpp receptors are autonomously required for cell proliferation in the entire developing Drosophila wing. *Development*, *122*, 2261-2269.

Campbell, G. and Tomlinson, A. (1999). Transducing the Dpp morphogen gradient in the wing of Drosophila: regulation of Dpp targets by brinker. *Cell*, *96*, 553-562.

Casci, T., Vinos, J. and Freeman, M. (1999). Sprouty, an intracellular inhibitor of Ras signaling. *Cell*, *96*, 655-665.

Cong, F., Schweizer, L. and Varmus, H. (2004). Casein kinase Iepsilon modulates the signaling specificities of dishevelled. *Mol Cell Biol*, *24*, 2000-2011.

del Alamo Rodriguez, D., Terriente Felix, J. and Diaz-Benjumea, F. J. (2004). The role of the T-box gene optomotor-blind in patterning the Drosophila wing. *Dev Biol*, *268*, 481-492.

Diaz-Benjumea, F. J. and Hafen, E. (1994). The sevenless signalling cassette mediates Drosophila EGF receptor function during epidermal development. *Development*, *120*, 569-578.

Duman-Scheel, M., Weng, L., Xin, S. and Du, W. (2002). Hedgehog regulates cell growth and proliferation by inducing Cyclin D and Cyclin E. *Nature*, *417*, 299-304.

Gibson, M. C. and Perrimon, N. (2005). Extrusion and death of DPP/BMP-compromised epithelial cells in the developing Drosophila wing. *Science*, *307*, 1785-1789.

Glise, B., Bourbon, H. and Noselli, S. (1995). hemipterous encodes a novel Drosophila MAP kinase kinase, required for epithelial cell sheet movement. *Cell*, *83*, 451-461.

Gomez-Lazaro, M., Fernandez-Gomez, F. J. and Jordan, J. (2004). p53: twenty five years understanding the mechanism of genome protection. *J Physiol Biochem*, *60*, 287-307.

Gorfinkiel, N., Morata, G. and Guerrero, I. (1997). The homeobox gene Distal-less induces ventral appendage development in Drosophila. *Genes Dev*, *11*, 2259-2271.

Greenberg, R. A. (2005). Telomeres, crisis and cancer. *Curr Mol Med*, *5*, 213-218.

Harvey, K. F., Pfleger, C. M. and Hariharan, I. K. (2003). The Drosophila Mst ortholog, hippo, restricts growth and cell proliferation and promotes apoptosis. *Cell*, *114*, 457-467.

Hipfner, D. R. and Cohen, S. M. (2003). The Drosophila sterile-20 kinase slik controls cell proliferation and apoptosis during imaginal disc development. *PLoS Biol*, *1*, E35.

Homsy, J. G., Jasper, H., Peralta, X. G., Wu, H., Kiehart, D. P. and Bohmann, D. (2006). JNK signaling coordinates integrin and actin functions during Drosophila embryogenesis. *Dev Dyn*, *235*, 427-434.

Huh, J. R., Guo, M. and Hay, B. A. (2004). Compensatory proliferation induced by cell death in the Drosophila wing disc requires activity of the apical cell death caspase Dronc in a nonapoptotic role. *Curr Biol*, *14*, 1262-1266.

Igaki, T., Kanda, H., Yamamoto-Goto, Y., Kanuka, H., Kuranaga, E., Aigaki, T. and Miura, M. (2002). Eiger, a TNF superfamily ligand that triggers the Drosophila JNK pathway. *Embo J*, *21*, 3009-3018.

Ito, K., Awano, W., Suzuki, K., Hiromi, Y. and Yamamoto, D. (1997). The Drosophila mushroom body is a quadruple structure of clonal units each of which contains a virtually identical set of neurones and glial cells. *Development*, *124*, 761-771.

Johnston, L. A. and Sanders, A. L. (2003). Wingless promotes cell survival but constrains growth during Drosophila wing development. *Nat Cell Biol*, *5*, 827-833.

Johnstone, R. W., Ruefli, A. A. and Lowe, S. W. (2002). Apoptosis: a link between cancer genetics and chemotherapy. *Cell*, *108*, 153-164.

Kanda, H. and Miura, M. (2004). Regulatory roles of JNK in programmed cell death. *J Biochem (Tokyo)*, *136*, 1-6.

Longenecker, K. L., Roach, P. J. and Hurley, T. D. (1998). Crystallographic studies of casein kinase I delta toward a structural understanding of auto-inhibition. *Acta Crystallogr D Biol Crystallogr*, *54*, 473-475.

Martin-Blanco, E., Gampel, A., Ring, J., Virdee, K., Kirov, N., Tolkovsky, A. M. and Martinez-Arias, A. (1998). puckered encodes a phosphatase that mediates a feedback loop regulating JNK activity during dorsal closure in Drosophila. *Genes Dev*, *12*, 557-570.

Milan, M., Perez, L. and Cohen, S. M. (2002). Short-range cell interactions and cell survival in the Drosophila wing. *Dev Cell*, *2*, 797-805.

Morata, G. and Ripoll, P. (1975). Minutes: mutants of drosophila autonomously affecting cell division rate. *Dev Biol*, *42*, 211-221.

Moreno, E., Basler, K. and Morata, G. (2002a). Cells compete for decapentaplegic survival factor to prevent apoptosis in Drosophila wing development. *Nature*, *416*, 755-759.

Moreno, E., Yan, M. and Basler, K. (2002b). Evolution of TNF signaling mechanisms: JNK-dependent apoptosis triggered by Eiger, the Drosophila homolog of the TNF superfamily. *Curr Biol*, *12*, 1263-1268.

Orange, J. S. and Ballas, Z. K. (2006). Natural killer cells in human health and disease. *Clin Immunol*, *118*, 1-10.

Perez-Garijo, A., Martin, F. A. and Morata, G. (2004). Caspase inhibition during apoptosis causes abnormal signalling and developmental aberrations in Drosophila. *Development, 131*, 5591-5598.

Perez-Garijo, A., Martin, F. A., Struhl, G. and Morata, G. (2005). Dpp signaling and the induction of neoplastic tumors by caspase-inhibited apoptotic cells in Drosophila. *Proc Natl Acad Sci U S A, 102*, 17664-17669.

Price, M. A. and Kalderon, D. (2002). Proteolysis of the Hedgehog signaling effector Cubitus interruptus requires phosphorylation by Glycogen Synthase Kinase 3 and Casein Kinase 1. *Cell, 108*, 823-835.

Reich, A., Sapir, A. and Shilo, B. (1999). Sprouty is a general inhibitor of receptor tyrosine kinase signaling. *Development, 126*, 4139-4147.

Rivers, A., Gietzen, K. F., Vielhaber, E. and Virshup, D. M. (1998). Regulation of casein kinase I epsilon and casein kinase I delta by an in vivo futile phosphorylation cycle. *J Biol Chem, 273*, 15980-15984.

Rodriguez Dd Ddel, A., Terriente, J., Galindo, M. I., Couso, J. P. and Diaz-Benjumea, F. J. (2002). Different mechanisms initiate and maintain wingless expression in the Drosophila wing hinge. *Development, 129*, 3995-4004.

Rogulja, D. and Irvine, K. D. (2005). Regulation of cell proliferation by a morphogen gradient. *Cell, 123*, 449-461.

Ryoo, H. D., Gorenc, T. and Steller, H. (2004). Apoptotic cells can induce compensatory cell proliferation through the JNK and the Wingless signaling pathways. *Dev Cell, 7*, 491-501.

Shen, J. and Dahmann, C. (2005). Extrusion of cells with inappropriate Dpp signaling from Drosophila wing disc epithelia. *Science, 307*, 1789-1790.

Wu, S., Huang, J., Dong, J. and Pan, D. (2003). hippo encodes a Ste-20 family protein kinase that restricts cell proliferation and promotes apoptosis in conjunction with salvador and warts. *Cell, 114*, 445-456.

Xu, Y., Padiath, Q. S., Shapiro, R. E., Jones, C. R., Wu, S. C., Saigoh, N., Saigoh, K., Ptacek, L. J. and Fu, Y. H. (2005). Functional consequences of a CKIdelta mutation causing familial advanced sleep phase syndrome. *Nature, 434*, 640-644.

Yanagawa, S., Matsuda, Y., Lee, J. S., Matsubayashi, H., Sese, S., Kadowaki, T. and Ishimoto, A. (2002). Casein kinase I phosphorylates the Armadillo protein and induces its degradation in Drosophila. *Embo J, 21*, 1733-1742.

Zilian, O., Frei, E., Burke, R., Brentrup, D., Gutjahr, T., Bryant, P. J. and Noll, M. (1999). double-time is identical to discs overgrown, which is required for cell survival, proliferation and growth arrest in Drosophila imaginal discs. *Development, 126*, 5409-5420.

In: New Developments in Cell Apoptosis Research
Editor: Alan J. Corvin, pp. 213-239
ISBN: 1-60021-453-3

Chapter IX

Response to Cadmium Stress by Neoplastic and Immortalized Human Breast Cells: Evidence for Different Modulation of Gene Expression

Claudio Luparello[1] *, Rosalia Sirchia[1], Laura Paci[1], Vitale Miceli[1], Rosaria Vella[1], Rosaria Scudiero[2] and Francesca Trinchella[2]

[1]Dipartimento di Biologia Cellulare e dello Sviluppo, Università di Palermo, Viale delle Scienze, 90128 Palermo, Italia,

[2]Dipartimento delle Scienze Biologiche - Sezione di Biologia Evolutiva e Comparata, Università di Napoli Federico II, Via Mezzocannone 8, 80134 Napoli, Italia

Abstract

Recently, increasing interest has been focused on the elucidation of the cellular and molecular aspects of cadmium (Cd)-dependent regulation of gene expression and signal transduction pathways in different model system. From available literature data, it appears that the scenario is multifaceted since Cd evokes in cells a number of responses that involve not only death-signalling- but also cellular protective reactions against the toxicity. Concerning breast cancer, there are few studies on the role played by Cd on ER^- (estrogen receptor-negative) human breast cancer cells, that are expected to be insensitive to the metalllo-estrogenic effect exerted by Cd on ER^+ breast cancer cells. Therefore, we wanted to examine the implications and cellular effects of Cd treatment on ER^- MDA-MB231 cells, obtained from mammary adenocarcinoma and endowed with high malignant potential, and, in parallel, on immortalized non-tumoral HB2 breast epithelial

* Corresponding author: clupar@tin.it

cells for comparison. In particular, in the present study we selected the minor concentration of the metal/exposition time capable of inducing an approximately 50% decrease of cell number (IC_{50}), estimated actual cellular uptake of Cd by atomic absorption spectrophotometry, and analyzed the typology of the cytotoxic effect by the evaluation of i) the physical state of the DNA, ii) the membrane localization of phosphatydilserine, and, iii) the expression levels of stress response genes such as those coding for heat shock proteins and metallothioneins, and for factors and enzymes involved in the onset of apoptosis, by conventional and semi-quantitative multiplex PCR. Our results indicate that, although both cytotypes do not show some signs of "classical" apoptosis (oligonucleosomal fragmentation of DNA, annexin binding of plasmalemma), a very different scenario emerges from the evaluation of gene expression levels. Treated non-tumoral HB2 cells exhibited a pattern typical of intoxicated cells with up-regulation of *hsp27*, *hsp70* and *hsc70* only, down-regulation or absence of expression of other genes assayed (e.g *caspase-8* and *metallothioneins*), and accumulation of the spliced form of XBP1 mRNA. On the other hand, Cd treatment of neoplastic cells resulted in the up-regulation of some genes (e.g. *hsp27*, *metallothioneins*) that can be interpreted as an attempt to resist to the toxic effect of the metal, but also in the down-regulation of other genes (e.g other *hsps* and *Bcl-2*) that, occurring in parallel with the up-regulation of death-promoting genes, such as those coding for Bad and some caspases, might account for final cell death which shares some aspects with both oxidative stress and apoptosis-like program.

Introduction

Cadmium (Cd) is an underground mineral extracted as part of zinc deposits, which is widely used in the steel industry, in plastics and as a component of batteries. It is an industrial and environmental pollutant released as air contaminant from fertilisers and, more prominently, in the form of wastewater. Food is at present the main source of daily exposure of humans to Cd, and sometimes drinking water may contain high levels of this heavy metal; also inhalation of smoke from cigarettes may lead to elevated Cd concentrations [1, for review]. Cd is not essential for the human body and has no known useful biological functions and, unlike many metals such as zinc, selenium, magnesium, it is not involved in enzymatic processes. On the other hand, as it is chemically-close to zinc, Cd may bind with high affinity the zinc-binding domains of several metalloproteins, interfering with or preventing zinc-dependent cellular functions from being completed [2]. Cd, which has a very long biological half-life (about 25 years in humans), exerts its toxic effect on a number of tissues such as the liver, kidney and thymus [3]. At cellular and molecular levels, a number of recent reports have provided increasing information on some aspects of Cd-dependent regulation of gene expression and signal transduction pathways. Cd ions are taken up through voltage-sensitive calcium channels of the plasma membrane and accumulate intracellularly due to their binding to cytoplasmic and nuclear material. It is known that, at least, in some of the cell lines tested, low concentrations of Cd stimulate DNA synthesis, enhance the expression of several classes of genes and up-regulate intracellular signalling pathways and cell proliferation, whilst more elevated concentrations are inhibitory and cytotoxic [4-6], although the mechanisms underlying such opposite effects are poorly understood. On the other hand, it is widely demonstrated that the sensitivity to Cd varies from one cytotype to another, being cancer cells mostly hypersensitive to treatment with the metal. It can be argued that exposure to Cd evokes in cells a number of responses that involve not only death signalling reactions but also cellular protective reactions against the toxicity [7]. For example, Cd at low concentrations has been proven to stimulate the transcription of immediate early genes (*c-fos*, *c-jun* and *c-myc*), of the tumor suppressor *p53*, and of genes coding for the synthesis of protective molecules, including *metallothioneins* (*MTs*), *glutathione* and *heat shock proteins (hsp)* [e.g. 8-10]. Also the expression of *TIF3* and *TEF-1δ* was recently found up-regulated in Cd-transformed BALB/c-3T3 cells [11], whilst Shin et al. [12] demonstrated the induction of orphan nuclear receptor *Nur77* expression in lung cells, proposing a correlation between the up-regulation of this apoptosis-related protein and the occurrence of pulmonary toxicity in response to cadmium exposure. Thus, the list of Cd-responsive genes expressed in the different cellular models is still growing [5, 6, 13]. Therefore, it can be hypothesized that cell fate is dependent on which pathway, the protective or the suicidal, exceeds the other, taking into consideration that both function as a cellular defense in the early stages of Cd cytotoxicity. The apoptosis-promoting nature of Cd was demonstrated in experiments involving DNA fragmentation and chromatin condensation in kidney, blood and liver cells [7, 14-16], although the pathway(s) followed still remain uncertain. It has been reported that apoptosis induced by Cd, at least in human myeloid leukemia HL-60 cells, involves cytochrome c release from mitochondria, that, in turn, may cleave and process caspases thorough the activation of Apaf-1 [17]. In addition, a novel type of apoptosis induced by Cd in BA/F3A cells was described, which does not display DNA fragmentation nor up-regulation of reactive oxygen species, and is not affected by *Bcl-2* over-expression [18]. Thus, the molecular basis of the cytotoxic effect(s) exerted by Cd represents a field of research that still requires further and more detailed investigation.

Cd is classified in group I of carcinogens by the International Agency of Research on Cancer [19], and studies reported in the literature indicate that it may play a role in both the initiation of cancer, by activating oncogenes, and in the progression of cancer, by increasing the metastatic potential of existing cancer cells, through mechanisms yet to be fully elucidated but which are likely to involve multiple factors, such as interference with E-cadherin function, followed by disruption of the junctional complexes of epithelial cell sheets and triggering of β-catenin-mediated activation of oncogenes [20]. On the other hand, a number of data demonstrate a Cd-induced tumor suppression when administered at not overtly-toxic doses to tumor-cell bearing immuno-depleted mice [e.g. 21, 22].

As far as breast cancer is concerned, few data are available from the literature and most of them are restricted to the effect of metal administration to the estrogen receptor (ER)-positive cell line MCF-7. In particular, incubation with Cd was shown to stimulate the synthesis of hsp27 [8] and to impair p53 function by inducing phosphorylation of a serine residue and conformational changes in the protein [23, 24]. More interestingly, Garcia-Morales et al. [25] found that Cd exerts an estrogen mimetic effect on ER^+ MCF-7 cells, thereby stimulating the transcription of several estrogen-inducible genes, like *progesterone receptor*, *cathepsin D* and *pS2*, and the increase of cell growth rate; these effects were also observed "in vivo" by Johnson et al. [26]. Taking into account that Cd-breast cancer biological interactions have been poorly investigated, and that to date there are no detailed studies on the role played by Cd on ER-negative human mammary carcinoma cells, that are expected to be insensitive to the estradiol-like modulation by Cd, in the present study we wanted to examine the implications and cellular effects of Cd treatment on ER-negative MDA-MB231 cells, obtained from mammary adenocarcinoma and endowed with high malignant potential, and, in parallel, on immortalized non-tumoral HB2 breast epithelial cells. In particular, we selected the minor concentration of the metal/exposition time capable of inducing a more than 50% decrease of cell number (IC_{50}) and analyzed the typology of the cytotoxic effect by the evaluation of i) the physical state of the DNA, ii) the membrane localization of phosphatydilserine, and, iii) the expression levels of stress response genes such as those coding for hsps and MTs and of genes coding for factors and enzymes involved in the onset of apoptosis (see Tables 1 and 2). The behaviour of an immortalized non-tumoral breast epithelial cell line (HB2) in response to Cd administration was studied in parallel and compared to that of the neoplastic model system under study.

Table 1. Stress response genes studied in the present work

Gene product	Function
hsp27	(27 kDa- *h*eat *s*hock *p*rotein) chaperonin endowed with protective activity against oxidative stress [79].
hsp70/hsc70/ hsp75 grp78	(70/75 kDa-*h*eat *s*hock (*c*ognate) *p*rotein; 78 kDa *g*lucose-*r*egulated stress *p*rotein) members of a chaperonin family whose over-expression protects cells from stress-induced apoptosis [46]. Hsp/hsc70 are cytosolic and nuclear whereas hsp75 is localized in mitochondria and grp78 is endoplasmic reticulum-associated.
hsp90 α/β	(90 kDa-*h*eat *s*hock *p*rotein) isoforms of a cytosolic chaperonin involved in signal transduction, protein folding and cell proliferation, showing both pro- and anti-apoptotic effects [46, 80].
hsf-1	(*h*eat *s*hock *f*actor-1) transcription factor for heat shock genes activated by heat, and other forms of environmental and chemical stress [81].
hsf-2	(*h*eat *s*hock *f*actor-2) transcription factor for heat shock genes controlling development and differentiation-specific gene expression [82].
EDEM	(*E*ndoplasmic reticulum *d*egradation *e*nhancing α-*m*annosidase-like protein) acceptor of misfolded glycoproteins whose expression is induced by various types of endoplasmic reticulum stresses [35].
XBP1	(*X*-box *b*inding *p*rotein) transcription factor that activates the unfolded protein response only when it is translated from the spliced form of its mRNA [35].
MTIA-L, MTIIA, MTIII, MTIV	(metallothionein isoforms) 6-7-kDa metal binding proteins that are induced upon exposure to heavy metals and participate in heavy metal regulation and detoxification. MTs I and II are the most widely expressed and are regulated coordinately in all tissues, MT III is brain specific, MT IV is expressed mainly in squamous epithelium [83].

Table 2. Apoptosis-related genes studied in the present work

Gene product	Function
Bcl-2	anti-apoptotic 26/30 kDa-component of nuclear, mitochondrial and endoplasmic reticulum membrane; founding member of Bcl-2 super-family, binds Bax and Bak [84].
Bcl-X	member of Bcl-2 super-family; L form of 29-kDa inhibits apoptosis, S form of 21 kDa promotes apoptosis; binds Bax and Bak [85].
Bad	5 kDa-pro-apoptotic member of Bcl-2 super-family; binds Bcl-2 and Bcl-X_L [85].
Bak	23 kDa-pro-apoptotic member of Bcl-2 super-family; binds Bcl-2, Bcl-X and E1B 19K [85].
Bax-α	21 kDa-pro-apoptotic member of Bcl-2 super-family; binds Bcl-2 and Bcl-X_L [85].
Rip-1	(*R*eceptor-*i*nteracting *p*rotein-1) Stress-sensor protein of 74 kDa which activates transcription factors such as NF-κB or AP-1, or stimulates cell death [77].
Tiar	(*TIA*-1-*r*elated protein) pro-apoptotic 42 kDa-RNA-binding factor [86].
Cas	(*C*ellular *a*poptosis-*s*usceptibility protein) pro-apoptotic 100 kDa-human homologue of yeast chromosome-segregation protein (CSE-1) [87].
Waf-1	pro-apoptotic 21 kDa-protein which mediates the tumor-suppressor properties of p53 [88].
p53	pro-apoptotic 53 kDa-chromatin- and nuclear matrix-associated factor [89].
Dapk	(*D*eath-*a*ssociated *p*rotein *k*inase) pro-apoptotic 160 kDa-Ca^{++}-calmodulin-regulated serine/threonine kinase [90].
Caspase-2	(long apoptotic initiators [91]. form),-8,-9, -10
Caspase-3, -6, -7	apoptotic executioners [91].
Caspase-1, -4,	cytokine processors, involved in some forms of apoptotic death [91]. -5, -13
Caspase-14	small prodomain-caspase expressed in embryonic tissue and involved in terminal differentiation of keratinocytes and apoptosis of breast cancer cells [92].

Materials and Methods

Cell Cultures and Treatments

The MDA-MB231 breast cancer cell line was routinely grown in RPMI 1640 medium plus 10% foetal calf serum (FCS; Invitrogen, Carlsbad, CA/USA), 100 U/ml penicillin, 100 μg/ml streptomycin, and 2.5 mg/L amphotericin B, at 37°C in a 5% CO_2 atmosphere. The HB2 breast epithelial cell line (courtesy of Cancer Research UK; [27]) was routinely grown in the same culture conditions with the exception of high glucose DMEM as medium and 5 μg hydrocortisone/ml (Sigma, St.Louis, MO/USA) and 10 μg bovine insulin (Sigma)/ml as further supplements. Both cell lines were plated in 24-well plates at the concentration of 2.5 x 10^4 cells/well in culture medium containing 1, 5, 10, 50 or 100 μM $CdCl_2$ and grown for 24, 48, 72 or 96 h; at the end of the incubation, cells were trypsinized and counted in a Bürker chamber. The data obtained were used to generate a dose/time/response curve and to determine the IC_{50}. In light of the results obtained, in all the subsequent experiments MDA-MB231 and HB2 cells were plated in the presence of 5 (for MDA-MB231) or 50 (for HB2) μM $CdCl_2$ and grown for 96 (for MDA-MB231) or 24 h (for HB2).

Evaluation of Cadmium Uptake

Control and treated MDA-MB231 and HB2 cells were grown for the appropriate time, trypsinized, collected by low-speed centrifugation and weighted. Cadmium content of the cell pellets was *deterined* by a graphite-furnace-equipped atomic absorption spectrophotometer (Perkin Elmer) as described by Biagioli et al. [28].

DNA Extraction and Electrophoretic Analysis

The occurrence of DNA fragmentation was evaluated by electrophoretic analysis following two different experimental protocols. *Assay 1* (according to Luparello et al. [29]): at the end of the incubation, control and treated cells were scraped from the plates, lysed in 100 μl of 45 mM Tris-borate, pH 8, containing 0.25% Nonidet P-40 substitute (Fluka, Buchs/CH) and 10 mM EDTA, and treated for 1 h at 50°C with 1 mg RNase A/ml (Fluka) and then for 1 h at 50°C with 1 mg Proteinase K/ml (Fluka). as described by. *Assay 2* (adapted from Roccheri et al. [30]): at the end of incubation, trypsinized control and treated cells recovered by centrifugation were lysed in 500 μl of 10 mM Tris-HCl, pH 7.5, containing 10 mM EDTA and 0.2% Triton X-100, for 15 min. at 4°C, centrifuged at 12,000 rpm for 10 min. at 4°C to eliminate cell debris, and the lysates extracted using the High Pure PCR Product purification kit (Roche, Mannheim/D) according to manufacturer's instructions.

In order to improve band resolution in a low-agarose-concentration gel, all samples were submitted to voltage gradient-gel electrophoresis (VGGE) in ethidium bromide-containing 1.2% agarose gel with TAE buffer for 5 h at 60 V [30, 31].

Microscopic Analysis of Phosphatdylserine (PS) Externalization

To examine the possible occurrence of PS translocation from the inner to the outer leaflet of the plasma membrane, one of the hallmarks of apoptosis, we used the Annexin V-EGFP Apoptosis Detection kit (Biovision, Mountain View, CA/USA) according to the manufacturer's instructions. Cells were plated in 6-well dishes and treated with $CdCl_2$ as described in the preceding paragraph; at the end of the incubation, the preparations were observed under an Olympus BX50 microscope equipped with a CoolSNAP-Pro Color digital camera (Media Cybernetics, Silver Springs, MD/USA) and images generated by either EGFP-annexin V or propidium iodide, when present, recorded using Image Pro Plus software (Media Cybernetics). Untreated cells were tested in parallel.

RNA Extraction and Reverse Transcription

Isolation of total RNA from monolayers of control and treated MDA-MB231 and HB2 cells grown in flasks was carried out with TRIzol reagent (Invitrogen), according to manufacturer's instructions. Before the reverse transcription, 1-2 μg of total RNA were treated with RQ1 RNase-free DNase (Promega, Madison,WI/USA). The cDNAs were synthesized in the presence of random 6-mer primers, using SuperScript II reverse transcriptase (Invitrogen).

Conventional and Semi-quantitative "Multiplex" (SM) Polymerase Chain Reaction (PCR)

PCR analysis was performed using 2.5 μM of appropriate sense and antisense primers, 1 U RedTaq DNA polymerase (Sigma)/μl, 200 μM each of dNTPs, and 1 μl of the cDNA template obtained from total RNA. All the primers used were from [29, 32, 33] except for those for metallothioneins [34], EDEM and XBP1 [35]. The thermal cycle used was a denaturation step of 94°C for 3 min., followed by 35 cycles of 94°C for 30 sec., the appropriate annealing temperature for 30 sec., and 72°C for 30 sec. A final extension of the product was performed for 5 min. at 72°C. PCR products were analysed by 2% agarose gel electrophoresis and visualized by ethidium bromide staining under UV light. RT-PCR quantitative measurement of the splicing of XBP1 transcription factor mRNA and evaluation of percentage activation of XBP1 was accomplished according to Shang [35]; the intensities of the bands corresponding to the spliced and the unspliced forms were evaluated with SigmaScan software (SPSS, USA).

SM-PCR was carried out as previously reported [29, 32, 33, 36, 37]. Essentially, the cDNA species of interest was co-amplified with cDNA from 18S rRNA over a range of cycles, followed by 2% agarose electrophoresis and ethidium bromide stain. Cycle profile was a denaturation step of 94.5°C for 3 min., followed by cycles of 94°C for 30 sec., 57°C for 30 sec., 72°C for 30 sec. and a final extension of the product for 5 min. at 72°C. Cycles were limited to the minimum necessary for detection. The intensities of the bands of interest, evaluated with SigmaScan software, were normalized for those of 18S, and plotted as a function of cycle number. Exponential regression equations fitted to the curves were used to calculate the number of cycles necessary to reach a normalized intensity threshold value = 1 for each sample. The relative difference in abundance between two samples was taken as 2^n where n is the difference between the numbers of cycles required by the samples to reach the threshold. Two different RNA preparations from each experimental condition were pooled to make more significant the differences between the expression levels, if any.

Results

In a first set of experiments, we checked the effect of the incubation of tumoral and immortalized cells with increasing concentrations of Cd for different times. The panel in Figure 1 shows the dose- and time-response curves of $CdCl_2$ on the viability of MDA-MB231 and HB2 cells; the graphs indicate a general higher sensitivity of non-tumoral HB2 cells to metal treatment. On the basis of the results obtained and the evaluation of the IC_{50} of the metal upon treatment for 24, 48, 72 and 96 h, we opted for the incubation with 5 μM $CdCl_2$ for 96 h (for MDA-MB231 cells) and with 50 μM $CdCl_2$ for 24 h (for HB2 cells) in the following cellular and molecular assays. The uptake of Cd by MDA-MB231 and HB2 cells in the different experimental conditions was measured by absorption spectrophotometry; the results obtained showed that the average metal uptake by HB2 cells cultured in the presence of 50 μM $CdCl_2$ was 164.6 pmoles/mg of cells already at 24 h from incubation, remaining constant thereafter, whilst for MDA-MB231 cells cultured in the presence of 5 μM $CdCl_2$ the average amount of incorporated Cd was found to be 21.3, 146.8, 144.11 and 166 pmoles/mg of cells at 24, 48, 72 and 96 h from incubation, respectively. Thus, despite the differing times of incubation and concentrations of exogenously-added metal, Cd uptake by the two cytotypes in the conditions chosen on the basis of IC_{50} was comparable.

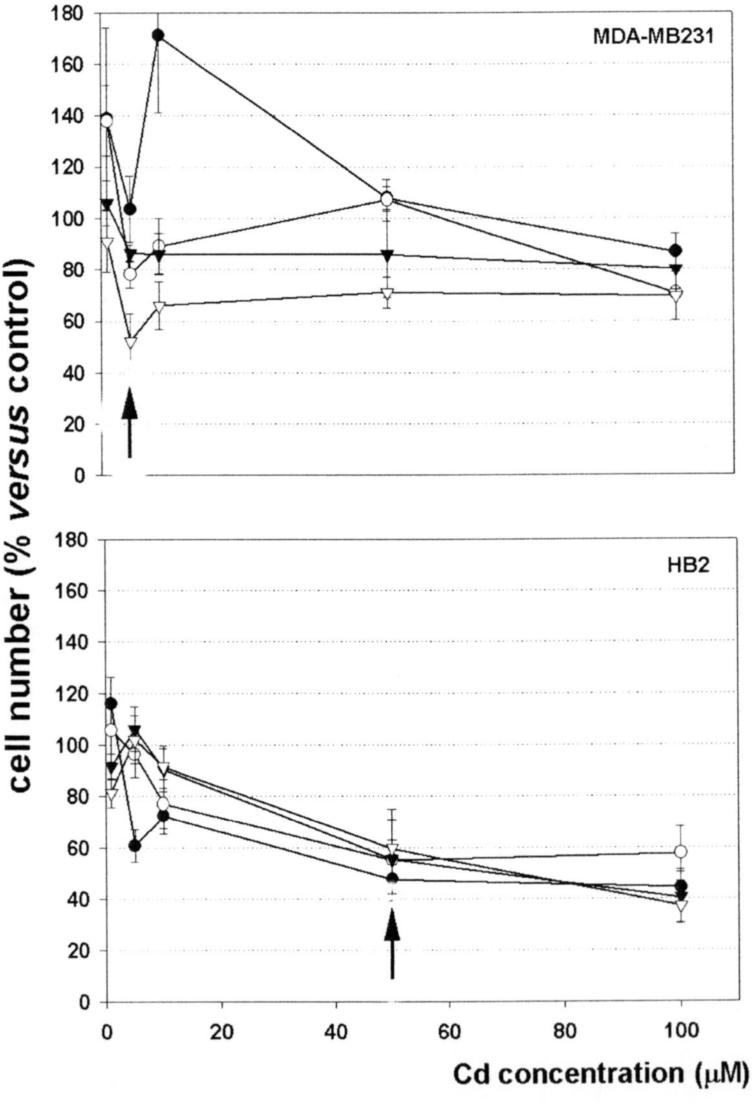

Figure 1. Cell number determination upon $CdCl_2$ treatment for various time courses. Cells were treated with different concentrations (1-100 μM) of $CdCl_2$ for 24 (▽), 48 (○), 72 (▼) and 96 h (●). Data are presented as means of ± s.e.m. for six replicate measurements. The percentage cell number is expressed relative to control, which is defined as 100%. The arrows point to the IC_{50} of $CdCl_2$ and the times of incubation chosen for the following experiments.

In order to check whether the Cd-promoted death event of the two cell lines under study displayed peculiar signs of "classical" apoptosis, we examined whether exposure to Cd resulted in oligonucleosomal fragmentation of DNA and/or externalization of phosphatydilserine. As shown in Figure 2, analysis by VGGE of the DNA extracted in two different ways demonstrated that incubation of both cell lines with $CdCl_2$ for different time lapses did not result in a "ladder"-like electrophoretic pattern indicative of the occurrence of oligonucleosomal fragmentation. We then submitted MDA-MB231 and HB2 cells to Cd treatment and subsequent phosphatydilserine detection assay by labeled annexin binding. The panel of micrographs in Figure 3 clearly indicates that no "flip-flop" of the phospholipid occurred in treated cells since no fluorescence was observable on cell surface; as expected about 50% of cells examined were positive for propidium iodide counterstaining, thus further indicating that, at the times and Cd concentrations chosen for the two cell lines, about half of the cell population present was severely damaged.

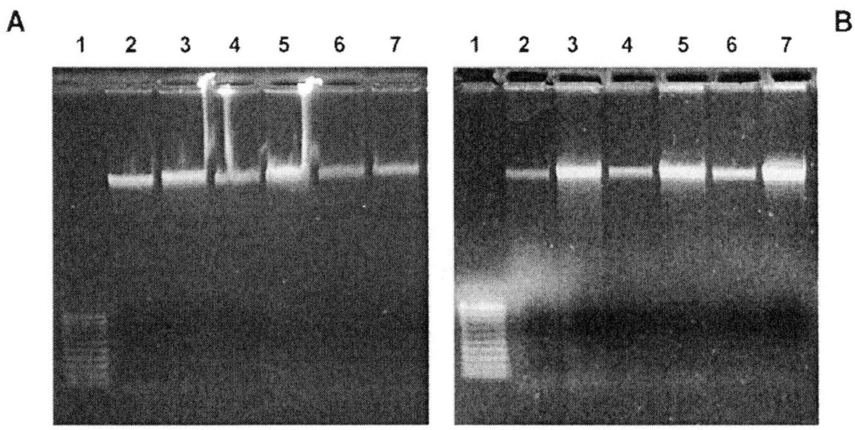

Figure 2. Representative VGGE of the DNA isolated from HB2 (A) and MDA-MB231 cells (B), cultured in control conditions (2, 4, 6) and in the presence of Cd (3, 5, 7) for 24 (2, 3), 72 (4, 5) and 96 h (6, 7). Lane 1 = 100 bp size marker.

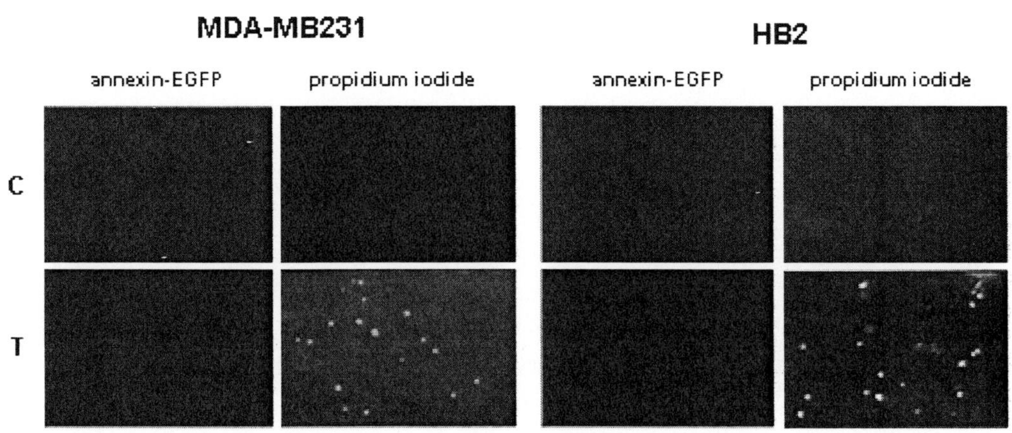

Figure 3. Representative fluorescence micrographs of control (C) and Cd-treated (T) tumoral and immortalized cells, stained with Annexin V-EGFP and propidium iodide.

In a following set of experiments, cDNA preparations from control and Cd-treated MDA-MB231 and HB2 cells were submitted to conventional- and SM-PCR in the presence of primers specific for the cDNAs of various stress response genes, MT isoforms, apoptosis-agonist and –antagonist factors and caspases. The panels in figure 4, representative of three independent experiments, show the results of conventional PCR assay. A positive signal was found for all the stress response genes in both cell lines with or without exposition to Cd, except for *grp78* that was absent in preparations from control and treated HB2 cells, and for *hsp90β* that was expressed only in untreated HB2 cells. Concerning MT expression by MDA-MB231 cells, amplification bands were observed for *MTIA*, *MTIE*, *MTIF*, *MTIG* and *MTIIA* in control and treated samples, whilst a positive signal for *MTIL* was evident only after exposition to Cd. Interestingly, although control HB2 cells appeared positive for *MTIE*, *MTIF*, *MTIG*, *MTIL* and *MTIIA*, no amplification band for any MT isoform was found in cDNA preparations from Cd-treated samples. Triggering of the unfolded protein response (UPR), conceivably promoted by accumulation of unfolded or misfolded proteins in the endoplasmic reticulum (ER), was checked by analysis of *EDEM* transcription and of the accumulation of the spliced form of XBP1 mRNA, which, being a substrate of UPR transducer IRE1, an ER-transmembrane protein kinase/RNase, can be used to monitor enzyme activation [35]. As shown in the panel, *EDEM* amplicons were found in all the experimental conditions tested; in addition, no difference between treated and control MDA-MB231 cells in the extent of XBP1 splicing was observed (mean % = 48.8 and 47.5, respectively), whilst, on the other hand, Cd treatment of HB2 cells resulted in an increase of the percentage of the spliced form of XBP1 from about 30.3 to about 39.2%, suggesting the occurrence of stress at the ER level in the non-tumoral cytotype. Concerning the expression of apoptosis factors and caspases, samples of the two cell lines, both control and treated, gave positive signals for *Tiar*, *Cas*, *p53*, *Dapk* and *Waf-1* among the modulators, and for *caspase-6* and *-8*; moreover, amplicons for *Bcl-2, Bad* and *caspase-1*, *-2*, *-3*, *-7* and *-9* were obtained from preparations of both control and treated MDA-MB231 cells, whilst *RIP-1* amplification band was observed only in treated MDA-MB231 samples. Expression of *caspase-5* was found in control MDA-MB231 cells only, whilst positive signals for *Bad* and *caspase-2*, *-5* and *-7* were obtained from cDNA preparations of control HB2 cells only. No amplification bands were observed for *MTIB*, *MTIH*, *MTIII*, *MTIV*, *Bcl-X_L*, *Bcl-X_S*, *caspase-4*, *-10*, *-13* and *-14* in preparations from control and treated MDA-MB231 and HB2 cells, at least under the experimental conditions used. The cDNA preparations were then submitted to triplicate SM-PCR to compare the expression levels of those genes whose amplicons were detected in preparations of both control and treated cells, selected on the basis of previous results. As shown in Figure 5, we found that incubation of HB2 cells with 50 μM $CdCl_2$ for 24 h was able to promote the up-regulation of only *hsp27* (+2.14 folds), *hsp70* (+12 folds), and prominently *hsc70* (+180 folds); the expression of *hsp90α* (-3.3 folds), *caspase-8* (-4.96 folds) and, more notably, *Dapk* (-23.3 folds) and *hsf1* (-40.8 folds) were down-regulated. All the standard errors of the mean (s.e.m) were less than 0.05. No significant difference was found for the expression levels of *hsf2*, *hsp75*, *EDEM*, *caspase-6*, *Cas*, *Tiar*, *p53* and *Waf-1* between control and Cd-treated HB2 cells (not shown). On the other hand, as shown in Figures 6 and 7, incubation of neoplastic MDA-MB231 cells with 5 μM $CdCl_2$ for 96 h was able to promote the up-regulation of *hsf1* (+7.5 folds), *hsp27* (+8.7 folds), *MTIA* (+2.34

folds), *MTIF* (+3.65 folds), *Waf-1* (+10.4 folds), *caspase-2* (+3 folds), *-7* (+15 folds), *-8* (+9.25 folds), *-9* (+4.7 folds), and, more prominently, *Dapk* (+55 folds), and *caspase-1* (+106 folds) and *-6* (+31.3 folds), whilst *MTIG* (-18.8 folds), *hsc70* (-4.9 folds), *hsp75* (-9.5 folds), *hsp90β* (-2.57 folds) , and, massively, *grp78* (-54.2 folds) and *Bcl-2* (-53 folds) were down-regulated. All the s.e.m were less than 0.05. No significant difference was found for the expression levels of *MTIE, MTIIA, hsf2, hsp70, hsp90α, EDEM, p53, Bad, Tiar* and *caspase-3* between control and Cd-treated MDA-MB231 cells (not shown).

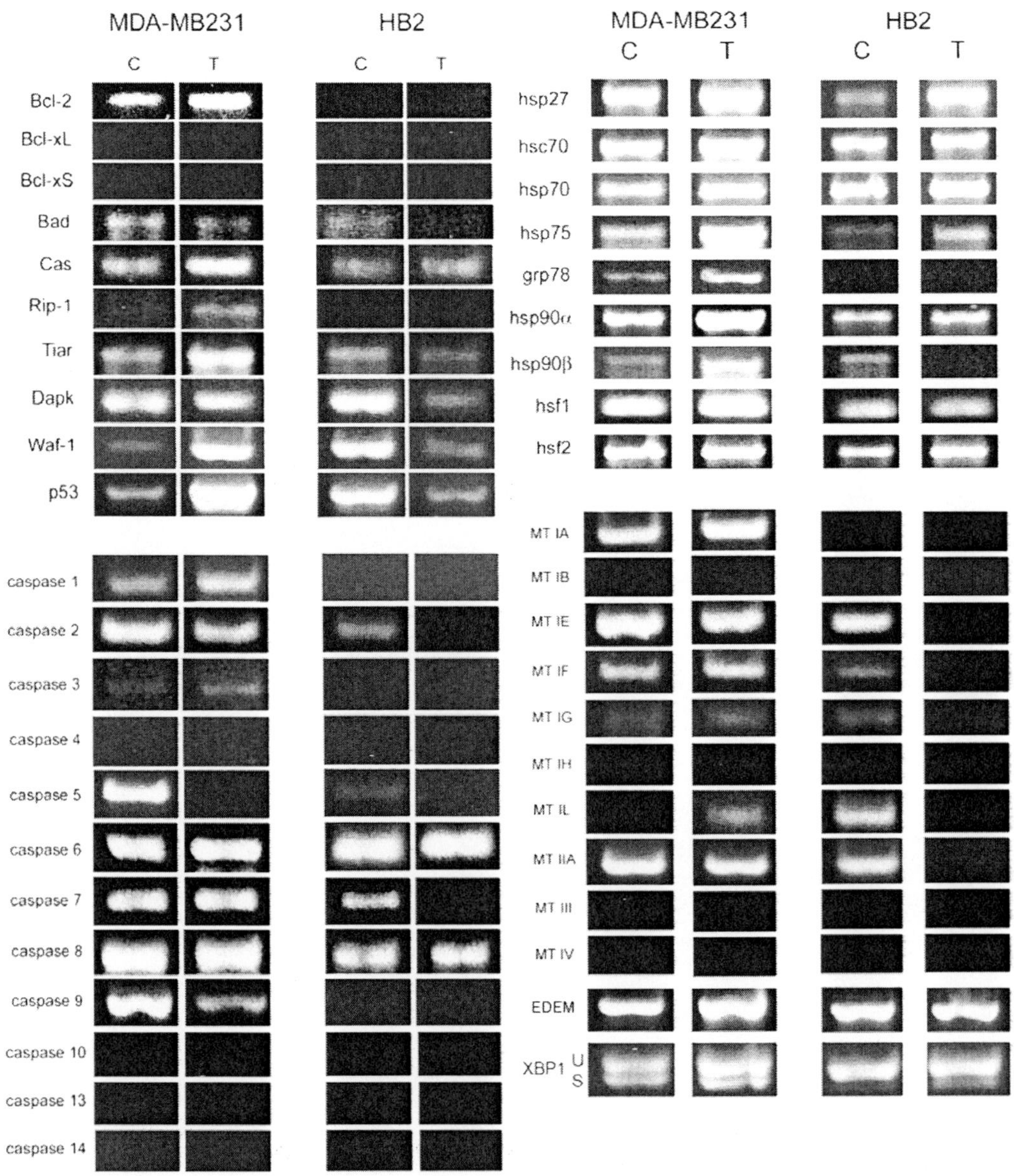

Figure 4. PCR analysis of expression of genes coding for stress response proteins, metallothioneins, apoptotis-related factors and caspases by control (C) and Cd-treated (T) HB2 and MDA-MB231 cells.

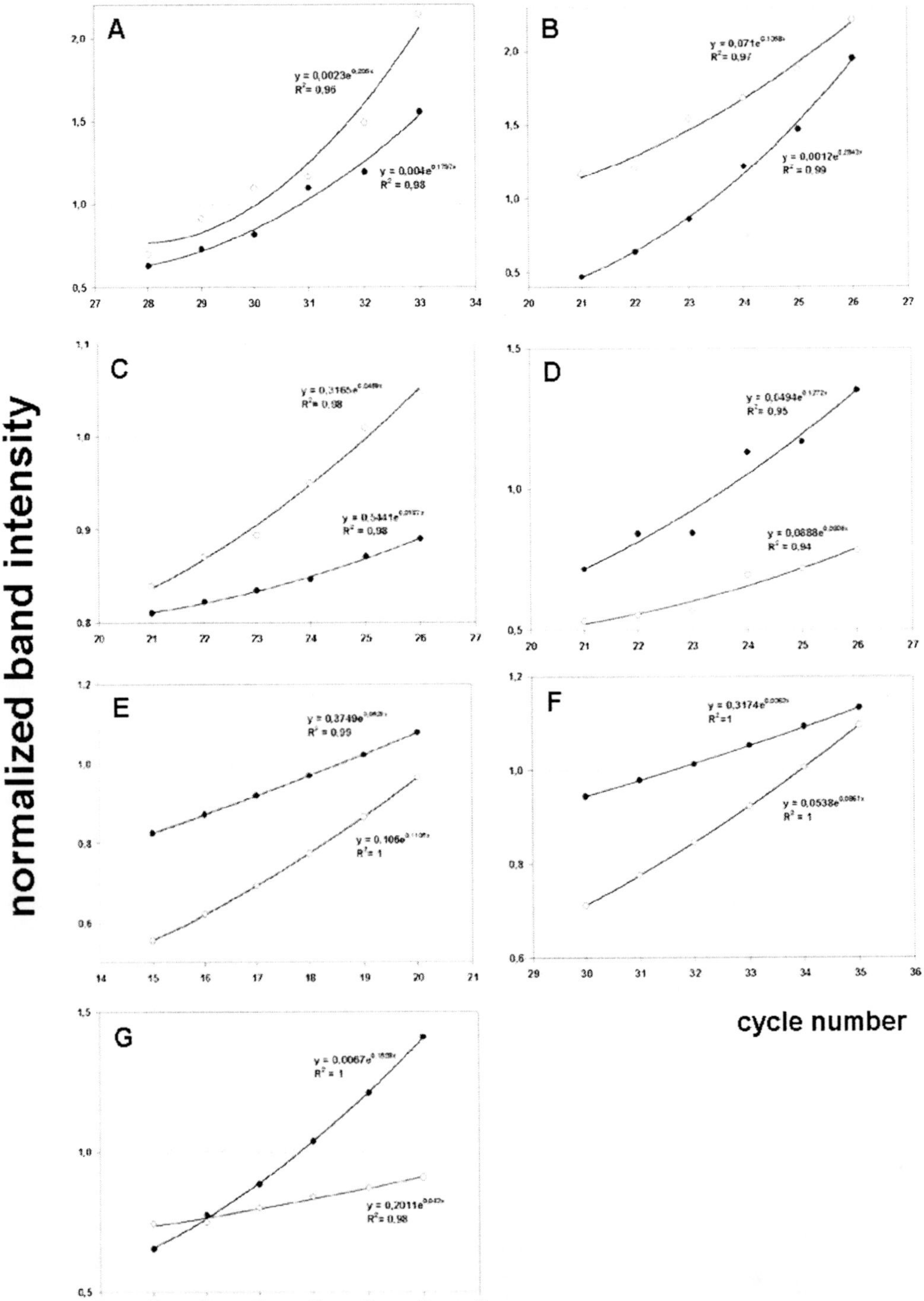

Figure 5. SM-PCR for hsp27 (A), hsp70 (B), hsc70 (C), hsf-1 (D), hsp90α (E), caspase-8 (F) and Dapk (G). Representative plots of normalized data vs. cycle number fit with an exponential curve for control (●) and Cd-treated (○) HB2 cells.

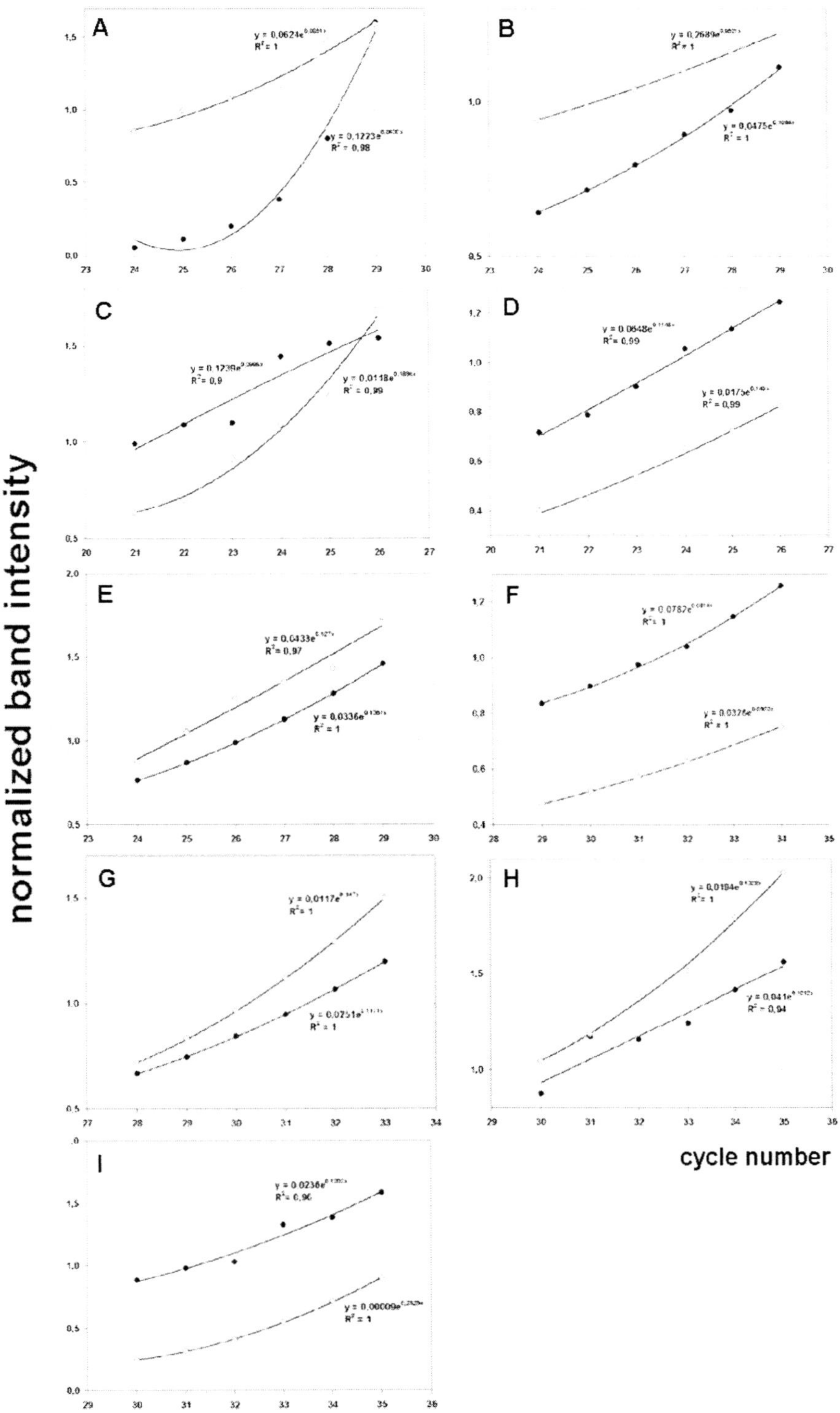

Figure 6. SM-PCR for hsp27 (A), hsf-1 (B), hsc70 (C), hsp75 (D), hsp90β (E), grp78 (F), MT IA (G), MT IF (H) and MT IG (I). Representative plots of normalized data vs. cycle number fit with an exponential curve for control (●) and Cd-treated (○) MDA-MB231 cells.

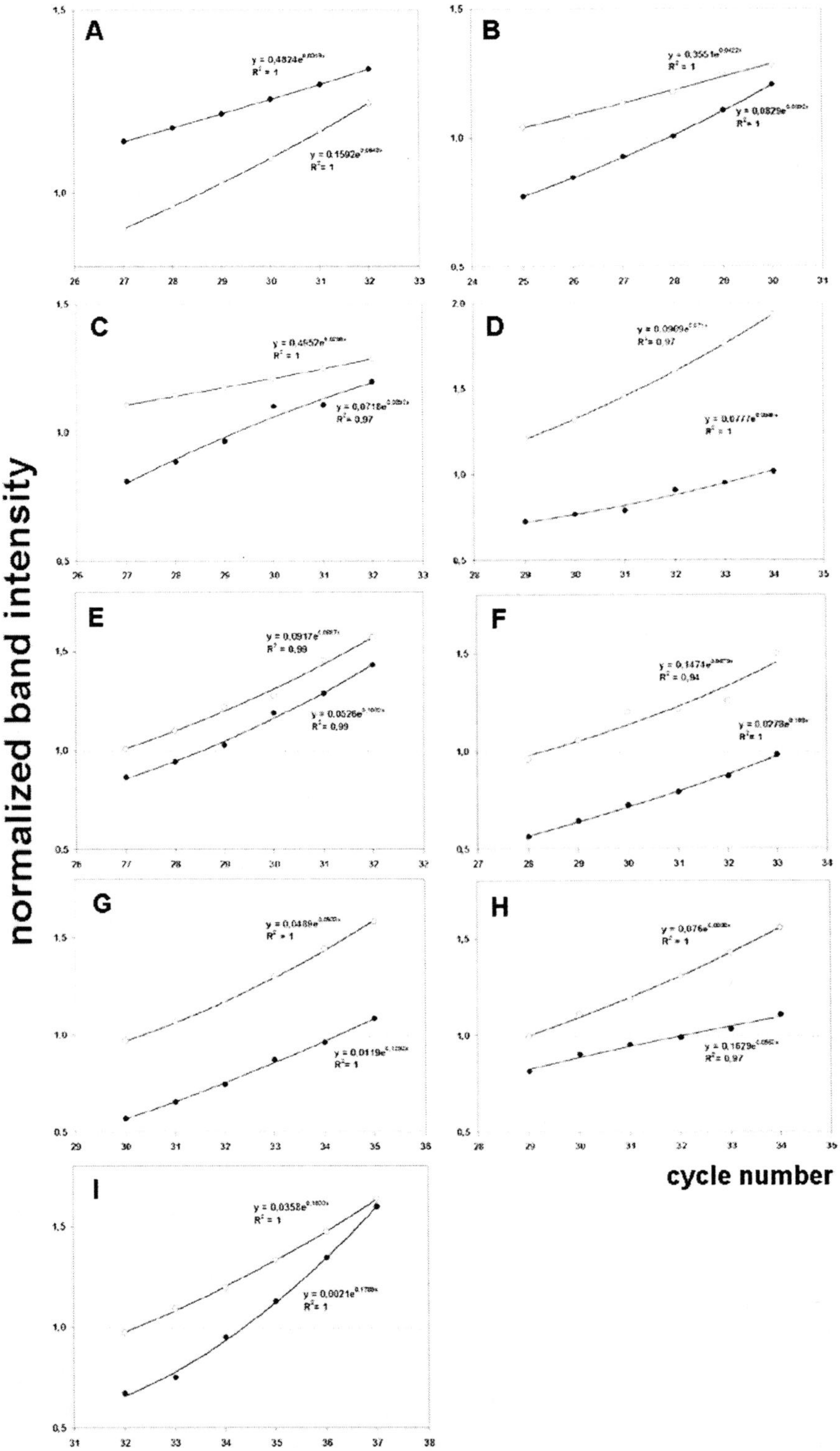

Figure 7. SM-PCR for Bcl-2 (A), Waf-1 (B), Dapk (C), caspase-1 (D), caspase-2 (E), caspase-6 (F), caspase-7 (G), caspase-8 (H) and caspase-9 (I). Representative plots of normalized data vs. cycle number fit with an exponential curve for control (•) and Cd-treated (○) MDA-MB231 cells.

In order to study the kinetics of Cd-dependent up- or down-regulation of gene expression, preparations of control and treated MDA-MB231 cells after 24, 48 and 72 h of incubation were submitted to SM-PCR for *hsf1*, *hsp27*, *hsc70*, *hsp75*, *grp78*, *hsp90β*, *MTIA*, *MTIF*, *MTIG*, *Bcl-2*, *Waf-1*, *Dapk*, and *caspase-1*, *-2*, *-6*, *-7*, *-8*, *-9*. As shown in Figure 8, up-regulation of *MTIA*, *MTIF*, *hsp27*, *hsf1*, *Waf-1*, *Dapk* and *caspases* are late events that take place between 72 and 96 h from cell plating in the presence of Cd, whereas down-regulation of *hsp90β*, *Bcl-2*, *hsc70* and *hsp75* can be considered as earlier Cd-dependent event, the first two occurring between 24 and 48 h of treatment, whilst the last two taking place subsequently, i.e between 48 and 72 h of treatment.

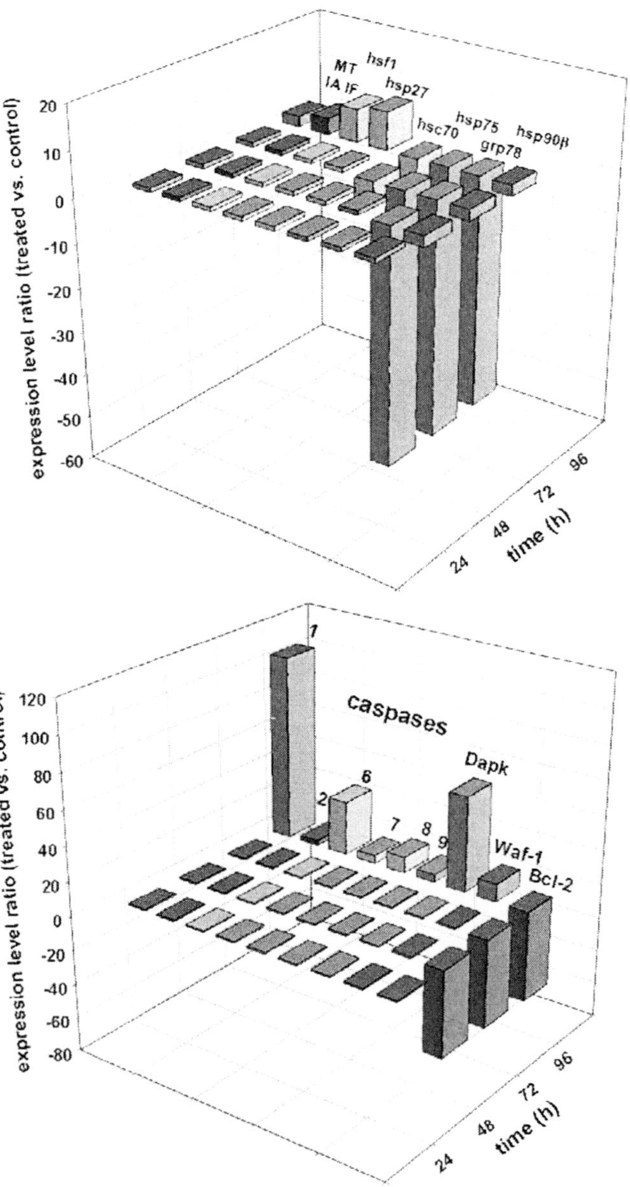

Figure 8. Histogram showing the time-dependent modulation of gene expression levels in Cd-treated MDA-MB231 cells *versus* controls, as resulted from SM-PCR assays.

Conclusion

It is now widely-accepted that Cd is able to exert various biological effects, both cytotoxic, which are counteracted by cells with the production of specific stress response proteins that, in turn, in some cases determine the switching-on of apoptotic mechanisms, and carcinogenetic, which are responsible of neoplastic transformation of cells. The complexity and diversity of the molecular events underlying cell-metal interactions are responsible of the data fragmentation and poor knowledge of the intracellular pathways involved in each experimental situation; nonetheless, significant indications are surely emerging from the evaluation of gene expression in the different cell model systems in the presence of Cd. Here, we have submitted to Cd treatment two cell lines from human breast epithelium, one neoplastic and the other immortalized and non-tumoral, and shown that, although the resulting cell loss was comparable and neither of the cytotypes displayed some signs of "classical" apoptosis, such as phosphatydilserine externalization and oligonucleosomal fragmentation of DNA, from the analysis of expression levels of stress response- and apoptosis-related genes a very different scenario could be disclosed, which deserves to be examined. As a general remark, taking into consideration the collective results from dose/time response curves and mRNA arrays, and other preliminary data showing the hyperpolarization of mitochondrial transmembrane potential (MTP) as well as the accumulation of reactive oxygen species (ROS) in Cd-treated MDA-MB231 cells, and, conversely, MTP collapse and absence of detectable ROS in Cd-treated HB2 cells (manuscript in preparation), we can assume that the immortalized non-neoplastic cytotype is more sensitive to metal damage and unable to activate defense mechanisms with respect to its tumoral counterpart. Preliminary communication to congress concerning the above-mentioned data may be downloaded from http://www.unipa.it/~dipbio/congresso2005/

Sirchia_Rinaldi_Luparello_dbcs_2005.pdf. Further studies will be required to define a precise model for the specific molecular mechanisms through which Cd accomplishes its cellular effects; however, the data here presented allow the following comments.

Expression of MTs

MTs are metal-binding proteins that are involved in cellular homeostatic control and regulation of trace elements, also acting as superoxide and hydroxyl radical scavengers thereby conferring protection against oxidative stresses. It is also postulated that MT can interact with and modulate the activity of other proteins controlling cell survival/death, such as p53 and NF-κB; in humans, ten different functional isoforms have been recognized [38]. As expected, MT III, which is brain-specific, and MT IV, limited to squamous epithelial cells, were not found in our experimental systems; on the other hand, the absence of MT IB message in cancer cells was consistent with data by Jin et al. [39], who found all breast tumor samples examined negative for this expression. The arrays of mRNA present in tumoral and immortalized cells were very similar, with the exception of MT IA which was found only in MDA-MB231 cell samples; interestingly, also ER^+ PMC42 breast cancer cells failed to express this MT isoform [40] and, therefore, our results further support the hypothesis that

MT IA transcription is likely to be related to ER status. In accordance with data by Fan and Cherian [41], Cd treatment of tumor cells resulted in the induction of MTs; in particular, we produced data at MT isoform level, demonstrating the up-regulation of *MT IA* and *IF*, and the switching-on of *MT IL*, whilst *MT IG* was conversely down-regulated. *MT IA* was found to be induced in Cd-, Zn- and Cu-treated human proximal tubule cells by exposure to those metal concentrations which elicited a loss of cell viability [42] and to protect embryonic kidney cells from damage caused by 10 μM Cd a bit more efficiently than MT II and MT III [43]. Thus, our results provide additional evidence that up-regulation of this particular isoform may be considered as a cellular defense reaction specifically set up against Cd intoxication. *MT IL* up-regulation was demonstrated in HeLa cells exposed to a non-lethal dose of $CdSO_4$ [44] whilst *MT IF* expression was found somehow related to the degree of differentiation of MDA-MB231 cells [45]; apart form these reports, the scarcity of literature data concerning human MT IF, IG and IL does not help us to make any assumption on the biological significance of the modulation of their expression in our experimental system, especially for what regards the prominent decrease of MT IG message. Nonetheless, their differential regulation may be conceivably related to the distinct biological features and roles that each of the isoforms is supposed to have acquired in humans following the duplication of the parental *MT I* gene; in addition, the specific involvement of some of the up-regulated MT isoforms in the protection against the oxidative stress arising from ROS production after Cd treatment of MDA-MB231 cells (manuscript in preparation) cannot be ruled out. Noteworthy, a completely-different response to Cd stress was observed with HB2 cells, where incubation with the heavy metal resulted in the disappearance of all the MT isoform amplicons present in the control samples. Obviously, the failure to express *MT* genes renders non-tumoral cells much less protected against injuries, thereby justifying the more pronounced sensitivity of this cytotype to Cd toxicity observed in the present work; moreover, since MTs are linked with cell proliferation [38], the absence of *MT* expression, more specifically *MT IIA*, could seemingly be considered as an event *per se* responsible, at least in part, of the growth arrest of HB2 cells when treated with $CdCl_2$.

Expression of Stress Response Genes

Cd administration to tumoral and immortalized breast cells generated a very different pattern of expression of *hsps* and the diversified occurrence of splicing events of XBP1 mRNA. Massive up-regulation of *hsc70/hsp70* was, in fact, the main effect observable in HB2 cells, coupled with the marked down-regulation of *hsf1* and the deficiency of *hsp90β* expression; minor effects were exerted on *hsp27* and *hsp90α*. Over-expression of the two members of the 70 kDa-hsp family is a typical stress response elicited by cells in the attempt to become more resistant to the effect of cytotoxic agents [46]; in particular, by analogy with the data of Su et al. [47] on heart myoblasts, *hsc70* considerable up-regulation could be interpreted as an auxiliary specific protective response against oxidative insult which sets up in place of that by MTs, whose expression appears to be completely inhibited after Cd exposition. It is conceivable that such positive effect on *hsc70/hsp70* expression is due to trimerization of pre-existent hsf-1 monomers, since our data indicate that after Cd treatment the expression of *hsf-1* by HB2 cells is drastically lowered. On the other hand, as reported by Yan et al. [48] for mouse kidney cells, down-regulation of *hsf-1* can be seemingly considered an additional cause of the observed mitochondrial loss of function (manuscript in preparation), since it can be correlated with increase in mitochondrial permeability transition pore opening. Also, the absence of detectable *hsp90β* expression and the more modest decrease of *hsp90α* transcription level may be ascribed to *hsf-1* down-regulation [48]; interestingly, a moderate down-regulation of α isoform of *hsp90* was also reported in Cd-treated human lung cells [49]. Although it is known that hsp90 disruption by specific hammerhead ribozyme compromises cellular integrity, probably by failure in the maintenance of cytoskeleton-related structural elements within cells [50], the biological significance of the diverse effect exerted by Cd on the expression of the two different hsp90 isoforms is yet to be determined. Differently to what found in MDA-MB231 cells, the Cd stress response of HB2 cells appears to be based also upon enhanced splicing rate of XBP1 message. Chemical conditions that alter the folding of proteins in the endoplamic reticulum (ER) promote cytoprotective measures against ER accumulation of unfolded proteins: in this context, Ire-1 endonuclease-mediated splicing of XBP-1 mRNA as well as induction of *EDEM*, critically controlled by XBP-1, and *grp78* (a.k.a *BiP*), only modestly regulated by XBP-1, are useful markers of ER stress and UPR [51, 52]. No amplicons for *grp78* were obtained from HB2 cell preparations, whilst expression levels of *EDEM* did not differ between control and treated samples; noteworthy, we found a 29% increase of XBP-1 splicing; thus, it is conceivable that in our model system Cd elicits the UPR, but that ER stress, *via* XBP-1, affects other target key regulators than *grp78* or *EDEM* [52], which remain to be identified. It must be also taken into account that ER is one of the intracellular key points of integration of pro-apoptotic signalling and damage sensing, taking part of an integrated system of cross-talk with other organelles which renders it an active contributor to the induction of mitochondrial permeabilization and caspase activation [53].

A more composite effect was exerted by Cd on the expression of stress response genes in neoplastic breast cells, resulting in the early down-regulation of some *hsp* genes, primarily *hsp90β* and *grp78* followed by *hsc70* and *hsp75*, and the late increase of *hsp27* and *hsf1* expression. The restraining effect on *grp78* expression, evident already after 48 h of Cd treatment, is the sole ER-stress related event recordable in MDA-MB231 cells, among those tested. Although most stresses were proven to induce *grp78* up-regulation [51], oxidative damage and ROS generation have been reported to selectively trigger significant decrease in grp78 mRNA levels [54-56]; on the other hand, glutathione peroxidase over-expression in Cd-resistant pneumocytes abrogated oxidative stress and induced *grp78* [57]. Our data are consistent with the finding of MTP hyperpolarization and ROS accumulation in Cd-treated MDA-MB231 cells, observed at 96h from incubation with the metal (manuscript in preparation); in addition, present results suggest that the oxidative insult initiates in earlier time, since substantial restraining of *grp78* expression is already evident after 48h. Also, ROS could be involved in the down-regulation of *hsp75* (a.k.a. *TRAP-1*), a mitochondrial hsp90-homologous stress sensor, as demonstrated by Masuda et al. [58] who also found that TRAP-1 might be involved in apoptosis *via* regulation of the release of cytochrome *c* from mitochondria. Interestingly, in this paper evidence was produced suggesting that Bcl-2 might be one of the regulators of expression of *TRAP-1*; this hypothesis is in accordance with our results indicating that *Bcl-2* down-regulation is an early Cd-dependent event preceding the decrease of *hsp75* level of expression. Decrease of *hsc70* expression levels is likely to contribute in lowering cell protection against ROS, in particular as regards their activity on membrane lipid peroxydation [47]. Moreover, Marchler and Wu [59] have produced evidence that in Drosophila SL2 cells the depletion of hsc70 leads to increased expression of *hsp26* mRNA and accumulation of hsp26 protein; interestingly, although MDA-MB231 cells are reported to be ER-negative, it is known that *hsp27* contains an imperfect estrogen-responsive element [60] and that Cd acts as a metallo-estrogen [26]. Also, Hoang et al. [61] demonstrated that elevated expression of *hsf-1* in prostate carcinoma cells resulted in the consistent increase of hsp27 levels, whereas hsp70 and -90 were not significantly influenced. In light of these literature data, one or more of these aspects could be involved in *hsp27* over-expression by MDA-MB231 cells. The increase in *hsp27* level of expression in the presence of Cd has been documented in some different model systems, such as human kidney cells [62, 63], pneumocytes [57], lens epithelial cells [64] and endothelial cells [65], and interpreted as an help for cells to cope with oxidative injuries based upon the ability of this stress response factor to up-regulate several key enzymes involved in the ROS-glutathione pathway and to promote the degradation of irreversibly-oxidized proteins by the ubiquitin-independent 20S proteasome [66]. Up-regulation of *hsp27* in MDA-MB231 cells has been shown to prevent doxorubicine-dependent apoptotic cell death, independently from Bcl-2 [67]; noteworthy, Kindas-Mugge et al. [68] reported the decrease of the proliferation rate of *hsp27*-overexpressing MDA-MB231 cells. It can be hypothesized that this restraining effect of hsp27 on growth behaviour could cooperate with the induction of cell death to determine the about 50% diminution of MDA-MB231 cell number we observed after 96h of incubation with 5μm $CdCl_2$.

Expression of Apoptosis-related Factors and Caspases

Preparations of cDNAs from HB2 and MDA-MB231 cells, cultured in the absence or presence of $CdCl_2$, were submitted to PCR analyses to evaluate the effect of the metal on a panel of apoptotic modulators and caspases at gene expression level. The tumoral cell line is listed among $p53^-$ cytotypes since it produces a mutated non-functional form of the protein [69]. Obviously, in accordance with Fan and Cherian [41], this can play a role in the resistance of MDA-MB231 cells to cell death when treated with Cd; nonetheless the analyses of *p53* expression levels in HB2 cells revealed no differences between control and treated samples. This result could suggest the independence of this protein from the more pronounced susceptibility of immortalized cells to Cd, whose molecular basis would rely upon other intracellular factors; on the other hand, as reported by Méplan et al. [23] for MCF-7 cells, Cd could not affect the level of p53 message, whilst acting on protein stabilization thereby inducing actual accumulation of p53. Towards HB2 cells, the effect exerted by Cd was limited and inhibitory, as from our results on *Dapk*, *Bad* and some *caspases*, whereas the only promoting influence on gene expression was the switching-on of *Bax-α* transcription. This latter event occurred also in Cd-treated MDA-MB231 cells, where the effect of Cd was more variegated, determining also the early down-regulation of *Bcl-2*, the increase of the expression levels of *Waf-1*, *Dapk*, and several *caspases*, the disappearance of caspase-5 and the appearance of Rip-1 mRNA. Up-regulation of *Bax-α* was reported also in Cd-treated human lung cells [70] and mouse embryo cells [71]; by analogy with MDA-MB231 cells, the latter showed also *Waf-1* induction and *Bcl-2* under-expression, although Bad transcripts were unaffected by the metal. In colon cancer cells, *Bad* was found to be down-regulated in conditions of oxygen deprivation [72]. Cellular responses to hypoxia include diverse processes such as switching from oxidative phosphorylation to anaerobic glycolysis; the decrease of Bad message in Cd-treated HB2 cells could therefore be related to the MTP collapse and mitochondrial loss of function which is selectively observed only in this cytotype (manuscript in preparation). Bax-α, once dimerized, has the ability to insert into mitochondrial membrane, increase membrane permeability and promote the release of cytochrome c in the cytoplasm; its up-regulation in both lines, coupled with the decreased *Bcl-2*/*Bax* ratio in MDA-MB231 cell line, strongly suggests the contribution of some sort of apoptosis signaling in Cd-promoted death of breast cells. Less clear is the opposite regulation of *Dapk* found in the two cytotypes under study. The depletion of this mRNA in HB2 cells could be part of the generalized repressing effect on gene expression as a result of metal intoxication; interestingly, *Dapk*-deficient mouse fibroblasts exhibited a significant decrease in the expression level of p53 and reduced sensitivity to apoptotic promotion [73]. Thus, the depletion of Dapk mRNA could be at least partly responsible of the absence of increase of p53 message level after Cd treatment of immortalized breast cells. Our data indicated that *Dapk* was conversely over-expressed in Cd-treated neoplastic cells; we have previously found that this gene was induced in two human breast cancer cell lines, 8701-BC and T-47D, in case of substrate-promoted apoptosis [29, 33]. It is known that Dapk has cytoskeletal effects, in that it triggers an uncoupling of the formation of stress fibers and focal adhesions, inhibiting integrin activity and cell attachment to substrate [74]; it is therefore conceivable that *Dapk* up-regulation may participate to MDA-MB231 loss of viability by inducing an anoikis-type cell death following cell detachment and rounding-up. Concerning the increase of *Waf-1* expression levels in MDA-MB231 cells, p53-independent mechanisms responsible of this up-regulation have been described for this cytotype, with contribution of diverse pathways such as those of protein kinase C and LKB1/STK11 kinase [e.g. 75, 76]. The appearance of Rip-1 amplicon in preparations from Cd-treated MDA-MB231 cells submitted to PCR assays is particularly interesting: Rip-1 is a member of a kinase family which act as sensors of cellular stress and crucial regulators of cell viability or death [77]. Jäättela and Tschopp [78] have reported that Rip-1 might be involved in ROS formation by activation of cytosolic phospholipase A2 and/or enhanced electron flow through the mitochondrial electron transport chain. Thus, in light of this latter case, selective upregulation of *Rip-1* in MDA-MB231 could seemingly be involved in the hyperpolarization of MTP, observed by JC-1 dye monitoring, and positivity to aminophenyl fluoresceine staining, a highly sensitive fluorescent probe for the detection of hypochlorites, hydroxyl radicals, and peroxynitrites, following exposition to the metal (manuscript in preparation). Concerning our studies on caspases, it must be taken into account that examining expression at the mRNA level could provide evidence of differences that do not necessarily correspond to changes in enzyme activity, whose evaluation awaits further investigation; nonetheless, our results give an indication of which *caspases* may be regulated during incubation of breast cells with Cd. In particular, we demonstrated the down-regulation of all the *caspases* tested, except for *caspase-6* whose expression level remained steady, in HB2 cells, strongly suggesting that the contribution of these enzymes to death of immortalized breast cells after 24 h of Cd treatment is unlikely. On the other hand, the tumoral counterpart exhibited concurrent transcriptional activation of six *caspases*, in particular of *caspase-1*, and *-6* to high degree, of *caspase-7* and *-8* to lesser degree, and of *caspase-2* and *-9* to low degree, whilst *caspase-3* level remained steady and *caspase-5* was down-regulated; this diversified patterns of transcription suggest that gene expression of these enzymes is under the control of distinct mechanisms, whose nature is still to be determined, and that these components of the apoptotic machinery can be involved in Cd-promoted MDA-MB231 cell death.

In conclusion, in this chapter we have presented a detailed analysis at gene expression level of the different response to Cd stress by two human breast epithelial cell lines, the one immortalized the other neoplastic. Immortalized HB2 cells appear to endure in a more intense manner the cytotoxic action of the metal and to fail committing suicide by apoptosis; conversely, tumoral MDA-MB231 cells seem to be capable to activate a certain number of defense mechanisms which lose their efficacy after about 72 h of incubation, when new intracellular pathways are switched-on conceivably directing cells to a kind of death which shares some aspects with "non-classical" apoptotic program. Additional experiments will be required for the identification of other differentially-expressed genes in the diverse experimental conditions, to get more insight into Cd-responsive genes in our "in vitro" model systems.

Acknowledgments

We thank Cancer Research (UK) for providing the HB2 cell line. This work has been supported by grants from MIUR (Cofin 2004; prot. 2004057282_001 and 2004057282_002) and the University of Palermo (ex60%).

References

[1] Jarup, L. (2003). Hazards of heavy metal contamination. *Br. Med. Bull.*, *68*,167-182.

[2] Hartwig, A. (2001). Zinc finger proteins as potential targets for toxic metal ions: differential effects on structure and function. *Antioxid. Redox Signal.*, *3*, 625-634.

[3] Goering, P. L., Waalkes, M. P., and Klaassen, C. D. (1995). Toxicology of cadmium. In R. A. Goyer, and M. G. Cherian (Eds.), *Toxicology of Metals: Biochemical Aspects. Handbook of Experimental Pharmacology* (Vol. 115, pp. 189–213). New York, USA: Springer-Verlag.

[4] Beyersmann, D. and Hechtenberg, S. (1997). Cadmium, gene regulation, and cellular signalling in mammalian cells. *Toxicol. Appl. Pharmacol.*, *144*, 247-261.

[5] Koizumi, S. and Yamada, H. (2003). DNA microarray analysis of altered gene expression in cadmium-exposed human cells. *J. Occup. Health*, *45*, 331-334.

[6] Waisberg, M., Joseph, P., Hale, B. and Beyersmann, D. (2003). Molecular and cellular mechanisms of cadmium carcinogenesis. *Toxicology*, *192*, 95-117.

[7] Ishido, M. and Kunimoto, M. (2001). Regulation of cell fate by cadmium and zinc. *J. Health Sci.*, *47*, 9-13.

[8] Regazzi, R., Eppenberger, U. and Fabbro, D. (1988). The 27,000 daltons stress proteins are phosphorylated by protein kinase C during the tumor promoter-mediated growth inhibition of human mammary carcinoma cells. *Biochem. Biophys. Res. Commun.*, *152*, 62-68.

[9] Tchounwou, P. B., Ishaque, A. B. and Schneider, J. (2001). Cytotoxicity and transcriptional activation of stress genes in human liver carcinoma cells (HepG2) exposed to cadmium chloride. *Mol. Cell. Biochem.*, *222*, 21-28.

[10] Garrett, S. H., Phillips, V., Somji, S., Sens, M. A., Dutta, R., Park, S., Kim, D. and Sens, D. A. (2002). Transient induction of metallothionein isoform 3 (MT-3), c-fos, c-jun and c-myc in human proximal tubule cells exposed to cadmium. *Toxicol. Lett.*, *126*, 69-80.

[11] Joseph, P., Lei, Y. X. and Ong, T. M. (2004). Up-regulation of expression of translation factors - a novel molecular mechanism for cadmium carcinogenesis. *Mol. Cell. Biochem.*, *255*, 93-101.

[12] Shin, H. J., Lee, B. H., Yeo, M. G., Oh, S. H., Park, J. D., Park, K. K., Chung, J. H., Moon, C. K. and Lee, M. O. (2004). Induction of orphan nuclear receptor Nur77 gene expression and its role in cadmium-induced apoptosis in lung. *Carcinogenesis*, *25*, 1467-1475.

[13] Shin, H. J., Park, K. K., Lee, B. H., Moon, C. K. and Lee, M.O. (2003). Identification of genes that are induced after cadmium exposure by suppression subtractive hybridization. *Toxicology*, *191*, 121-131.

[14] Hamada, T., Tanimoto, A. and Sasaguri, Y. (1997) Apoptosis induced by cadmium. *Apoptosis*, *2*, 359-367.

[15] Li M., Kondo T., Zhao Q. L., Li F. J., Tanabe K., Arai Y., Zhou Z. C. and Kasuya, M. (2000). Apoptosis induced by cadmium in human lymphoma U937 cells through Ca2+-calpain and caspase-mitochondria- dependent pathways. *J. Biol. Chem.*, *275*, 39702-39709.

[16] Aydin, H. H., Celik, H. A., Deveci, R., Terzioglu, E., Karacali, S., Mete, N., Akarca, U. and Batur, Y. (2003). Characterization of the cellular response during apoptosis induction in cadmium-treated HepG2 human hepatoma cells. *Biol. Trace Elem. Res.*, *95*, 139-153.

[17] Kondoh, M., Ogasawara, S., Araragi, S., Higashimoto, M. and Sato, M. (2001). Cytochrome C release from mitochondria induced by cadmium. *J. Health Sci.*, *47*, 78-82.

[18] Zhai, Q. W., Ji, H. B., Zheng, Z. C., Sun, L. Y. and Liu, X. Y. (2000). A novel type of apoptosis induced by cadmium in BA/F3 beta cells. *Sheng Wu Hua Xue Yu Sheng Wu Wu Li Xue Bao (Shanghai)*, *32*, 77-80.

[19] IARC (1993). *Chemicals, groups of chemicals, complex mixtures, physical and biological agents and exposure circumstances to be evaluated in future.* Lyon, France: IARC Monographs, Report of an ad-hoc Working Group.

[20] Pearson, C. A. and Prozialeck, W. C. (2001). E-cadherin, beta-catenin and cadmium carcinogenesis. *Med. Hypotheses*, *56*, 573-581.

[21] Waalkes, M. P., Diwan, B. A., Rehm, S., Ward, J. M., Moussa, M., Cherian, M. G. and Goyer, R. A. (1996). Down-regulation of metallothionein expression in human and murine hepatocellular tumors: association with the tumor-necrotizing and antineoplastic effects of cadmium in mice. *J. Pharmacol. Exp. Ther.*, *277*, 1026-1033.

[22] Waalkes, M. P. and Diwan, B. A. (1999). Cadmium-induced inhibition of the growth and metastasis of human lung carcinoma xenografts: role of apoptosis. *Carcinogenesis*, *20*, 65-70.

[23] Méplan, C., Mann, K. and Hainaut, P. (1999). Cadmium induces conformational modifications of wild-type p53 and suppresses p53 response to DNA damage in cultured cells. *J. Biol. Chem.*, *274*, 31663-31670.

[24] Matsuoka, M. and Igisu, H. (2001). Cadmium induces phosphorylation of p53 at serine 15 in MCF-7 cells. *Biochem. Biophys. Res. Commun.*, *282*, 1120-1125.

[25] Garcia-Morales, P., Saceda, M., Kenney, N., Kim, N., Salomon, D. S., Gottardis, M. M., Solomon, H. B., Sholler, P. F., Jordan, V. C. and Martin, M. B. (1994). Effect of cadmium on estrogen receptor levels and estrogen-induced responses in human breast cancer cells. *J. Biol. Chem.*, *269*, 16896-16901.

[26] Johnson, M. D., Kenney, N., Stoica, A., Hilakivi-Clarke, L., Singh, B., Chepko, G., Clarke, R., Sholler, P. F., Lirio, A. A., Foss, C., Reiter, R., Trock, B., Paik, S. and Martin, M. B. (2003). Cadmium mimics the in vivo effects of estrogen in the uterus and mammary gland. *Nat. Med.*, *9*, 1081-1084.

[27] Berdichevsky, F., Alford, D., D'Souza, B. and Taylor-Papadimitriou, J. (1994). Branching morphogenesis of human mammary epithelial cells in collagen gels. *J. Cell Sci.*, *107*, 3557-3368.

[28] Biagioli, M., Pinton, P., Scudiero, R., Ragghianti, M., Bucci, S. and Rizzuto, R. (2005). Aequorin chimeras as valuable tool in the measurement of Ca2+ concentration during cadmium injury. *Toxicology*, *208*, 389-398.

[29] Luparello, C., David, F., Campisi, G. and Sirchia, R. (2003). T47-D cells and type V collagen: a model for the study of apoptotic gene expression by breast cancer cells. *Biol. Chem.*, *384*, 965-75.

[30] Roccheri, M. C., Barbata, G., Cardinale, F., Tipa, C., Bosco, L., Oliva, O. A., Cascino, D. and Giudice, G. (1997). Apoptosis in sea urchin embryos. *Biochem. Biophys. Res. Commun.*, *240*, 359-366.

[31] Izzo, V., Asaro, M. R., Forte, G. I. and Barbieri, R. (2000). Use of voltage gradient gel electrophoresis in apoptotic DNA analysis. *J. Chromatogr. A*, *890*, 371-374.

[32] Luparello, C., Sirchia, R. and Pupello, D. (2003). PTHrP [67-86] regulates the expression of stress proteins in breast cancer cells inducing modifications in urokinase-plasminogen activator and MMP-1 expression. *J. Cell Sci.*, *116*, 2421-2430.

[33] Luparello, C. and Sirchia, R. (2005). Type V collagen regulates the expression of apoptotic and stress response genes by breast cancer cells. *J. Cell. Physiol.*, *202*, 411-421.

[34] Oppermann, B., Zhang, W., Magabo, K. and Kantorow, M. (2001). Identification and spatial analysis of metallothioneins expressed by the adult human lens. *Invest. Ophthalmol. Vis. Sci.*, *42*, 188-93.

[35] Shang, J. (2005). Quantitative measurement of events in the mammalian unfolded protein response. *Methods*, *35*, 390-394.

[36] Spencer, W. E. and Christensen, M. J. (1999). Multiplex relative RT-PCR method for verification of differential gene expression. *Biotechniques*, *27*, 1044-1046.

[37] Sirchia, R., Ciacciofera, V. and Luparello, C. (2003). Tumor cell-collagen interactions: identification and semi-quantitative evaluation of selectively-expressed genes by combination of differential display- and multiplex-PCR. *Biol. Proced. Online*, *5*, 222-227.

[38] Jin, R., Huang, J., Tan, P. H. and Bay, B. H. (2004). Clinicopathological significance of metallothioneins in breast cancer. *Pathol. Oncol. Res.*, *10*, 74-79

[39] Jin, R., Chow, V. T., Tan, P. H., Dheen, S. T., Duan, W. and Bay, B.H. (2002). Metallothionein 2A expression is associated with cell proliferation in breast cancer. *Carcinogenesis*, *23*, 81-86

[40] Barnes, N. L., Ackland, M. L. and Cornish, E. J. (2000). Metallothionein isoform expression by breast cancer cells. *Int. J. Biochem. Cell Biol.*, *32*, 895-903.

[41] Fan, L. Z. and Cherian, M. G. (2002) Potential role of p53 on metallothionein induction in human epithelial breast cancer cells. *Brit. J. Cancer*, *87*, 1019-1026.

[42] Garrett, S. H., Somji, S., Todd, J. H., Sens, M. A. and Sens, D. A. (1998). Differential expression of human metallothionein isoform I mRNA in human proximal tubule cells exposed to metals. *Environ. Health Perspect.*, *106*, 825-832.

[43] Li, J., Liu, Y. and Ru, B. (2005). Effect of metallothionein on cell viability and its interaction with cadmium and zinc in HEK293 cells. *Cell Biol. Int.*, *29*, 843-848.

[44] Yamada, H. and Koizumi, S. (2002). DNA microarray analysis of human gene expression induced by a non-lethal dose of cadmium. *Industrial Health*, *40*, 159–166.

[45] Jin, R., Bay, B. H., Chow, V. T. and Tan, P. H. (2001). Metallothionein 1F mRNA expression correlated with histological grade in breast carcinoma. *Breast Cancer Res. Treat.*, *66*, 265-272.

[46] Garrido, C., Gurbuxani, S., Ravagnan, L. and Kroemer, G. (2001). Heat shock proteins: endogenous modulators of apoptoic cell death. *Biochem. Biophys. Res. Commun.*, *296*, 433-442.

[47] Su, C. Y., Chong, K. Y., Edelstein, K., Lille, S., Khardori, R. and Lai, C. C. (1999). Constitutive hsp70 attenuates hydrogen peroxide-induced membrane lipid peroxidation. *Biochem. Biophys. Res. Commun.*, *265*, 279-284.

[48] Yan, L. J., Rajasekaran, N. S., Sathyanarayanan, S. and Benjamin, I. J. (2005). Mouse HSF1 disruption perturbs redox state and increases mitochondrial oxidative stress in kidney. *Antioxid. Redox Signal.*, *7*, 465-471.

[49] Andrew, A. S., Warren, A. J., Barchowsky, A., Temple, K. A., Klei, L., Soucy, N. V., O'Hara, K. A. and Hamilton, J. W. (2003). Genomic and proteomic profiling of responses to toxic metals in human lung cells. *Environ. Health Perspect.*, *111*, 825-835.

[50] Sreedhar, A. S., Mihály, K., Pató, B., Schnaider, T., Steták, A., Kis-Petik, K., Fidy, J., Simonics, T., Maráz, A. and Csermely, P. (2003). Hsp90 inhibition accelerates cell lysis. Anti-Hsp90 ribozyme reveals a complex mechanism of Hsp90 inhibitors involving both superoxide- and Hsp90-dependent events. *J. Biol. Chem.*, *278*, 35231-35240.

[51] Ma, Y. and Hendershot, L. M. (2002). The mammalian endoplasmic reticulum as a sensor for cellular stress. *Cell Stress Chaperones*, *7*, 222–229.

[52] Lee, A. H., Iwakoshi, N. N. and Glimcher, L. H. (2003). XBP-1 regulates a subset of endoplasmic reticulum resident chaperone genes in the unfolded protein response. *Mol. Cell. Biol.*, *23*, 7448-7459.

[53] Ferri, K. F. and Kroemer, G. (2001). Organelle-specific initiation of cell death pathways. *Nat. Cell Biol.*, *3*, E255-63.

[54] Paschen, W., Mengesdorf, T., Althausen, S. and Hotop, S. (2001). Peroxidative stress selectively down-regulates the neuronal stress response activated under conditions of endoplasmic reticulum dysfunction. *J. Neurochem.*, *76*, 1916-1924.

[55] Shimoke, K., Kudo, M. and Ikeuchi, T. (2003). MPTP-induced reactive oxygen species promote cell death through a gradual activation of caspase-3 without expression of GRP78/Bip as a preventive measure against ER stress in PC12 cells. *Life Sci.*, *73*, 581-593.

[56] Dey, A., Kessova, I. G. and Cederbaum, A. I. (2006). Decreased protein and mRNA expression of ER stress proteins GRP78 and GRP94 in HepG2 cells over-expressing CYP2E1. *Arch. Biochem. Biophys.*, *447*, 155-166.

[57] Croute, F., Beau, B., Murat, J. C., Vincent, C., Komatsu, H., Obata, F. and Soleilhavoup, J. P. (2005). Expression of stress-related genes in a cadmium-resistant A549 human cell line. *J. Toxicol. Environ. Health A*, *68*, 703-718

[58] Masuda, Y., Shima, G., Aiuchi, T., Horie, M., Hori, K., Nakajo, S., Kajimoto, S., Shibayama-Imazu, T. and Nakaya, K. (2004). Involvement of tumor necrosis factor receptor-associated protein 1 (TRAP1) in apoptosis induced by beta-hydroxyisovalerylshikonin. *J. Biol. Chem.*, *279*, 42503-42515.

[59] Marchler, G. and Wu, C. (2001). Modulation of Drosophila heat shock transcription factor activity by the molecular chaperone DROJ1. *EMBO J.*, *20*, 499-509.

[60] Oesterreich, S., Lee, A. V., Sullivan, T. M., Samuel, S. K., Davie, J. R. and Fuqua, S. A. (1997). Novel nuclear matrix protein HET binds to and influences activity of the HSP27 promoter in human breast cancer cells. *J. Cell. Biochem.*, *67*, 275-286.

[61] Hoang, A. T., Huang, J., Rudra-Ganguly, N., Zheng, J., Powell, W. C., Rabindran, S. K., Wu, C. and Roy-Burman, P. (2000). A novel association between the human heat shock transcription factor 1 (HSF1) and prostate adenocarcinoma. *Am. J. Pathol.*, *156*, 857-864.

[62] Somji, S., Sens, D. A., Garrett, S. H., Sens, M. A. and Todd, J. H. (1999). Heat shock protein 27 expression in human proximal tubule cells exposed to lethal and sublethal concentrations of $CdCl_2$. *Environ. Health Perspect.*, *107*, 545-552.

[63] Bonham, R. T., Fine, M. R., Pollock, F. M. and Shelden, E. A. (2003). Hsp27, Hsp70, and metallothionein in MDCK and LLC-PK1 renal epithelial cells: effects of prolonged exposure to cadmium. *Toxicol. Appl. Pharmacol.*, *191*, 63-73.

[64] Hawse, J. R., Cumming, J. R., Oppermann, B., Sheets, N. L., Reddy, V. N. and Kantorow, M. (2003). Activation of metallothioneins and alpha-crystallin/sHSPs in human lens epithelial cells by specific metals and the metal content of aging clear human lenses. *Invest. Ophthalmol. Vis. Sci.*, *44*, 672-679.

[65] Wagner, M., Hermanns, I., Bittinger, F. and Kirkpatrick, C. J. (1999). Induction of stress proteins in human endothelial cells by heavy metal ions and heat shock. *Am. J. Physiol. Lung Cell. Mol. Physiol.*, *277*, L1026-L1033.

[66] Arrigo, A. P., Firdaus, W. J., Mellier, G., Moulin, M., Paul, C., Diaz-Latoud, C., Kretz-Remy, C. (2005). Cytotoxic effects induced by oxidative stress in cultured mammalian cells and protection provided by Hsp27 expression. *Methods*, *35*, 126-138.

[67] Hansen, R. K., Parra, I., Lemieux, P., Oesterreich, S., Hilsenbeck, S. G. and Fuqua, S. A. (1999). Hsp27 overexpression inhibits doxorubicin-induced apoptosis in human breast cancer cells. *Breast Cancer Res. Treat.*, *56*, 187-196.

[68] Kindas-Mugge, I., Micksche, M. and Trautinger, F. (1998). Modification of growth in small heat shock (hsp27) gene transfected breast carcinoma. *Anticancer Res.*, *18*, 413-417.

[69] Runnebaum, I. B., Nagarajan, M., Bowman, M., Soto, D. and Sukumar, S. (1991). Mutations in p53 as potential molecular markers for human breast cancer. *Proc. Natl. Acad. Sci. USA*, *88*, 10657-10661.

[70] Lag, M., Westly, S., Lerstad, T., Bjornsrud, C., Refnes, M. and Schwarze, P. E. (2002). Cadmium-induced apoptosis of primary epithelial cell lungs: involvement of Bax and p53, but not oxidative stress. *Cell Biol. Toxicol.*, *18*, 29-42.

[71] Fernandez, E. L., Gustafson, A. L., Andersson, M., Hellman, B. and Dencker, L. (2003). Cadmium-induced changes in apoptotic gene expression levels and DNA damage in mouse embryos are blocked by zinc. *Toxicol. Sci.*, *76*, 162-170.

[72] Erler, J. T., Cawthorne, C. J., Williams, K. J., Koritzinsky, M., Wouters, B. G., Wilson, C., Miller, C., Demonacos, C., Stratford, I. J. and Dive, C. Hypoxia-mediated down-regulation of Bid and Bax in tumors occurs via hypoxia-inducible factor 1-dependent and -independent mechanisms and contributes to drug resistance. *Mol. Cell. Biol.*, *24*, 2875-89.

[73] Raveh, T., Droguett, G., Horwitz, M. S., DePinho, R. A. and Kimchi, A. (2001). DAP kinase activates a p19ARF/p53-mediated apoptotic checkpoint to suppress oncogenic transformation. *Nat. Cell Biol.*, *3*, 1-7.

[74] Kuo, J. C., Lin, J. R., Staddon, J. M., Hosoya, H. and Chen, R. H. (2003). Uncoordinated regulation of stress fibers and focal adhesions by DAP kinase. *J. Cell Sci.*, *116*, 4777-4790.

[75] Akashi, M., Osawa, Y., Koeffler, H. P. and Hachiya, M. (1999). p21WAF1 expression by an activator of protein kinase C is regulated mainly at the post-transcriptional level in cells lacking p53: important role of RNA stabilization. *Biochem. J.*, *337*, 607-616.

[76] Setogawa, T., Shinozaki-Yabana, S., Masuda, T., Matsuura, K. and Akiyama, T. (2006). The tumor suppressor LKB1 induces p21 expression in collaboration with LMO4, GATA-6, and Ldb1. *Biochem. Biophys. Res. Commun.*, *343*, 1186-1190.

[77] Meylan, E. and Tschopp, J. (2005) The RIP kinases: crucial integrators of cellular stress. *Trends Biochem. Sci.*, *30*, 151-159.

[78] Jäättela, M. and Tschopp, J. (2003). Caspase-independent cell death in T lymphocytes. *Nat Immunol.*, *4*, 416-423.

[79] Jäättelä, M. (1999). Escaping cell death: survival proteins in cancer. *Exp. Cell Res.*, *248*, 30-43.

[80] Jolly, C. and Morimoto, R. I. (2000). Role of the heat shock response and molecular chaperones in oncogenesis and cell death. *J. Natl. Cancer Inst.*, *92*, 1564-1572.

[81] Cotto, J. J. and Morimoto, R. I. (1999). Stress-induced activation of the heat-shock response: cell and molecular biology of heat-shock factors. *Biochem. Soc. Symp.*, *64*, 105-118.

[82] Pirkkala, L., Nykanen, P. and Sistonen, L. (2001). Roles of the heat shock transcription factors in regulation of the heat shock response and beyond. *FASEB J.*, *15*, 1118-1131.

[83] Vašák, M. and Hasler, D. W. (2000). Metallothioneins: new functional and structural insights. *Curr. Opin. Chem. Biol.*, *4*, 177-183.

[84] Newton, K. and Strasser, A. (1998). The Bcl-2 family and cell death regulation. *Curr. Opin. Genet. Develop.*, *8*, 68-75.

[85] White, E. (1996). Life, death and the pursuit of apoptosis. *Genes Develop.*, *10*, 1-15.

[86] Kawakami, A., Tian, Q., Duan, X., Streuli, M., Schlossman, S. F. and Anderson, P. (1992). Identification and functional characterization of a TIA-1-related nucleolysin. *Proc. Natl. Acad. Sci. USA*, *89*, 8681-8685.

[87] Brinkmann, U. (1998). CAS, the human homologue of the yeast chromosome-segeregation gene CSE-1, in proliferation, apoptosis and cancer. *Am. J. Hum. Genet.*, *62*, 509-513.

[88] Özçelik, M., Mousses, S. and Andrulis, I. L. (1995). Low levels of expression of an inhibitor of cyclin-dependent kinases (CIP1/WAF1) in primary breast carcinomas with p53 mutations. *Clin. Cancer Res.*, *1*, 907-912.

[89] Cotter, T. G. *and* McCarthy, J. (1994). Genes and apoptosis. *Biochem. Soc. Trans.*, *22*, 591-593.

[90] Cohen, O. and Kimchi, A. (2001). DAP-kinase: from functional gene cloning to establishment of its role in apoptosis. *Cell Death Differ.*, *8*, 6-15.

[91] Wolf, B. B. and Green, D. R. (1999). Suicidal tendencies: apoptotic cell death by caspase family of proteinases. *J. Biol. Chem.*, *274*, 20049-20052.

[92] Hu, S., Snipas, S.J., Vincenz, C., Salvesen, G. and Dixit, V. M. (1998). Caspase-14 is a novel developmentally regulated protease. *J. Biol. Chem.*, *273*, 29648-29653.

In: New Developments in Cell Apoptosis Research
Editor: Alan J. Corvin, pp. 241-272
ISBN: 1-60021-453-3

Chapter X

Photosensitizers as Novel Compounds in Apoptosis Research

***N. K. Mak*[a]*, M. L. Lung*[b]*, C .K. Chang*[c] *and K. N. Leung*[d]

[a]Department of Biology, Hong Kong Baptist University
[b]Department of Biology and [c]Department of Chemistry,
The Hong Kong University of Science and Technology
[d]Department of Biochemistry, The Chinese University of Hong Kong

Abstract

Death is an intricate process in all living cellular organisms. Apoptosis, also known as programmed cell death, is a normal physiological process occurring in all eukaryotic cells. This normal cell death process has been known for over half a century. Mechanisms leading to apoptotic cell death have also been studied extensively. Induction of tumour cell apoptosis becomes a new research direction for the treatment of malignant disease. In the past decade, photodynamic therapy (PDT) has been developed as a tumour therapy. The treatment relies on the accumulation of photosensitizers in the tumour mass. Tumour cell death, triggered by irradiating the tumour mass with visible light, is one of the major mechanisms leading to tumour eradication. Generation of singlet oxygen and induction of oxidative stress are the major intracellular events after PDT. In view of the short half-life of reactive oxygen species, the localization of photosensitizers in various subcellular compartments, such as the plasma membrane, mitochondria, lysosomes, Golgi body, and endoplasmic reticulum has been implicated in the induction of various modes of cell death. Photosensitizers targeting mitochondria received particular interest as mitochondria are the key organelles in the mitochondria-mediated apoptotic pathway. Other cellular events, such as activation of transcription factors and intracellular stress signaling pathways and changes of Ca^{2+} and lipid metabolism are also seen in PDT treated cells. This review focuses on the understanding of PDT-induced tumour cell apoptosis.

Abbreviations

AIF	apoptosis-inducing factor
ALA	5-aminolevulinate
ANT	adenine nucleotide translocator
Apaf-1	apoptotic protease-activating factor-1
APP	aminopyro-pheophorbide
Bcl-2	B-cell leukemia/lymphoma-2
Bid	BH3-interacting domain death agonist
BPD-MA	benzoporphyrin derivative monoacid ring A
CAD	caspase-activated DNase
CARD	caspase-activation recruitment domain
CHOP	C/EBP homologous protein
CPO	9-capronyloxytetrakis (methoxyethyl) porphycene
ER	endoplasmic reticulum
ERK	extracellular signal-regulated kinase
FADD	Fas-associated protein with death domain
FasL	Fas ligand
HIF	hypoxia inducible factor
HMME	hematoporphyrin monomethyl ether
Hp	hematoporphyrin
HpD	hematoporphyrin derivative
HVD	hydroxyethylvinyl-deuteroporphyrin
JNK	c-Jun N-terminal kinase
LMP	lysosomal membrane permeabilization
MMP	mitochondrial membrane permeabilization
MAPK	mitogen-activated protein kinase
MBD	methylene blue derivative
NF-kB	nuclear factor-kappa B
OMM	outer mitochondrial membrane
PBR	peripheral benzodiazepine receptor
Pc4	silicon phthalocyanine
PDT	photodynamic therapy
PMEE	pyropheophorbide-a methylester
Pp	protoporphyrin
PTPC	permeability transition pore complex
RB	rose bengal
ROS	reactive oxygen species
SAPK	stress activated kinase
tBid	truncated carboxy-terminal fragment of Bid
TNF	tumour necrosis factor
VBBO	Victoria Blue BO
VDAC	voltage-dependent anion channel
ZnPc	zinc (II) phthalocyanine

Introduction

Under normal conditions, cell growth and differentiation are tightly controlled by a number of factors. These include the extrinsic factors such as cytokines and hormones, and the intrinsic factors such as cell cycle regulators and transcription factors. Tumour cells are generally considered as a group of cells with genetic defects in the expression and production of the cell growth regulators. Tumour progression is frequently linked to defects in the controlling mechanisms of apoptosis (or programmed cell death) in the tumour cells. In the past few decades, targeted induction of tumour cell death is one of the strategies in the control of tumour cell growth. In addition to radiotherapy, many mechanism-based therapeutic agents were developed and used to induce tumour cell death. Several forms of cell death have been described. These include apoptosis, autophagy, mitotic catastrophe, necrosis, and senescence [1]. The classification of these different types of cell death are based on the morphological changes of the plasma membrane and nucleus, the involvement of caspases in the cell death process, and also the appearance of apoptotic bodies. Apoptosis is a distinct form of cell death, as apoptosis is generally considered as part of the normal physiological events. It is now understood that apoptosis is an important process to eliminate the transitory organs and tissues during the early stage of embryo development. It is also important in tissue remodeling and organ development. During the development of antigen-specific lymphocytes, apoptosis is involved in the removal of auto-reactive lymphocytes.

Apart from the induction of tumour cell apoptosis, recent research has also focused on the induction of non-apoptotic cell death pathways in tumour cells. This is especially the case for the tumour cells that cannot undergo the apoptotic type of cell death. This review will focus on the induction of apoptotic tumour cell death by a group of light-sensitive photosensitizers commonly used in photodynamic therapy (PDT).

Photoactivation of Photosensitizers

Photosensitizers belong to a group of light absorbing compounds. The photosensitizers used in PDT usually consist of heterocyclic ring structure for light capture. When the photosensitizer is irradiated with light of an appropriate wavelength, the photosensitizer is activated to the excited singlet state (S1). From the excited state, the photosensitizer may proceed in many different ways to reach the ground state. These include the radiationless transition to the ground state; fluorescence, intersystem crossing, phosphorescence, energy transfer, and charge transfer to other molecules (Figure 1). During radiationless transfer to the ground state, excess energy is released to the aqueous environment in the form of heat. The process usually occurs within a nanosecond to picoseconds and the thermal effect on the biomolecules is generally considered not significant. This form of energy release has little therapeutic value in cancer treatment. In a biological system, many biomolecules show various degrees of fluorescence. Thus, photosensitizers with high fluorescence yields are usually used in photodetection of tumours.

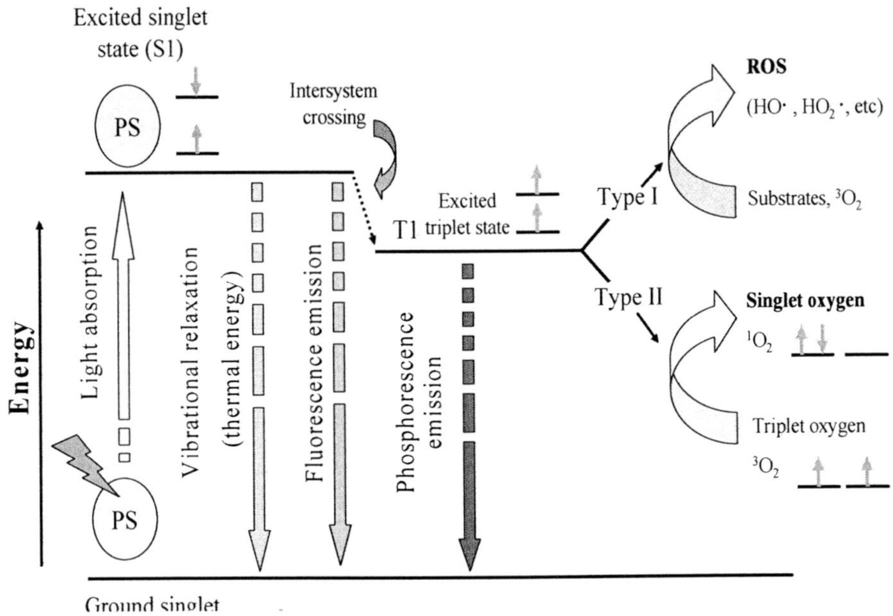

Figure 1. Photoactivation of photosensitizers. Photosensitizer at ground state (So) is activated to S1 state. On return to ground state, energy is released and transferred to other substrates (Type I reaction) or normal oxygen (Type II reaction). "Singlet" and "triplet" refers to the degeneracy (energy level) of the electron. ↑ and ↓ depicts the direction of spin of electron. ↑ ↑, same direction; ↑ ↓, opposite direction.

In intersystem crossing, the S1-state photosensitizer is converted to a triplet state (T1) with two electrons of the same spin in an orbital. Photosensitizers at T1 stage may have a lifetime of several seconds to less than a millisecond. Hence, T1-state photosensitizers with a long lifetime will have a greater chance to interact with surrounding biomolecules. In PDT, one of the mechanisms of cytotoxic killing of malignant cells is through the photodynamic action of the excited photosensitizer at T1 stage. Theoretically, the T1 state photosensitizer can interact with surrounding biomolecules through two different mechanisms, namely Type I and Type II reactions. Type I reaction involves the transfer of electrons or hydrogen between T1-state photosensitizer and the surrounding biomolecules. Reactive free radical intermediates of T1 photosensitizer, formed during the reaction, will react with oxygen and produce cytotoxic reactive oxygen species (ROS) such as superoxide and hydroxyl radicals. Type II reaction occurs when the T1 state photosensitizer and the ground state oxygen ($^3O_2$) molecule are in close proximity. Energy is transferred directly from the T1 state photosensitizer to the $^3O_2$. If $^3O_2$ absorbs enough energy to reverse the spin of one of the two unpaired electrons (Figure 1), the ground state oxygen is then activated to become the highly reactive singlet state oxygen ($^1O_2$). In a cellular environment, $^1O_2$ has a lifetime of less than 0.04 micro-second [2]. It has also been estimated that $^1O_2$ can only diffuse to a distance of less than 0.02 μm. Hence, the localization / site of photoactivation of photosensitizer is believed to be the target site of photodamage. After the energy transfer, T1 photosensitizer is returned to the ground state. The process of energy transfer and production of ROS or $^1O_2$ is then repeated and repeated until the photosensitizer is degraded (photo-bleached). The resulting oxidative stress will finally lead to necrosis or apoptosis of tumour cells.

Phosphorescence is distinguished from fluorescence. The light absorbing photosensitizer is excited to a metastable state, and the energy is released very slowly and continuously in the form of glowing light. Recent study shows that singlet oxygen can be detected from a single cell by measuring the weak 1270 nm phosphorescence upon photoactivation of the photosensitizer 5,10,15,20-tetrakis(N-methyl-4-pyridyl)-21H,23H-porphine (TMPyP) [3].

Photosensitizing Agents Used in Photodynamic Therapy

Application of photosensitizers in cancer research has a history of over one hundred years. The cytotoxic effect of a combination of light and certain light absorbing molecules was initially observed by a German medical student Oscar Raab. He found that a combination of light and acridine red was lethal to paramecium [4]. The term "photodynamic action" was introduced in 1907 by Von Tappeiner and Jodlbauer when they observed the requirement of oxygen in the treatment of skin cancer with light and eosin [5]. Subsequent to these studies, Hausmann observed the photocytotoxic effect of hematoporphyrin (Hp), an iron-containing porphyrin prepared by heating dried blood with sulfuric acid, on red cells [6]. Further chemical analysis revealed that Hp was a mixture of porphyrins [7]. After acetylation and neutralization, a preparation, called hematoporphyrin derivative (HpD) with a higher photocytotoxic activity than Hp was produced. The application of HpD in tumour eradication in tumour-bearing animals was demonstrated by Dougherty and co-workers in 1975 [8]. Successful application of HpD and argon dye laser in the treatment of patients with skin tumour was subsequently reported in 1978 [9]. Although HpD is more phototoxic than Hp, it is chemically impure. HpD is still a mixture containing Hp, hydroxyethylvinyl-deuteroporphyrin (HVD), protoporphyrin (Pp), and an oligomeric fraction containing ester, ether and carbon-carbon linkages of Hp. Subsequent studies found that Photofrin, the chromatographic fraction enriched with oligomeric porphyrins, is the most active fraction of HpD [10,11].

Before the development of the second generation photosensitizers, HpD and Photofrin were the major types of photosensitizer. In the past decade, most of the photosensitizers used in PDT are heterocyclic compounds with the light absorption ring structure similar to porphyrin or chlorophylls (Figure 2) [12,13]. Porphyrins are naturally occurring conjugated tetrapyrroles with various side chain substitutions. The absorption maximum of porphyrins ranged from 620 nm to 640 nm [14]. Reduction of one of the double bonds in the pyrrole group resulted in the formation of chlorins. The absorption maximum is also shifted to about 650 - 690 nm [12]. Chlorophyll-a is the substituted derivative of chlorin. Bacteriochlorins are derived from the chlorins through the further reduction of another pyrrole group. The absorption of bacteriochlorins is further shifted to the red region of about 780 nm. In addition to these naturally occurring tetrapyrroles, non-natural synthetic cyclic tetrapyrroles, such as porphycenes [15,16] and sapphyrins [17], and other structural classes of photosensitizers (e.g. Rose Bengal and hypericin from St John's Wort) have also been developed and evaluated for killing of cancer cells. Because of the variation in hydrophobic and ionic properties of the substituted side-chain in these parent compounds, the photosensitizers may localize in different compartments of the tumour cells. As mentioned earlier the half-life of the singlet oxygen is very short and the site of localization / photoactivation of the photosensitizers is the primary target site of damage. Thus, targeting of the photosensitizer to mitochondria, endoplasmic reticulum (ER), and/or lysosomes may trigger the process of apoptosis. The events of cell damage caused by the photosensitizer can be studied systematically. However, redistribution of photosensitizers in cells after photoactivation has also been reported. Hence, multiple subcellular sites may be affected by the photosensitizers. The effects of subcellular localization / photoactivation of photosensitizers on the molecular events and the induction of cell death will be discussed below.

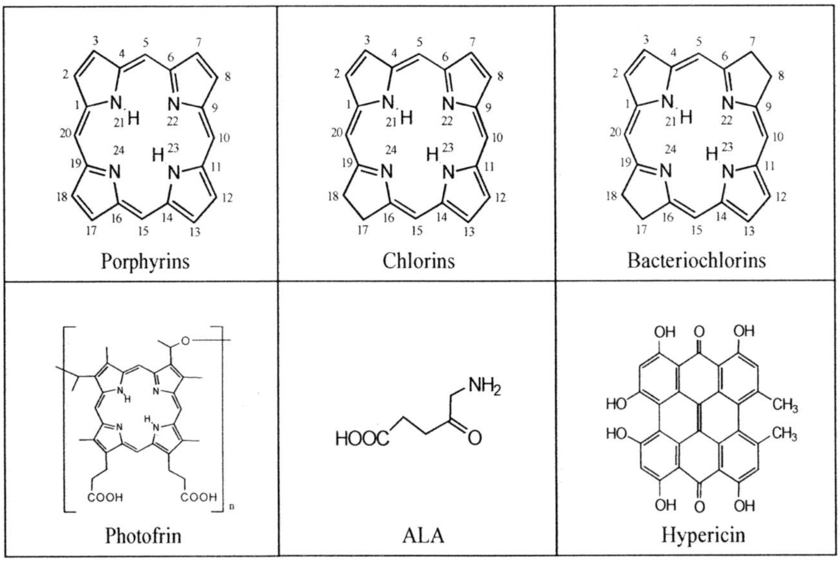

Figure 2. Basic chemical structure of selected photosensitizers. Reduction of one of the double bonds in the pyrrole group (C17 – C18) of porphyrins resulted in the formation of chlorins. Further reduction of another pyrrole group (C7 – C8) resulted in the formation of bacteriochlorins. ALA is a precursor of the photosensitizer protoprophyrin IX (PpIX).

Unlike other chemotherapeutic drugs, photosensitizers have many characteristics that make them suitable for the study of tumour cell apoptosis. The use of photosensitizers in the research of apoptosis is advantageous for several reasons: (i) It is a mechanism-based reaction. The activation process relies on the external source of light. In many cases, induction of tumour apoptosis can be achieved through the induction of oxidative stress and generation of ROS and singlet oxygen under a specific photodynamic condition. (ii) Many photosensitizers show distinct patterns of subcellular localization. This will facilitate the study to determine the importance of organelles in the cell death pathways. (iii) Localization of the photosensitizer can easily be monitored in the target cell or target tissue. The detection can easily be achieved by labeling the cell with fluorescence organelle-specific probes and determining the localization of the photosensitizer by overlapping the fluorescence images of the organelle-specific probes and the photosensitizer using confocal microscopy. (iv) Targeting of the photosensitizer to various organelles can be achieved by chemical modification of the functional group of the photosensitizer. This may improve the efficacy of the drug. (v) The process of cell death can only be initiated after light irradiation with an appropriate wavelength of light. The killing process is restricted to the target tissues exposed to light. This will limit the non-specific damage of normal tissue. (vi) It is possible to use certain prodrugs, such as ALA, for the purpose of both PDT and fluorescence photodetection of malignant cells.

An Overview on the Mechanisms of Cell Death

Cell death is a vital cellular process for development and homeostasis of multicellular organisms. It can be broadly classified into five different types based on the distinct morphological and biochemical features found in dying cells [1,18]: these include apoptosis, necrosis, autophagy, mitotic catastrophe and senescence. Apoptosis and autophagy are generally regarded to be "programmed cell death" due to the fact that these cell death processes are under strict genetic control [19,20]. On the other hand, necrosis and mitotic catastrophe are generally considered to be passive cellular responses to massive insults. However, recent findings suggest that these two forms of cell death may also be genetically controlled [21,22]. Senescence is a process of cellular aging and is a gene-directed program involving the shortening of chromosome ends (telomeres) and the activation of tumour suppression signaling pathways. Other modes of cell death have also been reported, including necroptosis, paraptosis, pyroptosis, and slow cell death, whose biochemical and morphological features differ somewhat from the five major modes of cell death mentioned above [18,23].

The term "apoptosis" was first coined by Kerr and co-workers in 1972 to describe controlled cell depletion occurring during normal development which could be distinguished from necrotic cell death resulting from acute tissue injury [24]. Apoptosis is an active, regulated form of cell death, which is a complex biochemical process involving multiple signaling pathways. This form of cell death has been conserved throughout evolution and plays pivotal roles in embryonic development, tissue homeostasis and remodeling, immune regulation, and has been implicated in various pathological states [25,26]. The morphological changes that occur during apoptosis include chromatin condensation, cytoplasmic shrinkage, plasma membrane blebbing, and nuclear membrane breakdown; whereas the molecular changes that arise during apoptosis involve inter-nucleosomal DNA cleavage and externalization of phosphatidylserine on the outer layer of the plasma membrane. Moreover, there are also apoptotic bodies formed during apoptosis [27], which are rapidly engulfed by neighboring phagocytic cells such as macrophages. "Necrosis" is defined as passive cell death triggered by diverse stimuli such as extreme temperatures, infections, toxins, and hypotonic conditions [28]. Traditionally necrosis is regarded as a form of cell death that is uncontrolled and pathological. However, more recent studies indicate that it is a regulated event that may have a significant role in developmental, physiological, and pathological conditions [18,29]. Biochemically it is characterized by depletion of cellular energy such as ATP, damage to membrane lipids, and loss of functions of ion pumps and channels. As a consequence, there is loss of membrane integrity, an increase in the influx of Ca^{2+} and a massive release of lysosomal and nuclear contents, which result in swelling and disruption of the cells. During necrosis, the release of cellular contents into the cellular environment will initiate an inflammatory process and cause necrosis and death of surrounding or neighboring cells [30].

In contrast to apoptosis and necrosis, "autophagy" is not synonymous with cell death [18]. It is an adaptive process in response to metabolic stresses such as nutrient depletion. Autophagy is exemplified by the degradation of intracellular proteins and organelles in the lysosomal pathway through which their constituents can be reused by the cell [31]. During the induction of autophagy, vesicles known as autophagosomes are formed by encapsulation of cytoplasmic components, intracellular organelles, and proteins in a double-layered, membrane-bound structure of endoplasmic reticulum. Autophagosomes then fuse with lysosomes, causing degradation of the autophagosomal contents by lysosomal enzymes [26]. Autophagy occurs in all eukaryotic cells and is an evolutionarily conserved pro-survival mechanism in which cellular components can be degraded for energy production during nutrient stress conditions, although excessive autophagy under some conditions can lead to cell death. Autophagic cell death is a nonapoptotic form of programmed cell death whose morphological features are quite distinct from apoptosis and necrosis [32]. Unlike apoptotic cells, there is an early degradation of organelles in the autophagic cells, but their cytoskeletal elements are preserved until the late stages. Moreover, in autophagic cell death, there is virtually no inflammatory response and caspase activation and DNA fragmentation occur (if at all) only at very late stages.

"Mitotic catastrophe" is another type of cell death caused by aberrant mitosis, in which there is an improper segregation of chromosomes during sister chromatid separation. Generally speaking, mitotic catastrophe may result from defects in cell cycle checkpoints or DNA damage [21]. It is also a highly conserved cellular mechanism in response to stress. It can be induced by chemotherapeutic agents that interfere with microtubule assembly and disassembly, thus, disrupting the mitotic spindle formation. Morphologically mitotic catastrophe is characterized by the formation of multinucleate giant cells that contain uncondensed chromosomes, which is quite distinct from apoptosis, necrosis, or autophagy [33]. On the other hand, "senescence" can be regarded as a form of "living cell death", since senescent cells undergo permanent growth arrest and loss of replication potential despite the fact that the integrity of their plasma membrane is maintained [1]. Replicative senescence is caused by DNA damage response triggered by the shortening of telomeres as a result of successive cell replication, whereas premature senescence is induced by DNA damage response activated by other cellular stresses, such as exposure to chemotherapeutic drugs, oncogenic or mitogenic signals [18]. Morphologically, senescent cells are larger in size with increased granularity; the cytoplasm is flattened and often vacuolated and distinct heterochromatic foci made up of transcriptionally inactive DNA are found in the nucleus. Biochemically, senescence is associated with changes in cellular metabolism and expression of the senescence-associated-galactosidase activity [1].

Among these five different types of cell death, apoptotic cell death has been studied most extensively due to the availability of many reagents and assay protocols for its detection, whereas examination of other modes of cell death has been hampered by the difficulties in quantitative measurement of cell death. In recent years, apoptosis is increasingly well-characterized both at the cellular and molecular levels. It can be divided into three phases:

initiation, execution and degradation [34]. The initiation phase can be triggered by a wide variety of apoptotic stimuli such as oxidative stress, DNA damage, ionizing radiation, chemotherapeutic drugs, nutrient deprivation and cytokines [25]. The execution or effector phase will lead to the activation of a number of proteases and nucleases, which participate in the degradation phase. Apoptosis is mediated by two central pathways, namely the "intrinsic or mitochondrial-mediated" pathway and the "extrinsic or death-receptor-mediated" pathway. These apoptotic pathways are driven by a family of highly specialized proteases known as caspases (cysteine-dependent aspartate-directed proteases) [26,35]. Two types of caspases can be distinguished; the upstream caspases are called initiator caspases which include caspases-8, -9, and -10. Their downstream targets are known as executioner or effector caspases, which include caspases-3, -6, and -7 [25]. Caspases-9 and -8 are initiators of the intrinsic and extrinsic pathways of apoptosis, respectively. The intrinsic pathway is characterized by dissipation of mitochondrial membrane potential, and release of mitochondrial cytochrome c leading to subsequent activation of initiator caspase-9. The extrinsic pathway is characterized by the activation of death receptors such as Fas and the resultant activation of initiator caspase-8. In fact, the two signaling cascades may sometimes be entwined. For example, caspase-9 and caspase-8 from both pathways can cleave and activate procaspase-3 into activated effector caspase-3 [36,37]. Activated caspase-3 then cleaves specific substrates including poly (ADP-ribose) polymerase (PARP), resulting in morphological and biochemical changes that characterize apoptotic cell death [38]. The intrinsic pathway is triggered by apoptotic stimuli, which induce mitochondrial membrane permeabilization (MMP) and leads to the release of cytochrome c from the mitochondrial intermembrane space. In the presence of ATP or dATP, cytochrome c binds to an adaptor protein, the apoptotic protease-activating factor-1 (Apaf-1), which recruits procaspase-9 to form an apoptosome [39]. Apaf-1 recruits procaspase-9 via its N-terminal caspase-activation recruitment domain (CARD). In its quiescent state, Apaf-1 is a compact molecule. In the presence of ATP or dATP, cytochrome c displaces the CARD domain, allowing the compact structure of Apaf-1 to stretch out into a more linear molecule to form an apoptosome with procaspase-9 [27]. This leads to autocatalytic activation of procaspase-9 to caspase-9. Activated caspase-9 then cleaves and activates the downstream procaspase-3, which is also recruited into the apoptosome [36]. On the other hand, in the extrinsic pathway, caspase-8 is activated when the death receptor Fas is activated by ligation of death receptors in the plasma membrane [37]. The Fas ligand (FasL) induces trimerization of Fas and the trimerized cytoplasmic region of Fas then transduces the death signal. When Fas is activated, an adaptor molecule called Fas-associated protein with death domain (FADD) is recruited to Fas and binds to Fas via the death domain at its C terminus [40]. The N terminus of FADD, known as the death-effector domain (DED), is responsible for recruiting procaspase-8. As a result, the death-inducing signaling complex (DISC) is formed, which leads to oligomerization, self-cleavage, and activation of procaspase-8 [37]. Activated caspase-8 then cleaves and activates downstream effector caspases such as caspases-3 and -7. In fact, caspase-8 also functions as a direct conduit between the extrinsic and intrinsic pathways by cleaving and activating Bid, a pro-apoptotic member of the Bcl-2 family. In response to FasL or TNF-α, activated caspase-8 cleaves Bid into a truncated carboxy-terminal fragment (tBid) that translocates from the cytosol to the mitochondria and inserts into the outer mitochondrial membrane (OMM) [41].

Oligomers of the tBid can trigger permeabilization of the OMM and the subsequent release of cytochrome c, which mediates activation of caspases such as caspases-9 and -3 in the intrinsic pathway.

During the execution phase of apoptosis, effector caspases cleave and inactivate certain vital cellular proteins including DNA repair enzymes, major cytoskeletal proteins of the cytoplasm and nucleus, and a number of protein kinases [42]. Effector caspases are known to play a dual role as executioners that cleave key proteins to promote death activities, and as regulators that turn off the cell-survival pathways [39]. One such death-promoting factor is a nuclease called caspase-activated DNase (CAD) [43]. CAD is ubiquitously expressed in various tissues and is normally inactivated by binding to its inhibitor iCAD. During apoptosis, iCAD is cleaved by effector caspases-3 and -7, leading to the release of active CAD, which then cleaves chromosomal DNA and produces nucleosomal DNA fragments. Moreover, CAD can induce the collapse and hypercondensation of the chromatin against the nuclear periphery [43].

Although caspases are known to act as central effectors of apoptotic cell death, recent evidence has shown that apoptosis can also occur in the absence of caspases in a number of *in vivo* cell death models [44,45] and that inhibition of caspases failed to block apoptosis in cultured cells in response to a variety of apoptotic stimuli [46-48], suggesting that caspase-independent mechanisms might exist as backups to regulate apoptosis [25]. In fact, non-caspase proteases such as granzymes, cathepsins and calpains have been reported to act as effector molecules of apoptosis in a number of *in vitro* studies [49-52]. Moreover, in addition to the mitochondria, other subcellular organelles and structures, including the endoplasmic reticulum, Golgi apparatus, and lysosomes, were found to generate caspase-independent signaling intermediates which might be involved in the effector and degradation phases of apoptosis [53]. In the following sections, the mechanisms by which photosensitizers may lead to the apoptotic cell death will be discussed.

Mechanisms of Photosensitizer-Induced Cell Death

A great deal of experimental evidence showed that PDT control of tumour growth occurs via multiple mechanisms. These include (1) the direct tumour cell killing through the induction of apoptosis, necrosis, and / or autophagy, (2) acting on the vascular bed in the tumour mass, and (3) modulation of immune response through the production of cytokines and inflammatory cytokines during photoactivation. For the direct induction of tumour cell death, subcellular localization / photoactivation of photosensitizers is an important issue in PDT. The chemical nature of photosensitizers may determine their intracellular localization. Lipophilic photosensitizers tend to bind to membranes, water-soluble photosensitizers are frequently found in lysosomes, and cationic photosensitizers are found in mitochondria and nuclei. In view of the short diffusing distance of singlet oxygen, it is generally believed that the site of photosensitizer localization / photoactivation is the target of photodamage. This may also explain the low direct DNA damaging effect of many organelle-localizing photosensitizers. This section summarizes the localization and effects of some selected photosensitizers on various organelles. It should also be noted that the primary sites of localization of certain photosensitizers may not be as important as their secondary sites in inducing cell death [54].

Mitochondria as a Major Target Leading to Apoptotic Cell Death

Mitochondria are organelles with two separated layers of membrane. The outer layer is smooth and continuous, while the inner layer is folded inside and forms the tubular structure (cristae). It is now clear that mitochondria are not only involved in energy metabolism, but also in the modulation of intracellular calcium signaling and the cell death process [55-57]. The space between the outer and inner membranes is filled with enzymes responsible for the energy production. The pro-apoptotic protein cytochrome c is mainly stored in the cristal lumen [58], while apoptosis-inducing factor (AIF), pro-caspases-2, -3, and 9 are found in the intermembrane space and cristal lumen [59-61]. Hence, the mitochondrial dynamics are important in the regulation of cellular apoptosis [62].

In view of the biological significance of mitochondria in controlling the life and death of a cell, designing mitochondrion-targeted cytotoxic drugs has been suggested as a novel strategy for cancer therapy. This is aimed at overcoming apoptosis-resistant tumour cells due to disruption of the p53 or death receptor pathway [63]. Various mechanism-based compounds have been developed and used to target mitochondria or mitochondria-mediated pathway for the purpose of tumour cell killing [64,65]. Among these compounds, the cationic drugs receive special interest. The preferentially accumulation of cationic lipophilic molecules in the mitochondria is due to the presence of an electrochemical gradient across the mitochondrial membrane. For example, a normal mitochondrial membrane of an intact hepatocyte may have a membrane potential ($\triangle\psi m$) of 161mV [66]. It has been estimated that every 61.5mV increase in $\triangle\psi m$ may lead to a 10-fold increase in the concentration of cationic molecules in the mitochondrial matrix [63]. As a result of this, a hundred- to a thousand-fold higher concentration of these cationic drugs may accumulate in the mitochondrial matrix rather than in the cytosol. Examples of cationic mitochondria-localizing photosensitizers are rhodamine-123 [67] and methylene blue derivative (MBD) [68]. In a recent study of rhodamine-123 and its monobromo to tetrabromo derivatives, Lacerda et al. demonstrated that the selective phototoxicity toward tumour cells is highly modulated by the lipophilic/hydrophilic characteristics of the photosensitizer [69]. These authors also proposed that the lipophilic/hydrophilic characteristics of the photosensitizers designed for photochemotherapy of neoplastic disease should fall within just about two orders of magnitude above the value of rhodamine-123. Photosensitizers of other chemical properties have also been found to localize at mitochondria (Table 1). However, the mechanisms of accumulation of these photosensitizers have not been fully studied.

Table 1. Site of localization / target of some selected photosensitizers

Site	Photosensitizer	Cell model	Reference
Plasma membrane	Photofrin *	A431 cells	[137]
	Dithiaporphyrins 1 and 2	R3230AC cells	[138]
	Chloroaluminum phthalocyanine (AlPc)	Chinese hamster ovary cells	[139]
	MC540	M1 and JCS myeloid leukemia cells	[140]
	Aminopyropheophorbide (APP)	HCT-116 colon cancer cells	[135]
Mitochondria	Zn-BC-AM	HONE-1 NPC cells	[88]
	Porphycene derivative PS6A	HONE-1 NPC cells	[16]
	ALA/protoporphyrin IX (PpIX)	COH-BR1 breast cancer cells	[141]
		NIH 3T3 cells	[142]
		MCF-7 breast cancer cells	[143]
		HL-60 cells	[144]
		Rat neonatal myocardial cells	[145]
	Hypericin	CNE-2 NPC cells	[146]
	MC540	M1 and JCS myeloid leukemia cells	[140]
	Pc4	Mouse LY-R lymphoma cells	[147]
	mTHPC	CNE-2 and HK-1 NPC cells	[148]
	Tin etiopurpurin	P388 murine leukemia cells	[111]
	Victoria Blue BO (VBBO)	Chinese hamster ovary cells (C-1)	[149]
	Methylene blue derivative (MBD)	V79 Chinese hamster fibroblasts	[150]
	Rhodamine-123	MGH-U1 human carcinoma cells	[151]
Endoplasmic reticulum	Azadipyrromethene (ADPM)01 and ADPM06	HeLa cells	[152]
	Zn-BC-AM	HONE-1 NPC cells	[88]
	mTHPC	MCF-7 cells	[85]
	ALA	HL-60 cells	[144]
	Hypericin (in perinuclear cytoplasmic area)	Adenocarcinoma cells WiDr, glioblastoma D54Mg	[153]
	Tolyporphin	EMT-6 cells	[154]

Table 1. Site of localization / target of some selected photosensitizers (Continued)

Site	Photosensitizer	Cell model	Reference
Lysosome	Hypericin	CNE-2 and TW0-1 NPC cells	[146]
	Pc4	Mouse LY-R lymphoma cells	[147]
		Rat R1 fibroblasts	[155]
	Hypocrellins A and B	NPC cells	[156]
	Dihydroxychlorins	Jurkat cells	[157]
	ATX-S10(Na)	Human malignant melanoma cells	[113]
	NPe6	Murine hepatoma cells	[158]
	Porphycene derivative PS6A	CNE-2 cells	[16]
	Tin etiopurpurin	P388 murine leukemia cells	[111]
	Mono-L-aspartyl chlorin e6	Potorous tridactylus (PTK2) cells	[159], [160]
Golgi body	Photofrin	A431 cells	[137]
	Pc4	Mouse LY-R lymphoma cells	[147]
	mTHPC	MCF-7 cells	[85]
	Zinc (II) phthalocyanine (ZnPc)	4R transformed fibroblast	[100]
Cytoplasm	Protoporphyrin IX (PpIX)	LLC cells	[161]
	ALA (in peri-nuclear cytoplasm)	Bovine pulmonary artery endothelium CPAE and rat kidney cell PTK2 cells	[145]
	Hypericin (in perinuclear cytoplasmic area)	Adenocarcinoma cells WiDr, glioblastoma D54Mg	[153]
Nucleus	Chlorin 6(6) conjugated with nucleus directed peptide	CHO cells	[162]

* With a short incubation time, photofrin localizes at the plasma membrane. With a long incubation time, it localizes to other intracellular membrane system [163].

Although many of these photosensitizers can be found in mitochondria, the exact molecular target of these photosensitizers has not been fully investigated. The mitochondrial permeability transition pore complex (PTPC) consists of the mitochondrial outer membrane voltage-dependent anion channel (VDAC), the peripheral benzodiazepine receptor (PBR), and the mitochondrial inner membrane adenine nucleotide translocator (ANT). VDACs are pore-forming proteins located at the outer mitochondria membrane. Pore formation by homo-oligomerization of VDAC or hetero-oligomerization of VDAC with pro-apoptotic proteins such as Bak or Bax has been suggested to be involved in controlling the release of apoptogenic proteins from mitochondria [70]. Opening of PTPC is known to cause membrane permeabilization, collapse of membrane potential, and initiation of mitochondria-mediated apoptotic events. For this reason, it is logical to deduce that PTPC may be the target of photodamage on mitochondria. Using the lymphoma cell model, light-activated Verteporfin was found to cause a dissipation of the mitochondrial inner transmembrane potential which is followed by tumour cell apoptosis [71]. The process of cell death could be abolished by transfection with the anti-apoptotic *Bcl*-2 gene. Using a cell-free system, these authors further demonstrated that Verteoporfin photodamages ANT-containing liposomes [71]. In another study, hexaminolevulinate (HAL)-PDT appears to target the PBR in Reh human leukemia cells [72]. All of these studies suggested that components of PTPC could be targets for some of the photosensitizers.

It is now clear that proteins of the Bcl-2 family play an important role in regulating the mitochondria-mediated apoptotic cell death pathway. The anti-apoptotic proteins such as Bcl-2 and Bcl-xL are known to stabilize the mitochondrial membrane and prevent mitochondria damage, while the pro-apoptotic proteins Bax and Bak are involved in inducing cell death via mitochondrial membrane permeabilization [73]. Using the mitochondrial photosensitizer Pc4, Xue demonstrated that Pc4 caused the disappearance of the anti-apoptotic Bcl-2 protein in MCF-7 cells [74]. Photochemical damage of Bax, Bad, VDAC, or the ANT was not seen. Subsequently, these authors further found that Bcl-xL localized to mitochondria was more sensitive to photodamage than was Bcl-xL in the cytosol [75]. In another study, photodamage of Bcl-2 was also seen in Pc4 PDT-induced MCF-7 cell apoptosis [76]. Using GFP- or Xpress-tagged Bcl-2, Usuda et al. further found that the PDT-induced disappearance of the Bcl-2 was accompanied by the appearance of cross-linked Bcl-2 protein. These authors further hypothesized that the region between the BH1 and BH2 domains, corresponding to the α5- and α6-helices of Bcl-2, is an important region for forming the photodamageable structure [77]. In this model system, both Bcl-2 and Bcl-xL appear to be the targets of Pc4 PDT. The effect of PDT on Bcl-2 is not limited to Pc4. Amphipathic photosensitizer, disulfonated diphthalimidomethyl phthalocyanine zinc (ZnPcS2P2), also down-regulated Bcl-2 protein levels in PDT-treated HL60 cells [78], indicating that Bcl-2 is a target of PDT with these photosensitizers. Taken together, both PTPC and Bcl-2 are the mitochondrial targets for certain photosensitizers. This may explain the general observation that photosensitizers localized at the mitochondria are very efficient in triggering apoptosis. In addition to the two strategies previously used to target mitochondria, namely the use of thio cross-linker for ANT and the use of ligands of the PBR that can interact with ANT [63], PDT-induced damage of Bcl-2 proteins may be considered as the third strategy to overcome Bcl-2-mediated cytoprotection.

Endoplasmic Reticulum (ER) as an Apoptotic Signaling Initiator

ER consists of a membranous system interconnected with many other membrane-bound organelles, such as the outer membrane of the nucleus and the Golgi body. ER is the organelle responsible for protein synthesis, protein post-translational modification, and protein folding. The process of protein folding is controlled by ER resident proteins and chaperones. ER is also the major intracellular store of calcium (Ca^{2+}). There was an estimation that the concentrations of intramembrane Ca^{2+} ranged from 2 to 5 mM [79].

Hence, ER plays an important role in conveying intracellular signal to other cellular compartments and maintaining a proper function of ER is required for the cell to carry out normal functions. Any disturbance, known as ER stress, of ER may result in inhibition of protein synthesis and glycosylation, accumulation of unfolded protein, interruption of protein trafficking to other organelles, and the release of Ca^{2+} into the cytosol. Oxidative stress is one of the stimuli that can induce ER stress [80,81]. Several distinct responses of a cell to ER stress have been identified [82-84]. These include the up-regulation of expression and production of ER chaperones for the prevention of protein folding, degradation of misfolded proteins by the proteasome, blocking of the translation through phosphorylation of eIF2α, and the induction of apoptosis if the damage is beyond repair.

The involvement of ER in PDT-induced apoptosis has also been studied in the past. Teiten and co-workers demonstrated that ER and the Golgi apparatus are the preferential sites for mTHPC to accumulate. These organelles are also the primary intracellular sites of damage by mTHPC-mediated PDT in MCF-7 cells, as the accumulation of mTHPC in mitochondria and lysosomes was weak [85,86]. In this model, mitochondria are not directly affected by mTHPC/PDT, and it is only involved at a later stage of the cell death process via an indirect mechanism. The involvement of ER-derived Ca^{2+} in PDT-induced cell death has been reported. Verteporfin-PDT was found to rapidly empty Ca^{2+} from ER in HeLa cells [87]. In a separate study, Zn-BC-AM PDT was found to enhance the expression of ER chaperones Bip/Grp78 and Grp94, and the ER resident lectin-like chaperone calnexin (CNX). Furthermore, buffering of intracellular Ca^{2+} released from ER resulted in a significant reduction of Zn-BC-AM PDT-induced cell death [88]. Similarly, photodamage of ER and disruption of Ca^{2+} homeostasis were observed in hypericin PDT-treated cells [89]. Further study using the ER localizing photosensitizer porphyrin showed that porphyrin PDT induced the expression of the ER stress-inducible transcription factor C/EBP homologous protein (CHOP) and the 78 kDa glucose regulated protein (Grp-78). When compared with chop -/- cells, porphyrin PDT caused an enhanced apoptotic cell death in chop +/+ cells [90]. In a recent study with the ER photosensitizer CPO (9-capronyloxytetrakis (methoxyethyl) porphycene, CPO-PDT induces both apoptosis and autophagy in L1210 leukemia cells after ER photodamage. The process of autophagy could be inhibited by PI3K inhibitors [91]. However, detailed mechanism of CPO PDT-induced ER stress leading to apoptosis and autophagy was still unknown, although caspase-12 has been implicated to play a role in apoptosis under ER stress [92,93]. However, in the colon cancer cell model, PMEE (Pyropheophorbide-a methylester) was found to localize in the ER/Golgi system and lysosomes, but caspase-12 was not involved in PMEE/PDT-induced apoptosis [94]. The discrepancy in the observation of the actions of caspase-12 is probably due to the fact that in human cells caspase-12 is mainly involved in the maturation of cytokines, whereas in mouse cells it is involved in apoptosis induced by ER stress [95]. Taken together, PDT-induced Ca^{2+} release from the ER may confer cell sensitivity to mitochondria-mediated apoptotic cell death.

Golgi Complex as an Additional Stress Sensor

The Golgi complex, consisting of polarized stacks of flattened membranous structure, is usually in close association with the rough ER. It has been known for a long time that Golgi complex is mainly involved in protein processing and protein transportation. The additional role of Golgi complex in sensing of intracellular stress and in initiation of apoptosis has recently been reviewed [96]. Like caspases-8, -9, and -10, caspase-2 is a pro-apoptotic caspase, but the localization of the caspase-2 is distinct from caspases-8 and -9. Caspase-2 is localized predominantly at the Golgi complex and nucleus [97]. Activation of caspase-2 is achieved through dimerization and followed by autocatalytic cleavage [98]. Upon induction of apoptosis by staurosporine, activated caspase-2 cleaves the membrane component golgin-160 [97]. The cleavage of golgin-160 at the Golgi appears to be essential for the development of apoptosis. These authors further hypothesized that the Golgi complex transduces pro-apoptotic signals through caspase-2. The role of golgin-160 in transducing signals arising from ER was also demonstrated by Maag et al. using cell lines stably expressing caspase-resistant golgin-160 [99].

In the mechanistic study of zinc (II) phthalocyanine (ZnPc) PDT, Fabris et al. observed that the Golgi complex was the primary target of photodamage by ZnPc PDT [100]. In a more recent study, caspase-2 activation in the Golgi region was observed immediately after ZnPc photodynamic treatment of A549 cells [101]. The morphology of this organelle also starts to disrupt a few minutes after photoactivation. Results from these studies are in agreement with the recent view on the stress sensing function of Golgi complex and the role of caspase-2 in transmission of apoptotic signals originating from the Golgi complex [102].

Lysosome as a Target for the Future Cancer Therapy

Lysosomes are membrane-bound vesicles containing a number of lysosomal enzymes capable of breaking down a wide range of exogenous and endogenous macromolecules. Hence, the functions of lysosomes are involved in recycling of intracellular molecules, and the breaking down of exogenous antigens for antigen presentation. Cathepsins, a group of well-characterized endopeptidases with varying substrate specificities, have recently been implicated to play an important role in normal and malignant cell development. Cathepsins are divided into 3 groups, namely cysteine proteases (cathepsin B, C, H, F, K, L, O, S, V, W, and X/Z), aspartate proteases (cathepsin D and E), and serine protease (cathepsin G). The cysteine proteases are separately encoded by 11 genes and the expression levels vary among different tissues [103]. Although both cathepsin D and cathepsin E are aspartate proteases, cathepsin E is mainly localized in endosomes and its function is primarily involved in antigen presentation by the Class II MHC system [104,105]. In addition to the roles in cancer progression and metastasis, cathepsin D appears to function as a key mediator in the lysosome-mediated-apoptotic cell death pathway [106].

There is growing evidence that the lysosome is important in initiating the lysosome-mitochondria cell death pathway. The pathway can be initiated by a variety of stimuli, such as through the stimulation of TNF receptor [107], lysosomotropic agents [108], and oxidative stress [109]. Recent studies indicate that lysosomal membrane permeabilization (LMP) is an early event preceding the release of capthespins into the cytosol. The downstream events of LMP include the increase in MMP and release of AIF, which is followed by apoptosis [110]. In the study of PDT, many photosensitizers are found to localize at lysosomes (Table 1). Early studies showed that photodamage of lysosomes and mitochondria by tin etiopurpurin PDT resulted in a rapid apoptotic response [111]. Similarly, photosensitizers that localize at lysosomes and other organelles were found to induce apoptosis [16,94]. The more direct linkage between lysosomes and mitochondria in PDT-induced apoptosis was recently demonstrated by Reniners and co-workers [112]. Using murine hepatoma cells, the lysosomal sensitizer NPe6 was found to localize to lysosomes. Upon photoactivation, lysosome damage was found to precede the cleavage/activation of Bid and cytochrome c. This study clearly supports the model of lysosome-mitochondria apoptotic pathway. The role of lysosomes in PDT-induced apoptosis was also supported in the study using the amphiphilic photosensitizer ATX-S10(Na). Nagata and co-workers found that lysosomes are the primary target site for accumulation of ATX-S10(Na). The apoptotic event was initiated through the damage of lysosome and followed by mitochondrial destabilization [113].

Although lysosomes are the primary targets for these sensitizers, the molecular targets on the lysosome have not been fully investigated. Direct damage of lysosome, due to the oxidative stress during photoactivation, may explain the release of cathepsins into the cytosol. This is supported by the other studies that lysosomal membrane can be destabilized by oxidative stress [114]. Using the H_2O_2-induced oxidative stress model, Antunes and co-workers also demonstrated that lysosomal rupture is an early upstream initiating event induced by oxidative stress [115]. Hence, the event of PDT-induced LMP in PDT-treated cells is similar to the induction of LMP by certain anti-cancer drugs [116]. Conversely, the apoptosis-inducing signal delivered to lysosomes may be derived from the cytosol. Kagedal and co-workers demonstrated that exposing the human fibroblasts to the apoptosis-inducing drug staurosporine resulted in the release of cathepsin D to the cytosol. Insertion of Bax into the lysosomal membrane is responsible for the induction of release of these enzymes [117]. Whether a similar mechanism is also involved in the PDT-treated cells remains to be determined.

Effects on the Intracellular Signaling Pathways

PDT has been shown to activate a wide range of intracellular signaling pathways. As most of the photosensitizers generate singlet oxygen during photoactivation, the signaling pathways known to be sensitive to oxidative stress are most likely to be affected by PDT. The mitogen-activated protein kinase (MAPK) signaling pathway is one of the most well characterized intracellular signaling pathways. The kinases in this signal transduction system sense the stress signals that arise both intracellularly or from outside of the cells. Examples for these stimuli include oxidative stress caused by chemotherapeutic drugs, UV-radiation, binding of inflammatory cytokines to cell surface receptors, and PDT. Extracellular signal regulated kinases (ERKs), the c-Jun N-terminal kinases (JNKs)/stress activated protein kinases (SAPKs), and the p38-MAPKs are the three most studied signaling pathways in this stress signaling system. In each case, the MAPKs are activated in response to activation of a dual-specificity MAPK kinase (MAPKK/MEK). The activating MAPKKs phosphorylate their target MAPKs on conserved threonine and tyrosine residues, which are separated by an intervening amino acid that is characteristic for each MAPK family. The upstream MAPKKs and MAPKK kinases (MAPKKK) are distinct for each pathway, but some crosstalks may be possible. ERKs are activated by mitogens (e.g. growth factor binding) and stress stimuli and are important for cell survival. JNKs and p38-MAPKs are activated by environmental stress and cellular stress (e.g. oxidative stress), and are, thus, involved in apoptosis.

In PDT, the effects of PDT-induced MAPK activation on cell death are controversial (Table 2). Previous studies showed that p38 and JNK were activated in cells after benzoporphyrin derivative monoacid ring A (BPD) - mediated PDT [118]. In this study, activation of ERK1/2 was not observed and reactive oxygen intermediates were shown to play a significant role in PDT-induced signaling. A similar observation was also made on ALA PDT-treated human skin cells [119]. In the animal study, an increase in phosphorylation of p38 MAPK was also observed in tumour-bearing nude mice after phthalocyanine Pc4 PDT [120]. However, the role of these stress pathways on tumour cell death was not examined in these studies. Using pathway-specific inhibitors, Xue and co-workers demonstrated that the p38 MAPK inhibitor blocked Pc4 PDT-induced apoptosis in leukemic lymphoblasts [121]. Inhibition of JNK also reduced PDT-induced caspase-3 activation [122]. However, in the case of hypericin PDT, HeLa cell transfection with the dominant negative inhibitors of the JNK or pretreatment with the p38 MAPK inhibitor PD169316 was found to have an enhancing effect in PDT-induced apoptosis [123]. In this model, JNK and p38 MAPK pathways appear to play an important role in cellular resistance against PDT-induced apoptosis with hypericin. Furthermore, using a PDT-sensitive GM38A cell line, Tong et al. demonstrated that transient ERK activation leads to cell death, while a sustained ERK activation protected the resistant LFS807 cells from Photofrin-mediated cytotoxicity [124].

As with all chemical inhibitors, there is a concern about the specificity of the “pathway-specific inhibitors” at high drug concentrations. Genetic approaches (e.g. by over-expression or by knocking out the gene functions) appear to be appropriate in dissecting the role of these pathways. Using both pathway-specific inhibitors and the genetic approaches, Hendrickx and co-workers recently demonstrated that over-expression of WT-p38 MAPK increased cellular resistance (HeLa cells) to hypericin-PDT induced-apoptosis [125]. These authors further proposed that a combination of PDT with pyridinyl imidazole inhibitors of p38 may improve the therapeutic efficacy of PDT.

Taken together, the response of the MAPK signaling pathways to PDT is complex. It is unlikely that there is a single universal mechanism for the response of cells to PDT [13]. The results derived from one tumour model may not be applied to another tumour model and the effect sometimes is cell type dependent. All of these signaling pathways, the lysosomes, ER, Golgi complex, and mitochondria are involved in PDT-induced apoptotic cell death. The temporal relationship between the MAPK signaling pathways and these organelles, especially the organelles with stress sensing function, is still unknown.

Table 2. Intracellular stress signaling pathways being studied in PDT

Photo-sensitizer	Pathway being studied	Implication in apoptosis	Cell model	Reference
Pc 4	p38 MAP kinase activation	Not determined	HCT116 cell line	[164]
	P38 activation	Triggering of apoptosis, but not confirmed	Colon cancer xenograft	[120]
	SAPK/JNK and p38/HOG	High level of p38 may poise the cells for rapid activation of apoptosis	LY-R and CHO cells	[121]
Hypericin	p38 MAP kinase activation	Protective role in PDT-treated cells	Mouse embryonic fibroblasts MEF	[165]
			HeLa and T24 bladder carcinoma	[125]
	JNK and p38 activation	Increase cellular resistance against PDT-induced apoptosis	HeLa cells	[123]
Photofrin	JNK1 and p38	No effect on PDT-treated cells	LFS cells	[166]
	JNK activation	Not determined	A431 carcinoma	[137]
	ERK1/2	Sustained ERK1/2 activation protects cells from Photofrin-mediated phototoxicity	LFS087 cells	[124]
ALA	JNK and p38	Not determined	Human skin fibroblasts	[167]
	JNK and p38	Not determined	Human skin cells	[119]
RB	ERK, JNK, p38	Associate apoptosis with JNK and p38 activation	Fibroblasts	[168]
	JNK activation	JNK is required for caspase-3 activation	A431 carcinoma	[122]
BPD-MA	SAPK and p38 HOG1	Not determined	Murine keratinocyte cell line	[118]

Ceramide in Apoptosis Signaling Pathway

Sphingolipid metabolites are second messengers involved in the signaling pathways activated by cytokines, growth factors, and cellular stress [126]. The sphingolipid metabolites, such as ceramide and sphingosine are considered to be endogenous mediators of apoptosis in response to cellular stress [127]. Ceramide can be produced from sphingomyelin or by *de novo* synthesis from palmitoyl CoA [128]. The ER appears to be the major site of *de novo* synthesis of ceramide. Earlier study on L5178Y lymphoma, U937 leukemia, CHO Chinese hamster ovary cells, and RIF-1 fibrosarcoma cells showed that the production of ceramide was rapidly increased in cells treated with Pc4 PDT [129,130]. The accumulation of ceramide is associated with PDT-induced apoptosis. Subsequent study using CHO-derived mutants deficient in the *de novo* sphingolipid synthesis enzyme serine palmitoyltransferase (SPT) showed that *de novo* generated ceramide and / or sphinganine initiates apoptosis, and the process is downstream of the release of cytochrome c from mitochondria [128]. Similarly, Dolgachev et al. also demonstrated the accumulation of *de novo* ceramide in Pc4 photosensitized Jurkat and CHO cells in the absence of SPT up-regulation [131]. By comparing the Pc4 PDT-induced apoptosis with other apoptosis-inducing agents such as the Bcl-2 inhibitor HA14-1, Separovic et al. found that the ceramide response to PDT in Jurkat cells is not influenced by other pro-apoptotic stimuli and the mechanism of *de novo* ceramide accumulation is an unique lipid response, as a result of oxidative stress, during PDT-induced apoptosis [132]. So far most of the studies on the ceramide-apoptosis signaling pathways center on Pc4 photosensitizer; the involvement of ceramide and other sphingolipid metabolites in PDT-induced apoptosis is still not fully investigated.

Effects on the Transcription Factors

Normal cell growth and differentiation are known to be controlled by transcription factors. Alteration of activities of transcription factors may lead to deregulation of gene expression and, hence, affect the cell development and survival. PDT has been shown to affect the activities of a number of transcription factors. These include NF-kB, AP-1, HIF-1α, CHOP, and E2F (Table 3). Although activation of transcription factors was reported in cells after PDT, the exact role of these transcription factors in PDT-induced apoptosis was not fully studied. C/EBP homologous protein (CHOP), also known as growth arrest- and DNA damage-inducible gene 153 (GADD153), is an ER stress inducible transcription factor involved in the development of apoptosis [133]. Using CHOP deficient (chop - / -) mouse embryonic fibroblasts (MEFs), Wong et al. clearly demonstrated that the increase in CHOP expression is associated with porphyrin PDT-induced apoptosis [90]. In the A431 epidermoid carcinoma cell model, Pc4 PDT results in a time-dependent accumulation of hypo-phosphorylated pRb proteins and also the down-regulation of protein expression of all five E2F (1-5) family transcription factors [134]. These two studies clearly demonstrated the role of CHOP and E2F in PDT-induced apoptotic cell death. Using the HCT-116 cell line expressing the dominant negative mutant of inhibitor-kappa B alpha, Matroule et al. demonstrated that NF-kB is involved in the anti-apoptotic response of HCT-116 cell to APP-mediated photosensitization [135]. It is now clear that NF-kB activation may have apoptotic or anti-apoptotic effects in cells [136]. Whether the anti-apoptotic response due to NF-kB activation is cell type or photosensitizer specific remains to be determined.

Table 3. PDT-induced activation of transcription factors

Photo-sensitizer	Cell model	Transcription factor being studied	Effect	Reference
Porphyrin	CHOP +/+ and CHOP -/- mouse embryonic fibroblast	Induces ER-stress inducible transcription factor CHOP	Increase apoptosis	[90]
Photofrin	Transplantable BA mouse mammary carcinoma	Induce expression of HIF-1α	Effect on apoptosis was not determined	[169]
	HeLa epithelial cells	AP-1	IL-6 secretion	[170]
	L1210 cells	Increase NF-kB binding activity	Increase NF-kB binding activity	[171]
HpD	Esophageal cancer cells from patients	HIF-1α and HIF-2α	As a predictors of sensitivity to PDT	[172]
ALA	Esophageal normal Het-1A and squamous cell carcinoma KYSE-70 and KYSE-450 cell lines	HIF-1α	Increase cellular resistance to PDT	[173]
Pc4	A431 cells	Decrease hyper-phosphorylated form of pRb and E2F1 - 5	Induce cell cycle G1 arrest and apoptosis	[134]
PPME	Colon cancer cells	Activate NF-kB	Counteract apoptosis	[94]
APP	HCY-116 colon cancer cells	NF-kB	Anti-apoptotic response	[135]

Conclusions

In view of the chemical nature of the photosensitizers, many of the photosensitizers used in PDT localize in various subcellular structures. Plasma membrane, ER, Golgi complex, lysosomes, and mitochondria are the targets for these photosensitizers (Figure 3). The stress signals generated from the damaged organelles may converge, via the mitochondria-dependent or mitochondria-independent pathways, to the effector caspase system for cell destruction. Mitochondria are generally considered as the major targets of PDT. The roles of the lipid messenger ceramide, the intracellular stress signaling pathways, and transcription factors in the regulation of PDT-induced apoptosis have not yet been fully elucidated. Understanding the signaling mechanisms in PDT may provide means to modulate the PDT effects at the molecular level and potentiate its anti-tumour effectiveness.

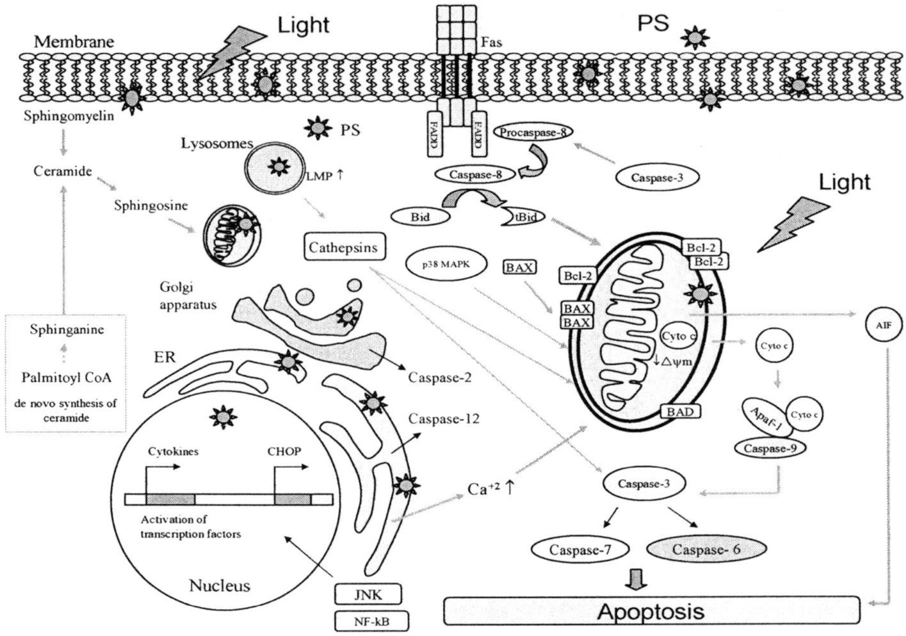

Figure 3. PDT-induced activation of cell death pathways. Photosensitizers (PS) localizes at various subcellular compartments. Activation signals, as a result of photoactivation of the PS, are conveyed from the site of activation to mitochondria and other cell death executioners for the initiation of cell death process.

Acknowledgements

This work was supported by the Research Grant Council of Hong Kong (UST/CA/03-04/2) and Faculty Development Grant of Hong Kong Baptist University (FRG/04-05/II-63).

References

[1] Okada H, Mak TW. Pathways of apoptotic and non-apoptotic death in tumour cells. *Nat Rev Cancer* 2004; *4*(8):592-603.

[2] Moan J, Berg K. The photodegradation of porphyrins in cells can be used to estimate the lifetime of singlet oxygen. *Photochem Photobiol* 1991; *53*(4):549-553.

[3] Snyder JW, Lambert JD, Ogilby PR. 5,10,15,20-tetrakis(N-methyl-4-pyridyl)-21H,23H-porphine (TMPyP) as a sensitizer for singlet oxygen imaging in cells: characterizing the irradiation-dependent behavior of TMPyP in a single cell. *Photochem Photobiol* 2006; *82*(1):177-184.

[4] Raab O. Uber die Wirkung fluoreszierender Stoffe auf Infusorien. *Zeitung Biol* 1900; *39*:524-546.

[5] Von Tappeiner H, Jodlbauer A. Die Sensibilisieren-de Wirkung fluorieszierender Substanzer. Gesammte Untersuchungen uber die photodynamische Erscheinung. F.C.W. Vogel, Leipzig, 1907.

[6] Hausmann W. Die sensibilisierende Wrikung des Hematoporphyrins. *Biochem Z* 1911; *30*:276-316.

[7] Schwartz S.K, Absolon K, Vermund H. Some relationships of porphyrins, x-rays, and tumours. *Univ Minn Med Bull* 1955; *27*:7-8.

[8] Dougherty TJ, Grindey GB, Fiel R, Weishaupt KR, Boyle DG. Photoradiation therapy. II. Cure of animal tumors with hematoporphyrin and light. *J Natl Cancer Inst* 1975; *55*(1):115-121.

[9] Dougherty TJ, Kaufman JE, Goldfarb A *et al.* Photoradiation therapy for the treatment of malignant tumors. *Cancer Res* 1978; *38*(8):2628-2635.

[10] Dougherty TJ. Photodynamic therapy (PDT) of malignant tumors. *Crit Rev Oncol Hematol* 1984; *2*(2):83-116.

[11] Little FM, Gomer CJ, Hyman S, Apuzzo ML. Observations in studies of quantitative kinetics of tritium labelled hematoporphyrin derivatives (HpDI and HpDII) in the normal and neoplastic rat brain model. *J Neurooncol* 1984; *2*(4):361-370.

[12] Nyman ES, Hynninen PH. Research advances in the use of tetrapyrrolic photosensitizers for photodynamic therapy. *J Photochem Photobiol B* 2004; *73*(1-2):1-28.

[13] Oleinick NL, Morris RL, Belichenko I. The role of apoptosis in response to photodynamic therapy: what, where, why, and how. *Photochem Photobiol Sci* 2002; *1*(1):1-21.

[14] Jori G. Photosensitized processes in vivo: proposed phototherapeutic applications. *Photochem Photobiol* 1990; *52*(2):439-443.

[15] Szeimies RM, Calzavara-Pinton P, Karrer S, Ortel B, Landthaler M. Topical photodynamic therapy in dermatology. *J Photochem Photobiol B* 1996; *36*(2):213-219.

[16] Mak NK, Kok TW, Wong RN *et al.* Photodynamic activities of sulfonamide derivatives of porphycene on nasopharyngeal carcinoma cells. *J Biomed Sci* 2003; *10*(4):418-429.

[17] Kral V, Davis J, Andrievsky A *et al.* Synthesis and biolocalization of water-soluble sapphyrins. *J Med Chem* 2002; *45*(5):1073-1078.

[18] Ricci MS, Zong WX. Chemotherapeutic approaches for targeting cell death pathways. *Oncologist* 2006; *11*(4):342-357.

[19] Danial NN, Korsmeyer SJ. Cell death: critical control points. *Cell* 2004; *116*(2):205-219.

[20] Lum JJ, DeBerardinis RJ, Thompson CB. Autophagy in metazoans: cell survival in the land of plenty. *Nat Rev Mol Cell Biol* 2005; *6*(6):439-448.

[21] Castedo M, Perfettini JL, Roumier T *et al.* Cell death by mitotic catastrophe: a molecular definition. *Oncogene* 2004; *23*(16):2825-2837.

[22] Zong WX, Thompson CB. Necrotic death as a cell fate. *Genes Dev* 2006; *20*(1):1-15.

[23] Fink SL, Cookson BT. Apoptosis, pyroptosis, and necrosis: mechanistic description of dead and dying eukaryotic cells. *Infect Immun* 2005; *73*(4):1907-1916.

[24] Kerr JF, Wyllie AH, Currie AR. Apoptosis: a basic biological phenomenon with wide-ranging implications in tissue kinetics. *Br J Cancer* 1972; *26*(4):239-257.

[25] Hail N, Jr., Carter BZ, Konopleva M, Andreeff M. Apoptosis effector mechanisms: A requiem performed in different keys. *Apoptosis* 2006; *11*(6):889-904.

[26] Jin Z, El Deiry WS. Overview of cell death signaling pathways. *Cancer Biol Ther* 2005; *4*(2):139-163.
[27] Jiang X, Wang X. Cytochrome C-mediated apoptosis. *Annu Rev Biochem* 2004; *73*:87-106.
[28] Majno G, Joris I. Apoptosis, oncosis, and necrosis. An overview of cell death. *Am J Pathol* 1995; *146*(1):3-15.
[29] Proskuryakov SY, Gabai VL, Konoplyannikov AG. Necrosis is an active and controlled form of programmed cell death. *Biochemistry (Mosc)* 2002; *67*(4):387-408.
[30] Waring P. Redox active calcium ion channels and cell death. *Arch Biochem Biophys* 2005; *434*(1):33-42.
[31] Klionsky DJ, Emr SD. Autophagy as a regulated pathway of cellular degradation. *Science* 2000; *290*(5497):1717-1721.
[32] Levine B, Yuan J. Autophagy in cell death: an innocent convict? *J Clin Invest* 2005; *115*(10):2679-2688.
[33] Takai H, Tominaga K, Motoyama N *et al.* Aberrant cell cycle checkpoint function and early embryonic death in Chk1(-/-) mice. *Genes Dev* 2000; *14*(12):1439-1447.
[34] Green D, Kroemer G. The central executioners of apoptosis: caspases or mitochondria? *Trends Cell Biol* 1998; *8*(7):267-271.
[35] Boatright KM, Salvesen GS. Mechanisms of caspase activation. *Curr Opin Cell Biol* 2003; *15*(6):725-731.
[36] Bratton SB, Walker G, Srinivasula SM *et al.* Recruitment, activation and retention of caspases-9 and -3 by Apaf-1 apoptosome and associated XIAP complexes. *EMBO J* 2001; *20*(5):998-1009.
[37] Nagata S. Fas ligand-induced apoptosis. *Annu Rev Genet* 1999; *33*:29-55.
[38] Hengartner MO. The biochemistry of apoptosis. *Nature* 2000; *407*(6805):770-776.
[39] Earnshaw WC, Martins LM, Kaufmann SH. Mammalian caspases: structure, activation, substrates, and functions during apoptosis. *Annu Rev Biochem* 1999; *68*:383-424.
[40] Boldin MP, Goncharov TM, Goltsev YV, Wallach D. Involvement of MACH, a novel MORT1/FADD-interacting protease, in Fas/APO-1- and TNF receptor-induced cell death. *Cell* 1996; *85*(6):803-815.
[41] Cheng EH, Sheiko TV, Fisher JK, Craigen WJ, Korsmeyer SJ. VDAC2 inhibits BAK activation and mitochondrial apoptosis. *Science* 2003; *301*(5632):513-517.
[42] Strasser A, O'Connor L, Dixit VM. Apoptosis signaling. *Annu Rev Biochem* 2000; *69*:217-245.
[43] Enari M, Sakahira H, Yokoyama H *et al.* A caspase-activated DNase that degrades DNA during apoptosis, and its inhibitor ICAD. *Nature* 1998; *391*(6662):43-50.
[44] Doonan F, Donovan M, Cotter TG. Caspase-independent photoreceptor apoptosis in mouse models of retinal degeneration. *J Neurosci* 2003; *23*(13):5723-5731.
[45] Lockshin RA, Zakeri Z. Caspase-independent cell death? *Oncogene* 2004; *23*(16):2766-2773.
[46] Alonso M, Tamasdan C, Miller DC, Newcomb EW. Flavopiridol induces apoptosis in glioma cell lines independent of retinoblastoma and p53 tumor suppressor pathway alterations by a caspase-independent pathway. *Mol Cancer Ther* 2003; *2*(2):139-150.

[47] Maianski NA, Roos D, Kuijpers TW. Tumor necrosis factor alpha induces a caspase-independent death pathway in human neutrophils. *Blood* 2003; *101*(5):1987-1995.

[48] Zhang XD, Gillespie SK, Hersey P. Staurosporine induces apoptosis of melanoma by both caspase-dependent and -independent apoptotic pathways. *Mol Cancer Ther* 2004; *3*(2):187-197.

[49] Heibein JA, Barry M, Motyka B, Bleackley RC. Granzyme B-induced loss of mitochondrial inner membrane potential (Delta Psi m) and cytochrome c release are caspase independent. *J Immunol* 1999; *163*(9):4683-4693.

[50] Takano J, Tomioka M, Tsubuki S *et al.* Calpain mediates excitotoxic DNA fragmentation via mitochondrial pathways in adult brains: evidence from calpastatin mutant mice. *J Biol Chem* 2005; *280*(16):16175-16184.

[51] Turk V, Turk B, Turk D. Lysosomal cysteine proteases: facts and opportunities. *EMBO J* 2001; *20*(17):4629-4633.

[52] Johnson DE. Noncaspase proteases in apoptosis. *Leukemia* 2000; *14*(9):1695-1703.

[53] Ferri KF, Kroemer G. Organelle-specific initiation of cell death pathways. *Nat Cell Biol* 2001; *3*(11):E255-E263.

[54] Wood SR, Holroyd JA, Brown SB. The subcellular localization of Zn(II) phthalocyanines and their redistribution on exposure to light. *Photochem Photobiol* 1997; *65*(3):397-402.

[55] Brdiczka DG, Zorov DB, Sheu SS. Mitochondrial contact sites: their role in energy metabolism and apoptosis. *Biochim Biophys Acta* 2006; *1762*(2):148-163.

[56] Kovacic P, Pozos RS, Somanathan R, Shangari N, O'Brien PJ. Mechanism of mitochondrial uncouplers, inhibitors, and toxins: focus on electron transfer, free radicals, and structure-activity relationships. *Curr Med Chem* 2005; *12*(22):2601-2623.

[57] Kuwana T, Newmeyer DD. Bcl-2-family proteins and the role of mitochondria in apoptosis. *Curr Opin Cell Biol* 2003; *15*(6):691-699.

[58] Bernardi P, Scorrano L, Colonna R, Petronilli V, Di Lisa F. Mitochondria and cell death. Mechanistic aspects and methodological issues. *Eur J Biochem* 1999; *264*(3):687-701.

[59] Mancini M, Nicholson DW, Roy S *et al.* The caspase-3 precursor has a cytosolic and mitochondrial distribution: implications for apoptotic signaling. *J Cell Biol* 1998; *140*(6):1485-1495.

[60] Susin SA, Lorenzo HK, Zamzami N *et al.* Molecular characterization of mitochondrial apoptosis-inducing factor. *Nature* 1999; *397*(6718):441-446.

[61] Susin SA, Lorenzo HK, Zamzami N *et al.* Mitochondrial release of caspase-2 and -9 during the apoptotic process. *J Exp Med* 1999; *189*(2):381-394.

[62] Heath-Engel HM, Shore GC. Mitochondrial membrane dynamics, cristae remodelling and apoptosis. *Biochim Biophys Acta* 2006; *1763*(5-6):549-560.

[63] Costantini P, Jacotot E, Decaudin D, Kroemer G. Mitochondrion as a novel target of anticancer chemotherapy. *J Natl Cancer Inst* 2000; *92*(13):1042-1053.

[64] Solary E, Bettaieb A, Dubrez-Daloz L, Corcos L. Mitochondria as a target for inducing death of malignant hematopoietic cells. *Leuk Lymphoma* 2003; *44*(4):563-574.

[65] Andre N, Rome A, Carre M. [Antimitochondrial agents: a new class of anticancer agents]. *Arch Pediatr* 2006; *13*(1):69-75.

[66] Hoek JB, Nicholls DG, Williamson JR. Determination of the mitochondrial protonmotive force in isolated hepatocytes. *J Biol Chem* 1980; *255*(4):1458-1464.

[67] Johnson LV, Walsh ML, Chen LB. Localization of mitochondria in living cells with rhodamine 123. *Proc Natl Acad Sci U S A* 1980; *77*(2):990-994.

[68] Ball DJ, Luo Y, Kessel D *et al.* The induction of apoptosis by a positively charged methylene blue derivative. *J Photochem Photobiol B* 1998; *42*(2):159-163.

[69] Lacerda SH, Abraham B, Stringfellow TC, Indig GL. Photophysical, photochemical, and tumor-selectivity properties of bromine derivatives of rhodamine-123. *Photochem Photobiol* 2005; *81*(6):1430-1438.

[70] Shoshan-Barmatz V, Israelson A, Brdiczka D, Sheu SS. The Voltage-Dependent Anion Channel (VDAC): Function in Intracellular Signalling, Cell Life and Cell Death. *Curr Pharm Des* 2006; *12*(18):2249-2270.

[71] Belzacq AS, Jacotot E, Vieira HL *et al.* Apoptosis induction by the photosensitizer verteporfin: identification of mitochondrial adenine nucleotide translocator as a critical target. *Cancer Res* 2001; *61*(4):1260-1264.

[72] Furre IE, Shahzidi S, Luksiene Z *et al.* Targeting PBR by hexaminolevulinate-mediated photodynamic therapy induces apoptosis through translocation of apoptosis-inducing factor in human leukemia cells. *Cancer Res* 2005; *65*(23):11051-11060.

[73] Kim R. Unknotting the roles of Bcl-2 and Bcl-xL in cell death. *Biochem Biophys Res Commun* 2005; *333*(2):336-343.

[74] Xue LY, Chiu SM, Oleinick NL. Photochemical destruction of the Bcl-2 oncoprotein during photodynamic therapy with the phthalocyanine photosensitizer Pc 4. *Oncogene* 2001; *20*(26):3420-3427.

[75] Xue LY, Chiu SM, Fiebig A, Andrews DW, Oleinick NL. Photodamage to multiple Bcl-xL isoforms by photodynamic therapy with the phthalocyanine photosensitizer Pc 4. *Oncogene* 2003; *22*(58):9197-9204.

[76] Usuda J, Azizuddin K, Chiu SM, Oleinick NL. Association between the photodynamic loss of Bcl-2 and the sensitivity to apoptosis caused by phthalocyanine photodynamic therapy. *Photochem Photobiol* 2003; *78*(1):1-8.

[77] Usuda J, Chiu SM, Murphy ES *et al.* Domain-dependent photodamage to Bcl-2. A membrane anchorage region is needed to form the target of phthalocyanine photosensitization. *J Biol Chem* 2003; *278*(3):2021-2029.

[78] Huang HF, Chen YZ, Wu Y. Mitochondria-dependent apoptosis induced by a novel amphipathic photochemotherapeutic agent ZnPcS2P2 in HL60 cells. *Acta Pharmacol Sin* 2005; *26*(9):1138-1144.

[79] Corbett EF, Michalak M. Calcium, a signaling molecule in the endoplasmic reticulum? *Trends Biochem Sci* 2000; *25*(7):307-311.

[80] Hayashi T, Saito A, Okuno S *et al.* Oxidative damage to the endoplasmic reticulum is implicated in ischemic neuronal cell death. *J Cereb Blood Flow Metab* 2003; *23*(10):1117-1128.

[81] Hayashi T, Saito A, Okuno S *et al.* Damage to the endoplasmic reticulum and activation of apoptotic machinery by oxidative stress in ischemic neurons. *J Cereb Blood Flow Metab* 2005; *25*(1):41-53.

[82] Lindholm D, Wootz H, Korhonen L. ER stress and neurodegenerative diseases. *Cell Death Differ* 2006; *13*(3):385-392.

[83] Oyadomari S, Araki E, Mori M. Endoplasmic reticulum stress-mediated apoptosis in pancreatic beta-cells. *Apoptosis* 2002; *7*(4):335-345.

[84] Sekine Y, Takeda K, Ichijo H. The ASK1-MAP kinase signaling in ER Stress and neurodegenerative diseases. *Curr Mol Med* 2006; *6*(1):87-97.

[85] Teiten MH, Marchal S, D'Hallewin MA, Guillemin F, Bezdetnaya L. Primary photodamage sites and mitochondrial events after Foscan photosensitization of MCF-7 human breast cancer cells. *Photochem Photobiol* 2003; *78*(1):9-14.

[86] Teiten MH, Bezdetnaya L, Morliere P, Santus R, Guillemin F. Endoplasmic reticulum and Golgi apparatus are the preferential sites of Foscan localisation in cultured tumour cells. *Br J Cancer* 2003; *88*(1):146-152.

[87] Granville DJ, Ruehlmann DO, Choy JC *et al.* Bcl-2 increases emptying of endoplasmic reticulum Ca2+ stores during photodynamic therapy-induced apoptosis. *Cell Calcium* 2001; *30*(5):343-350.

[88] Mak NK, Li KM, Leung WN *et al.* Involvement of both endoplasmic reticulum and mitochondria in photokilling of nasopharyngeal carcinoma cells by the photosensitizer Zn-BC-AM. *Biochem Pharmacol* 2004; *68*(12):2387-2396.

[89] Buytaert E, Callewaert G, Hendrickx N *et al.* Role of endoplasmic reticulum depletion and multidomain proapoptotic BAX and BAK proteins in shaping cell death after hypericin-mediated photodynamic therapy. *FASEB J* 2006; *20*(6):756-758.

[90] Wong S, Luna M, Ferrario A, Gomer CJ. CHOP activation by photodynamic therapy increases treatment induced photosensitization. *Lasers Surg Med* 2004; *35*(5):336-341.

[91] Kessel D, Vicente MG, Reiners JJ, Jr. Initiation of apoptosis and autophagy by photodynamic therapy. *Lasers Surg Med* 2006.

[92] Wootz H, Hansson I, Korhonen L, Napankangas U, Lindholm D. Caspase-12 cleavage and increased oxidative stress during motoneuron degeneration in transgenic mouse model of ALS. *Biochem Biophys Res Commun* 2004; *322*(1):281-286.

[93] Liu H, Baliga R. Endoplasmic reticulum stress-associated caspase 12 mediates cisplatin-induced LLC-PK1 cell apoptosis. *J Am Soc Nephrol* 2005; *16*(7):1985-1992.

[94] Matroule JY, Carthy CM, Granville DJ *et al.* Mechanism of colon cancer cell apoptosis mediated by pyropheophorbide-a methylester photosensitization. *Oncogene* 2001; *20*(30):4070-4084.

[95] Saleh M, Vaillancourt JP, Graham RK *et al.* Differential modulation of endotoxin responsiveness by human caspase-12 polymorphisms. *Nature* 2004; *429*(6987):75-79.

[96] Hicks SW, Machamer CE. Golgi structure in stress sensing and apoptosis. *Biochim Biophys Acta* 2005; *1744*(3):406-414.

[97] Mancini M, Machamer CE, Roy S *et al.* Caspase-2 is localized at the Golgi complex and cleaves golgin-160 during apoptosis. *J Cell Biol* 2000; *149*(3):603-612.

[98] Baliga BC, Read SH, Kumar S. The biochemical mechanism of caspase-2 activation. *Cell Death Differ* 2004; *11*(11):1234-1241.

[99] Maag RS, Mancini M, Rosen A, Machamer CE. Caspase-resistant Golgin-160 disrupts apoptosis induced by secretory pathway stress and ligation of death receptors. *Mol Biol Cell* 2005; *16*(6):3019-3027.

[100] Fabris C, Valduga G, Miotto G *et al.* Photosensitization with zinc (II) phthalocyanine as a switch in the decision between apoptosis and necrosis. *Cancer Res* 2001; *61*(20):7495-7500.

[101] Cristobal J, Stockert JC, Villanueva A *et al.* Caspase-2: a possible trigger of apoptosis induced in A-549 tumor cells by ZnPc photodynamic treatment. *Int J Oncol* 2006; *28*(5):1057-1063.

[102] Machamer CE. Golgi disassembly in apoptosis: cause or effect? *Trends Cell Biol* 2003; *13*(6):279-281.

[103] Rossi A, Deveraux Q, Turk B, Sali A. Comprehensive search for cysteine cathepsins in the human genome. *Biol Chem* 2004; *385*(5):363-372.

[104] Ostermann N, Gerhartz B, Worpenberg S, Trappe J, Eder J. Crystal structure of an activation intermediate of cathepsin E. *J Mol Biol* 2004; *342*(3):889-899.

[105] Nishioku T, Hashimoto K, Yamashita K *et al.* Involvement of cathepsin E in exogenous antigen processing in primary cultured murine microglia. *J Biol Chem* 2002; *277*(7):4816-4822.

[106] Liaudet-Coopman E, Beaujouin M, Derocq D *et al.* Cathepsin D: newly discovered functions of a long-standing aspartic protease in cancer and apoptosis. *Cancer Lett* 2005.

[107] Foghsgaard L, Wissing D, Mauch D *et al.* Cathepsin B acts as a dominant execution protease in tumor cell apoptosis induced by tumor necrosis factor. *J Cell Biol* 2001; *153*(5):999-1010.

[108] Boya P, Gonzalez-Polo RA, Poncet D *et al.* Mitochondrial membrane permeabilization is a critical step of lysosome-initiated apoptosis induced by hydroxychloroquine. *Oncogene* 2003; *22*(25):3927-3936.

[109] Kagedal K, Johansson U, Ollinger K. The lysosomal protease cathepsin D mediates apoptosis induced by oxidative stress. *FASEB J* 2001; *15*(9):1592-1594.

[110] Chen W, Li N, Chen T *et al.* The lysosome-associated apoptosis-inducing protein containing the pleckstrin homology (PH) and FYVE domains (LAPF), representative of a novel family of PH and FYVE domain-containing proteins, induces caspase-independent apoptosis via the lysosomal-mitochondrial pathway. *J Biol Chem* 2005; *280*(49):40985-40995.

[111] Kessel D, Luo Y, Deng Y, Chang CK. The role of subcellular localization in initiation of apoptosis by photodynamic therapy. *Photochem Photobiol* 1997; *65*(3):422-426.

[112] Reiners JJ, Jr., Caruso JA, Mathieu P *et al.* Release of cytochrome c and activation of pro-caspase-9 following lysosomal photodamage involves Bid cleavage. *Cell Death Differ* 2002; *9*(9):934-944.

[113] Nagata S, Obana A, Gohto Y, Nakajima S. Necrotic and apoptotic cell death of human malignant melanoma cells following photodynamic therapy using an amphiphilic photosensitizer, ATX-S10(Na). *Lasers Surg Med* 2003; *33*(1):64-70.

[114] Brunk UT, Neuzil J, Eaton JW. Lysosomal involvement in apoptosis. *Redox Rep* 2001; *6*(2):91-97.

[115] Antunes F, Cadenas E, Brunk UT. Apoptosis induced by exposure to a low steady-state concentration of H2O2 is a consequence of lysosomal rupture. *Biochem J* 2001; *356*(Pt 2):549-555.

[116] Erdal H, Berndtsson M, Castro J *et al.* Induction of lysosomal membrane permeabilization by compounds that activate p53-independent apoptosis. *Proc Natl Acad Sci U S A* 2005; *102*(1):192-197.

[117] Kagedal K, Johansson AC, Johansson U *et al.* Lysosomal membrane permeabilization during apoptosis--involvement of Bax? *Int J Exp Pathol* 2005; *86*(5):309-321.

[118] Tao J, Sanghera JS, Pelech SL, Wong G, Levy JG. Stimulation of stress-activated protein kinase and p38 HOG1 kinase in murine keratinocytes following photodynamic therapy with benzoporphyrin derivative. *J Biol Chem* 1996; *271*(43):27107-27115.

[119] Klotz LO, Fritsch C, Briviba K *et al.* Activation of JNK and p38 but not ERK MAP kinases in human skin cells by 5-aminolevulinate-photodynamic therapy. *Cancer Res* 1998; *58*(19):4297-4300.

[120] Whitacre CM, Feyes DK, Satoh T *et al.* Photodynamic therapy with the phthalocyanine photosensitizer Pc 4 of SW480 human colon cancer xenografts in athymic mice. *Clin Cancer Res* 2000; *6*(5):2021-2027.

[121] Xue L, He J, Oleinick NL. Promotion of photodynamic therapy-induced apoptosis by stress kinases. *Cell Death Differ* 1999; *6*(9):855-864.

[122] Chan WH, Wu HJ. Anti-apoptotic effects of curcumin on photosensitized human epidermal carcinoma A431 cells. *J Cell Biochem* 2004; *92*(1):200-212.

[123] Assefa Z, Vantieghem A, Declercq W *et al.* The activation of the c-Jun N-terminal kinase and p38 mitogen-activated protein kinase signaling pathways protects HeLa cells from apoptosis following photodynamic therapy with hypericin. *J Biol Chem* 1999; *274*(13):8788-8796.

[124] Tong Z, Singh G, Rainbow AJ. Sustained activation of the extracellular signal-regulated kinase pathway protects cells from photofrin-mediated photodynamic therapy. *Cancer Res* 2002; *62*(19):5528-5535.

[125] Hendrickx N, Volanti C, Moens U *et al.* Up-regulation of cyclooxygenase-2 and apoptosis resistance by p38 MAPK in hypericin-mediated photodynamic therapy of human cancer cells. *J Biol Chem* 2003; *278*(52):52231-52239.

[126] Sigal YJ, McDermott MI, Morris AJ. Integral membrane lipid phosphatases/phosphotransferases: common structure and diverse functions. *Biochem J* 2005; *387*(Pt 2):281-293.

[127] Cuvillier O. Sphingosine in apoptosis signaling. *Biochim Biophys Acta* 2002; *1585*(2-3):153-162.

[128] Wispriyono B, Schmelz E, Pelayo H, Hanada K, Separovic D. A role for the de novo sphingolipids in apoptosis of photosensitized cells. *Exp Cell Res* 2002; *279*(1):153-165.

[129] Separovic D, He J, Oleinick NL. Ceramide generation in response to photodynamic treatment of L5178Y mouse lymphoma cells. *Cancer Res* 1997; *57*(9):1717-1721.

[130] Separovic D, Mann KJ, Oleinick NL. Association of ceramide accumulation with photodynamic treatment-induced cell death. *Photochem Photobiol* 1998; *68*(1):101-109.

[131] Dolgachev V, Farooqui MS, Kulaeva OI *et al.* De novo ceramide accumulation due to inhibition of its conversion to complex sphingolipids in apoptotic photosensitized cells. *J Biol Chem* 2004; *279*(22):23238-23249.

[132] Separovic D, Wang S, Awad Maitah MY, Hanada K, Kessel D. Ceramide response post-photodamage is absent after treatment with HA14-1. *Biochem Biophys Res Commun* 2006.

[133] Oyadomari S, Mori M. Roles of CHOP/GADD153 in endoplasmic reticulum stress. *Cell Death Differ* 2004; *11*(4):381-389.

[134] Ahmad N, Gupta S, Mukhtar H. Involvement of retinoblastoma (Rb) and E2F transcription factors during photodynamic therapy of human epidermoid carcinoma cells A431. *Oncogene* 1999; *18*(10):1891-1896.

[135] Matroule JY, Hellin AC, Morliere P *et al.* Role of nuclear factor-kappa B in colon cancer cell apoptosis mediated by aminopyropheophorbide photosensitization. *Photochem Photobiol* 1999; *70*(4):540-548.

[136] Shishodia S, Aggarwal BB. Nuclear factor-kappaB activation: a question of life or death. *J Biochem Mol Biol* 2002; *35*(1):28-40.

[137] Hsieh YJ, Wu CC, Chang CJ, Yu JS. Subcellular localization of Photofrin determines the death phenotype of human epidermoid carcinoma A431 cells triggered by photodynamic therapy: when plasma membranes are the main targets. *J Cell Physiol* 2003; *194*(3):363-375.

[138] You Y, Gibson SL, Hilf R *et al.* Water soluble, core-modified porphyrins. 3. Synthesis, photophysical properties, and in vitro studies of photosensitization, uptake, and localization with carboxylic acid-substituted derivatives. *J Med Chem* 2003; *46*(17):3734-3747.

[139] Ben Hur E, Dubbelman TM, Van Steveninck J. Effect of fluoride on inhibition of plasma membrane functions in Chinese hamster ovary cells photosensitized by aluminum phthalocyanine. *Radiat Res* 1992; *131*(1):47-52.

[140] Chen JY, Cheung NH, Fung MC *et al.* Subcellular localization of merocyanine 540 (MC540) and induction of apoptosis in murine myeloid leukemia cells. *Photochem Photobiol* 2000; *72*(1):114-120.

[141] Kriska T, Korytowski W, Girotti AW. Hyperresistance to photosensitized lipid peroxidation and apoptotic killing in 5-aminolevulinate-treated tumor cells overexpressing mitochondrial GPX4. *Free Radic Biol Med* 2002; *33*(10):1389-1402.

[142] Tsai JC, Wu CL, Chien HF, Chen CT. Reorganization of cytoskeleton induced by 5-aminolevulinic acid-mediated photodynamic therapy and its correlation with mitochondrial dysfunction. *Lasers Surg Med* 2005; *36*(5):398-408.

[143] Tsai T, Hong RL, Tsai JC *et al.* Effect of 5-aminolevulinic acid-mediated photodynamic therapy on MCF-7 and MCF-7/ADR cells. *Lasers Surg Med* 2004; *34*(1):62-72.

[144] Grebenova D, Kuzelova K, Smetana K *et al.* Mitochondrial and endoplasmic reticulum stress-induced apoptotic pathways are activated by 5-aminolevulinic acid-based photodynamic therapy in HL60 leukemia cells. *J Photochem Photobiol B* 2003; *69*(2):71-85.

[145] Liang H, Shin DS, Lee YE *et al.* Subcellular phototoxicity of 5-aminolaevulinic acid (ALA). *Lasers Surg Med* 1998; *22*(1):14-24.

[146] Ali SM, Olivo M. Bio-distribution and subcellular localization of Hypericin and its role in PDT induced apoptosis in cancer cells. *Int J Oncol* 2002; *21*(3):531-540.

[147] Trivedi NS, Wang HW, Nieminen AL, Oleinick NL, Izatt JA. Quantitative analysis of Pc 4 localization in mouse lymphoma (LY-R) cells via double-label confocal fluorescence microscopy. *Photochem Photobiol* 2000; *71*(5):634-639.

[148] Yow CM, Chen JY, Mak NK, Cheung NH, Leung AW. Cellular uptake, subcellular localization and photodamaging effect of temoporfin (mTHPC) in nasopharyngeal carcinoma cells: comparison with hematoporphyrin derivative. *Cancer Lett* 2000; *157*(2):123-131.

[149] Morgan J, Whitaker JE, Oseroff AR. GRP78 induction by calcium ionophore potentiates photodynamic therapy using the mitochondrial targeting dye victoria blue BO. *Photochem Photobiol* 1998; *67*(1):155-164.

[150] Noodt BB, Rodal GH, Wainwright M *et al.* Apoptosis induction by different pathways with methylene blue derivative and light from mitochondrial sites in V79 cells. *Int J Cancer* 1998; *75*(6):941-948.

[151] Shea CR, Sherwood ME, Flotte TJ *et al.* Rhodamine 123 phototoxicity in laser-irradiated MGH-U1 human carcinoma cells studied in vitro by electron microscopy and confocal laser scanning microscopy. *Cancer Res* 1990; *50*(13):4167-4172.

[152] Gallagher WM, Allen LT, O'Shea C *et al.* A potent nonporphyrin class of photodynamic therapeutic agent: cellular localisation, cytotoxic potential and influence of hypoxia. *Br J Cancer* 2005; *92*(9):1702-1710.

[153] Uzdensky AB, Ma LW, Iani V *et al.* Intracellular localisation of hypericin in human glioblastoma and carcinoma cell lines. *Lasers Med Sci* 2001; *16*(4):276-283.

[154] Morliere P, Maziere JC, Santus R *et al.* Tolyporphin: a natural product from cyanobacteria with potent photosensitizing activity against tumor cells in vitro and in vivo. *Cancer Res* 1998; *58*(16):3571-3578.

[155] Sakharov DV, Elstak ED, Chernyak B, Wirtz KW. Prolonged lipid oxidation after photodynamic treatment. Study with oxidation-sensitive probe C11-BODIPY581/591. *FEBS Lett* 2005; *579*(5):1255-1260.

[156] Ali SM, Olivo M. Efficacy of hypocrellin pharmacokinetics in phototherapy. *Int J Oncol* 2002; *21*(6):1229-1237.

[157] Rancan F, Wiehe A, Nobel M *et al.* Influence of substitutions on asymmetric dihydroxychlorins with regard to intracellular uptake, subcellular localization and photosensitization of Jurkat cells. *J Photochem Photobiol B* 2005; *78*(1):17-28.

[158] Caruso JA, Mathieu PA, Joiakim A *et al.* Differential susceptibilities of murine hepatoma 1c1c7 and Tao cells to the lysosomal photosensitizer NPe6: influence of aryl hydrocarbon receptor on lysosomal fragility and protease contents. *Mol Pharmacol* 2004; *65*(4):1016-1028.

[159] Roberts WG, Berns MW. In vitro photosensitization I. Cellular uptake and subcellular localization of mono-L-aspartyl chlorin e6, chloro-aluminum sulfonated phthalocyanine, and photofrin II. *Lasers Surg Med* 1989; *9*(2):90-101.

[160] Roberts WG, Liaw LH, Berns MW. In vitro photosensitization II. An electron microscopy study of cellular destruction with mono-L-aspartyl chlorin e6 and photofrin II. *Lasers Surg Med* 1989; *9*(2):102-108.

[161] Nishiyama N, Stapert HR, Zhang GD *et al.* Light-harvesting ionic dendrimer porphyrins as new photosensitizers for photodynamic therapy. *Bioconjug Chem* 2003; *14*(1):58-66.

[162] Bisland SK, Singh D, Gariepy J. Potentiation of chlorin e6 photodynamic activity in vitro with peptide-based intracellular vehicles. *Bioconjug Chem* 1999; *10*(6):982-992.

[163] Morgan J, Potter WR, Oseroff AR. Comparison of photodynamic targets in a carcinoma cell line and its mitochondrial DNA-deficient derivative. *Photochem Photobiol* 2000; *71*(6):747-757.

[164] Chiu SM, Xue LY, Azizuddin K, Oleinick NL. Photodynamic therapy-induced death of HCT 116 cells: Apoptosis with or without Bax expression. *Apoptosis* 2005; *10*(6):1357-1368.

[165] Hendrickx N, Dewaele M, Buytaert E *et al.* Targeted inhibition of p38alpha MAPK suppresses tumor-associated endothelial cell migration in response to hypericin-based photodynamic therapy. *Biochem Biophys Res Commun* 2005; *337*(3):928-935.

[166] Tong Z, Singh G, Valerie K, Rainbow AJ. Activation of the stress-activated JNK and p38 MAP kinases in human cells by Photofrin-mediated photodynamic therapy. *J Photochem Photobiol B* 2003; *71*(1-3):77-85.

[167] Buchczyk DP, Klotz LO, Lang K, Fritsch C, Sies H. High efficiency of 5-aminolevulinate-photodynamic treatment using UVA irradiation. *Carcinogenesis* 2001; *22*(6):879-883.

[168] Schieke SM, von Montfort C, Buchczyk DP *et al.* Singlet oxygen-induced attenuation of growth factor signaling: possible role of ceramides. *Free Radic Res* 2004; *38*(7):729-737.

[169] Ferrario A, von Tiehl KF, Rucker N *et al.* Antiangiogenic treatment enhances photodynamic therapy responsiveness in a mouse mammary carcinoma. *Cancer Res* 2000; *60*(15):4066-4069.

[170] Kick G, Messer G, Goetz A, Plewig G, Kind P. Photodynamic therapy induces expression of interleukin 6 by activation of AP-1 but not NF-kappa B DNA binding. *Cancer Res* 1995; *55*(11):2373-2379.

[171] Ryter SW, Gomer CJ. Nuclear factor kappa B binding activity in mouse L1210 cells following photofrin II-mediated photosensitization. *Photochem Photobiol* 1993; *58*(5):753-756.

[172] Koukourakis MI, Giatromanolaki A, Skarlatos J *et al.* Hypoxia inducible factor (HIF-1a and HIF-2a) expression in early esophageal cancer and response to photodynamic therapy and radiotherapy. *Cancer Res* 2001; *61*(5):1830-1832.

[173] Ji Z, Yang G, Shahzidi S *et al.* Induction of hypoxia-inducible factor-1alpha overexpression by cobalt chloride enhances cellular resistance to photodynamic therapy. *Cancer Lett* 2006.

In: New Developments in Cell Apoptosis Research
Editor: Alan J. Corvin, pp. 273-287
ISBN: 1-60021-453-3

Chapter XI

Discrimination of Apoptotic Cells by Image Analysis

Maria Luiza S. Mello
Department of Cell Biology, Institute of Biology, UNICAMP,
13083-863 Campinas, SP, Brazil; E-mail: mlsmello@unicamp.br

Dedicated to Professor Benedicto De Campos Vidal
on the occasion of his 75th birthday

Abstract

Changes in chromatin higher packing state levels have been demonstrated for transformed and tumorigenic cells as well as expected for cells under advanced apoptosis stages. Image analysis is an efficient tool to recognize increased or decreased levels of chromatin compactness in interphase nuclei taking into account geometric, densitometric and textural parameters that describe image properties established by video or scanning microspectrophotometric systems. A previously described scatter diagram relating the percentage of nuclear areas covered by Feulgen-stained chromatin presenting absorbances above a selected threshold versus a dimensionless parameter that expresses how many times the average absorption of the condensed chromatin exceeds that of the entire nucleus is proposed here for discrimination of cells under advanced steps of apoptosis or programmed cell death. This proposition arose by assuming that the position acquired by points representing apoptotic cells in the mentioned diagram could be discriminated from that acquired by non-apoptotic cells, and considering the participation of condensed chromatin image patterns to define specific nuclear phenotypes. This study shows that a scatter diagram based on image analysis features is a useful tool to discriminate apoptotic cells in Feulgen-stained preparations.

Introduction

Chromatin is the complex of DNA, histones, nonhistone proteins, and RNA in the nucleus of the interphase cell. Chromatin packages into the cell nuclei, acquiring increased levels of hierarchical organization which participate actively in the regulation of gene transcription [Creston and Kadonga, 1993; van Driel and Otte, 1997; Cremer and Cremer, 2001; Alberts et al., 2002; Gilbert et al., 2005]. Based on chromatin texture, which reveals the distribution and organization of higher order chromatin packing states (supraorganization) and on nuclear size and shape, nuclear phenotypes can be defined at the light microscope [Vidal et al., 1973; Vidal, 1984; Marchevsky and Bartels, 1994; Mello et al., 1995, 2003; Oberholzer et al., 1996; Vidal et al., 1998]. Nuclear phenotypes may differ for different cells or for diverse physiological states of the same cell [Mello and Vidal, 2000].

Modern approaches for nuclear phenotype quantitative identification as related to the cell cycle, cell transformation and tumorigenesis and the cell death involve computer-aided automatic microspectrophotometry or video image analysis procedures.

Mathematical parameters studied with image analysis and processed statistically allow the description of geometric, densitometric and textural characteristics of the cell nuclei and of chromatin components adequately revealed by specific topochemical tests as is the case of the Feulgen reaction, specific for DNA. When associated to other data provided, for instance, by molecular biology, immunology, genetics, and pathology, image analysis findings acquire a powerful estimative reliability and may provide useful tools particularly for pattern recognition of cellular lesions and specific cell physiologies, and even for cytodiagnosis [Nicolini et al., 1982; Dawson et al., 1991; Marchevsky and Bartels, 1994; Kong and Ringer, 1995; Mello et al., 1995, 2003; Oberholzer et al., 1996; Wallet et al., 1996; Vidal et al., 1998; Mello and Vidal, 2000; Leyk et al., 2005; Knowles et al., 2006].

Chromatin Texture and the S_C% versus AAR Scatter Diagram

A scatter diagram relating the percentage of nuclear areas covered by Feulgen-stained chromatin (S_C%) presenting absorbances above a selected threshold (cutoff point) versus a dimensionless parameter that expresses how many times the average absorption of the condensed chromatin exceeds that of the entire nucleus (AAR) has been proposed by Vidal [1984] and applied to several subsequent studies. This scatter diagram representation permits to show a distribution of points corresponding to nuclei with specific phenotypes (Figure 1).

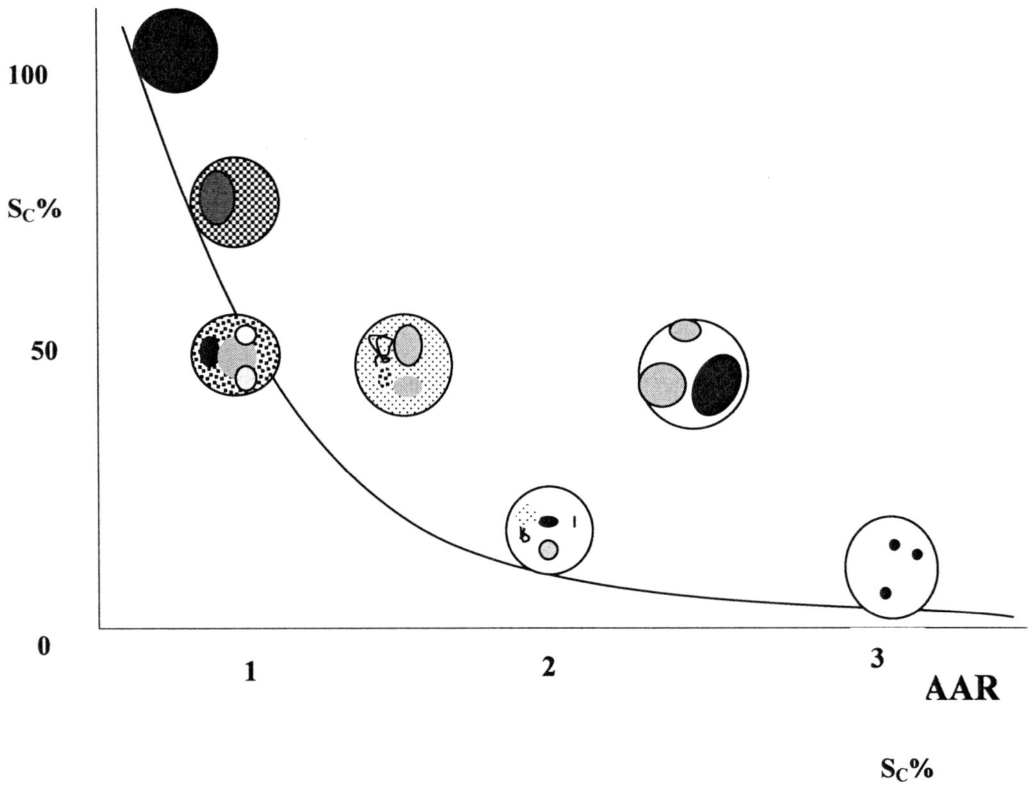

Figure 1. Scatter diagram representation for nuclear phenotypes studied in Feulgen-stained preparations, modified from Vidal [1984] and Vidal et al. [1998]. This representation relates S_C% (relative chromatin area) and AAR (average absorption ratio, that is, the textural contrast between condensed and non-condensed chromatin) parameters; the approximate position of each point (nucleus) is a function of the respective phenotypic image.

Polyploidy accented with aging is accompanied by an increase in the chromatin packing state and redistribution of the heterochromatin areas in the salivary gland cells of the rat [Vidal, 1984]. Because of the textural changes in the chromatin with increasing polyploidy, the ploidy degrees in this case could be associated with different nuclear phenotypes which by turn could be discriminated in the S_C% vs. AAR diagram [Vidal, 1984].

NIH 3T3 fibroblastic mouse cells transformed with the Ha-*ras* oncogene exhibit nuclear phenotypes with increased levels of chromatin condensation also distinguishable in the S_C% vs. AAR scatter diagram representation [Mello and Russo, 1990]. Furthermore, this diagram was especially valuable to discriminate changes in chromatin texture in non-tumorigenic persistent revertants of long terminal repeat (LTR)-activated c-Ha-*ras*-transformed NIH 3T3 cells and in cell lines obtained by transfection of these revertants with sense and antisense contructs of the lysyl oxidase (*Lox*) gene. Modulation of *ras* transformation by *Lox* expression affecting chromatin supraorganization could thus be demonstrated [Mello et al., 1995].

Immortal human breast epithelial cells, MCF-10F, which show characteristics of normal cells [Soule et al., 1990], when treated by the carcinogen benzo[a]pyrene (BP) gave rise to several cell lines in which the phenotypes associated with neoplastic transformation appear in a progressive fashion, [Calaf and Russo, 1993]. Differences in chromatin condensation characterize the transformed and full tumorigenic cells so generated (BP1 and BP1-E1 cells, respectively) [Vidal et al., 1998]. Transfection of the c-Ha-*ras* oncogene to transformed BP1 cells originated the BP1-T*ras* cell line which rapidly attained enhancement of the neoplastic changes initiated by exposure to BP, thus becoming as aggressive as the BP1-E1 cells in SCID mice [Calaf et al., 1995]; this was assumed to be favored by an enhanced genomic instability provided by the c-Ha-*ras* oncogene in the chemically transformed MCF-10F cells [Mello et al., 1999]. The differences in chromatin texture with the increasing transformation status in the MCF-10F cells were especially evident when the point position distribution corresponding to Feulgen-stained nuclei of the non-transformed MCF-10F cells, the BP1, BP1-E1, and BP1-T*ras* cells were compared in the S_C% vs. AAR scatter diagram [Vidal et al., 1998; Mello et al., 1999].

The S_C% vs. AAR representation also allowed concluding on different participation of DNA methylated CpG sequences in nuclei of stingless bees with low and high heterochromatin contents, as assessed by *Msp* I and *Hpa* II restriction enzyme digestion and image analysis [Mampumbu and Mello, 2006].

Apoptosis and Image Analysis

Programmed cell death and apoptosis are essential physiological mechanisms for selective elimination of cells by activation of a genetically determined cell suicide program [Horvitz et al., 1994; Jacobson et al., 1997]. These mechanisms play an integral part in a variety of biological events, including morphogenesis, the process of cell turnover, and removal of harmful cells; they are often involving death proteases called caspases and are regulated by several sets of genes, including the *bcl-2* family [Tsujimoto and Shimizu, 2000]. Programmed cell death and apoptosis are known to generally induce cell shrinkage, plasma membrane blebbing, an increase in chromatin condensation especially near the cell nuclear borders, DNA fragmentation, and ultimately breakdown of the cells into membrane-bound apoptotic bodies that are rapidly phagocytosed [Kerr et al., 1972; Wyllie et al., 1981; Majno and Joris, 1995].

The DNA fragmentation involved in cell death can be evaluated in different cell types through the TUNEL assay and single-cell gel electrophoresis (comet assay) [Singh et al., 1988; Wijsman et al., 1993; Darfier and Karaszkiewicz, 1995; Kong and Ringer, 1995; Klaude et al., 1996; Mello et al., 2000; Tice et al., 2000; Leyk et al., 2005; Miyamoto et al., 2005]. However, even in studies using *in situ* labeling of DNA strand breaks as a marker, there is a heavy reliance on morphological features for the identification of apoptotic nuclei. Chromatin condensation as associated to apoptosis can be visualized morphologically with the light or electron microscopy or with image analysis procedures [Wyllie et al., 1981; Majno and Joris, 1995; Diaz et al., 1997; Kuo et al., 1998; Maria et al., 2000; Yatouji et al., 2000; Mello et al., 2001]. Feulgen-stained preparations studied with image analysis or flow cytometry have been proposed to evaluate proliferation and apoptotic indices in glucocorticoid-treated rat thymocytes [Camby et al., 1995], to establish apoptosis-cell cycle relationships in a mammary epithelial cell line [Wallet et al., 1995], to associate levels of chromatin condensation with decrease in Feulgen-DNA amounts in V79 cells [Maria et al., 2000], and to reveal the DNA fragmentation associated with different Feulgen-DNA hydrolysis kinetics in nucleate erythrocytes of different snake species compared to the bullfrog erythrocytes [Miyamoto et al., 2005].

Considering that chromatin supraorganization and nuclear sizes and shapes generally change with programmed cell death and apoptosis phenomena, a representation that seeks for relationships between these properties established at the light microscopy level would probably identify the affected cells in a cellular population screened. As mentioned above, image analysis has been found to be an efficient tool to recognize increased or decreased levels of chromatin compactness in various cell types under different physiological or pathological conditions. It is thus expected that nuclei from cells under programmed cell death or apoptosis could be discriminated from other nuclear phenotypes in the S_C% vs. AAR scatter diagram due to the especial condensed chromatin patterns involved in these cell death forms [Kerr et al., 1972; Majno and Joris, 1995].

In the present study, apoptotic and non-apoptotic BP1-T*ras* cells and erythrocytes from bullfrogs under a developmental phase in which these cells are strongly committed to DNA fragmentation and cell death [Hasebe et al., 1996; Mello et al., 2000], were investigated by image analysis, using Vidal's [1984] scatter diagram representation. The distribution of points associated to the nuclear images of mentioned cells could be discriminated from that of other nuclear phenotypes in this diagram.

Materials and Methods

Cells

Benzo[a]pyrene (BP)-transformed human breast MCF-10F epithelial cells transfected with the c-Ha-*ras* oncogene (BP1-T*ras* cells) [Calaf et al., 1995], maintained in the Breast Cancer Research Laboratory of the Fox Chase Cancer Center at Philadelphia, and generously provided by Dr. Jose Russo, were used. The BP1-T*ras* cells were at passage 61. They were grown in Dulbecco's minimal essential medium/F-12 medium (1:1) supplemented with 5% equine serum, 0.1 μg/mL cholera-toxin, 10 μg/mL insulin, 100 units/mL penicillin, 100 μg/mL streptomycin. 2.5 μg/mL amphotericin B (Gibco Island, NY), 0.5 μg/mL hydrocortisone (Sigma, Saint Louis, MO), and 0.02 μg/mL epidermal growth factor (Collaborative Research Inc., Palo Alto, CA) as described previously [Soule et al., 1990; Calaf and Russo, 1993; Calaf et al., 1995]. The cells were cultivated on glass, a substrate which induces apoptotic ratios much higher than those in non-transformed MCF-10F cells or in the tumorigenic BP1-E cells receiving normal chromosomes 11 and 17 by microcell mediated-chromosome transfer [Mello et al., 2002, 2005].

Blood samples from young adult bullfrogs (*Rana catesbeiana*) obtained commercially were collected by cardiac puncture in presence of 0.24 M EDTA in 0.9% saline (2:10, v/v) [Mello et al., 2000]. Blood smears were then prepared.

Cell Fixation and Staining

The BP1-T*ras* cells and bullfrog erythrocytes were fixed in an absolute ethanol-acetic acid mixture (3:1) for 1 min, rinsed in 70% ethanol for 5 min, air dried at room temperature, and subjected to the Feulgen reaction (hydrolysis: 4 M HCl at 21°C for 75 min – BP1-T*ras* cells [Mello et al., 1999]; 4 M HCl at 25°C for 50 min – erythrocytes [Miyamoto et al., 2005]). The stained preparations were mounted in Canada balsam ($n_D = 1.54$).

Some bullfrog preparations were fixed in 1.75% paraformaldehyde for 15 min at room temperature, rinsed at tap water for 30-50 min, and processed for DNA fragmentation typical of programmed cell death using the terminal deoxynucleotidyl transferase (TdT)-mediated dUTP nick 3-end DNA labeling (TUNEL) technique as described by the manufacturer (Roche/Amersham, Mannheim, Germany). Since UTP was labeled with fluorescein, antifluorescein antibodies linked to horse raddish peroxidase were processed by treatment with 3,3'-diaminobenzidine (Fluka, Buchs, Switzerland) for 10 min at room temperature. Endogenous peroxidase was blocked with 3% H_2O_2 in absolute methanol for 30 min at room temperature. A negative control was done by omitting TdT. Counterstaining was performed in a methyl green (Merck, Darmstadt, Germany) solution.

Scanning Microspectrophotometry Image Analysis

Images were obtained with a Zeiss (Oberkochen, Germany) automatic scanning microspectrophotometer interfaced to a personal computer. The microscope used in this video image analysis procedure has construction characteristics which reduce glare (straylight) effects and diffraction errors. The operating conditions were: predominantly monodirectional scanning motion, Planapo objective 100/1.25; optovar 1.25; measuring diaphragm diameter, 0.10 mm; field diaphragm diameter, 0.20 mm; an LD-Epiplan 16/0.30 condenser; scanning spot size of 0.5 µm x 0.5 µm; halogen 100-W/12-V lamp; stabilized electronic power supply; Zeiss light modulator; a wavelength of 565 nm obtained with a Schott monochromator filter ruler; an R-928 photomultiplier; and a Pentium II microcomputer. Grid points (individual measuring points) showing absorbances no higher than 0.020 were considered to be background and were automatically removed from the nuclear image. The cutoff point of 0.200 was selected as the threshold level to evaluate the absorbing areas covered by condensed chromatin. The nuclei were measured individually. The image analysis parameters pertinent to this study were:

1. A_T, total integrated absorbance (= nuclear Feulgen-DNA values in arbitrary units).
2. A_C, integrated absorbance over the pre-selected cutoff which was the same for all the nuclei analyzed here. A_C corresponds to the condensed chromatin Feulgen-DNA values.
3. A_C%, condensed chromatin Feulgen-DNA values relative to the nuclear (whole chromatin) Feulgen-DNA values.
4. S_T, nuclear absorbing area in μm^2.

5. S_C, area in μm^2 covered with stained chromatin showing absorbances above the selected cutoff point.
6. S_C%, area covered with condensed chromatin relative to the nuclear area.
7. AAR = $(A_C/S_C)/(A_T/S_T)$ = average absorption ratio, a dimensionless parameter that expresses how many times the average absorption of the condensed chromatin exceeds that of the entire nucleus [Vidal et al., 1973].

The scatter diagram relating AAR to S_C%, as previously proposed by Vidal [1984] was then plotted.

Nuclei from BP1-T*ras* cells with typical morphological characteristics of apoptosis (n = 21) were purposedly chosen for measurements in separate from non-apoptotic nuclei (n = 100). One hundred bullfrog erythrocytes were measured.

Photomicrographs

The Feulgen-stained BP1-T*ras* cells were photographed directly from the video monitor in a Zeiss photomicroscope – Data Translation Inc. equipment, using the Global Lab Acquire Software™ (Marlboro, MA) for image acquisition, segmentation and featuring. In this case, the operational conditions involved a JVC TK-S240 monochrome CCD, Pol-Neofluar 25/0.60 objective, optovar 1.25, LD-Epiplan 16/0.30 condenser, λ = 550 nm, and 1 μm = 4.45 pixels. The bullfrog erythrocytes subjected to the Feulgen reaction and to the TUNEL assay were photographed in a Zeiss Axiophot II Universal microscope.

Statistics

All calculations and graphs were done using the Minitab 12™ statistical software for Windows (Minitab, State College, PA).

Results and Discussion

Presence of apoptotic cells in the Feulgen-stained BP1-T*ras* cell preparations was revealed (Figure 2a, b); it was probably elicited by the glass substrate on which the cells were cultivated, since presence of apoptosis in transformed and tumorigenic BP-treated MCF-10F cells cultivated on plastic is reduced [Mello et al., 2002, 2005].

The bullfrog erythrocytes analyzed here showed an intense positive response when subjected to the TUNEL assay (Figure 3a, b), and stained deeply with the Feulgen reaction (Figure 3c). In both cases the morphological images showing condensed chromatin typically apposed to the nuclear borders confirmed previous reports indicating that these cells are highly committed to programmed cell death [Hasebe et al., 1996; Mello et al., 2000; Miyamoto et al., 2005].

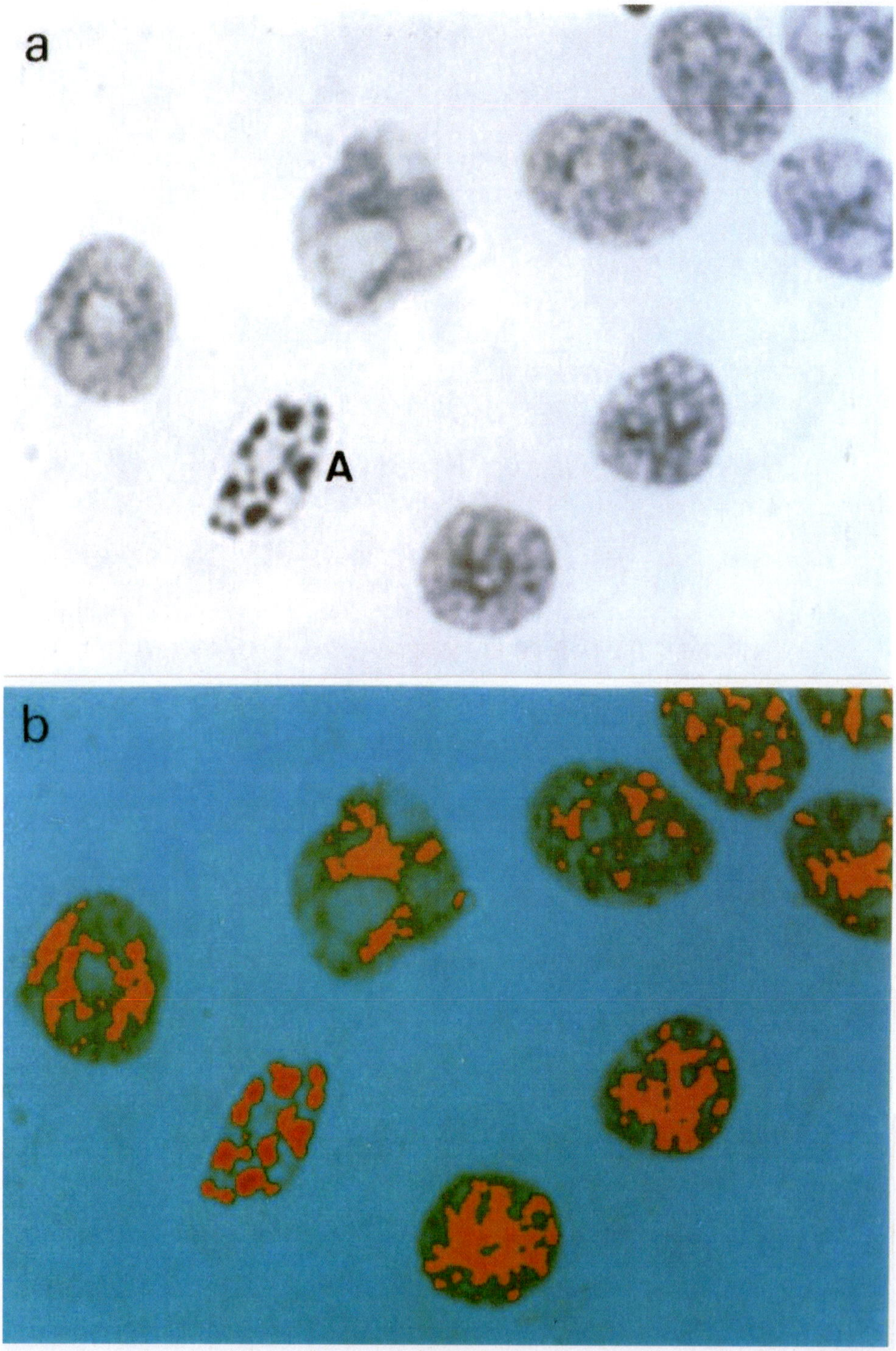

Figure 2. Nuclear images (a, b) of Feulgen-stained BP1-T*ras* cells photographed from the color monitor of a video image analyzer (Data Translation, Inc.). In the pseudocolorized image (b) differences in area covered by the condensed chromatin and its distribution are highlighted. The segmentation step used in the video image analyzer considered a threshold level corresponding to the absorbance cutoff of 0.200 used in the scanning microspectrophotometer. A, apoptotic nucleus.

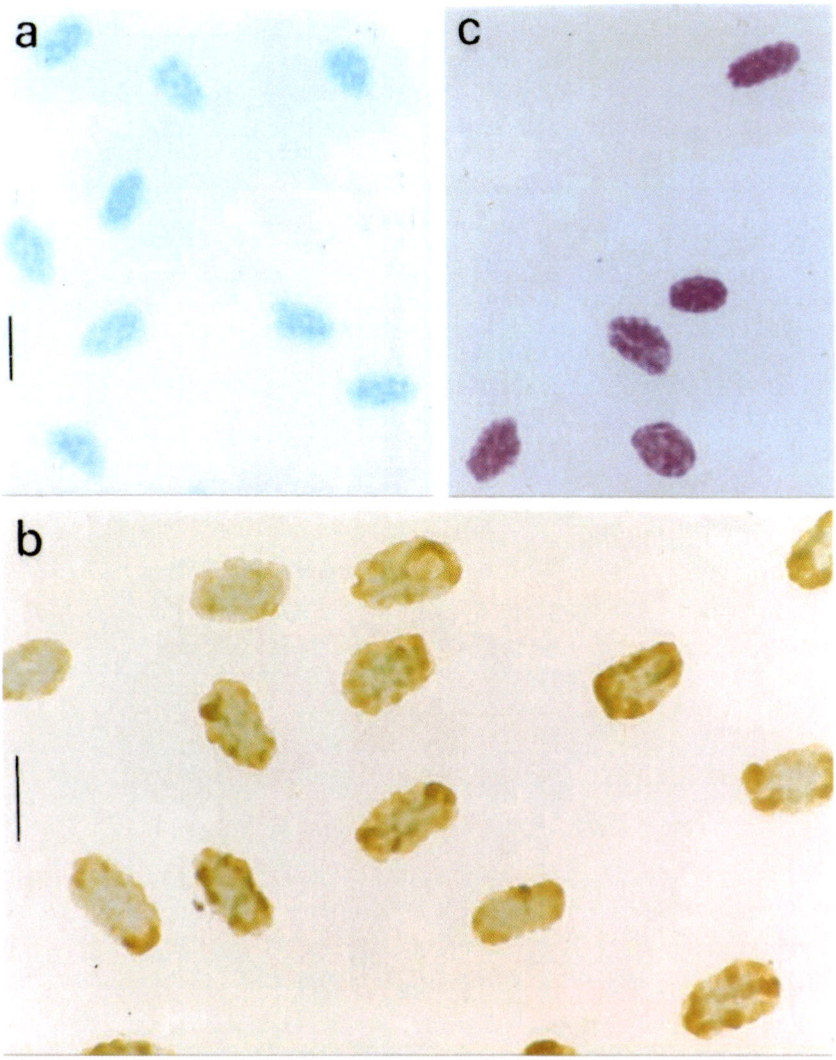

Figure 3. Bullfrog erythrocyte response in the TUNEL assay (a, b) and to the Feulgen reaction (c). a, TUNEL negative control in the absence of TdT. Bars = 10 μm.

The Feulgen-stained BP1-T*ras* cell nuclei represented by match of specific AAR-S_C% values in the scatter diagram are identified in Figure 4. An imaginary parabola developed by the distribution of points of the apoptotic nuclei in the S_C% vs. AAR representation is easily discriminated in this case from that regarding the distribution of the points for the non-apoptotic nuclei. The apoptotic nuclei attained higher S_C% values matching AAR values in the 1.5-3.5 range than the non-apoptotic nuclei. In this case condensed chromatin covered more than 30% of the nuclear area and the contrast between condensed and non-condensed chromatin as detected by the AAR parameter even attained the largest values yet reported. This is in agreement with the morphological aspects of the apoptotic nuclei in which highly-stained points were sharply distinguished from paler-stained points (Figs. 2a, b; 5). No intermediate staining characteristics were observed in nuclei where this chromatin condensation was typically occurring.

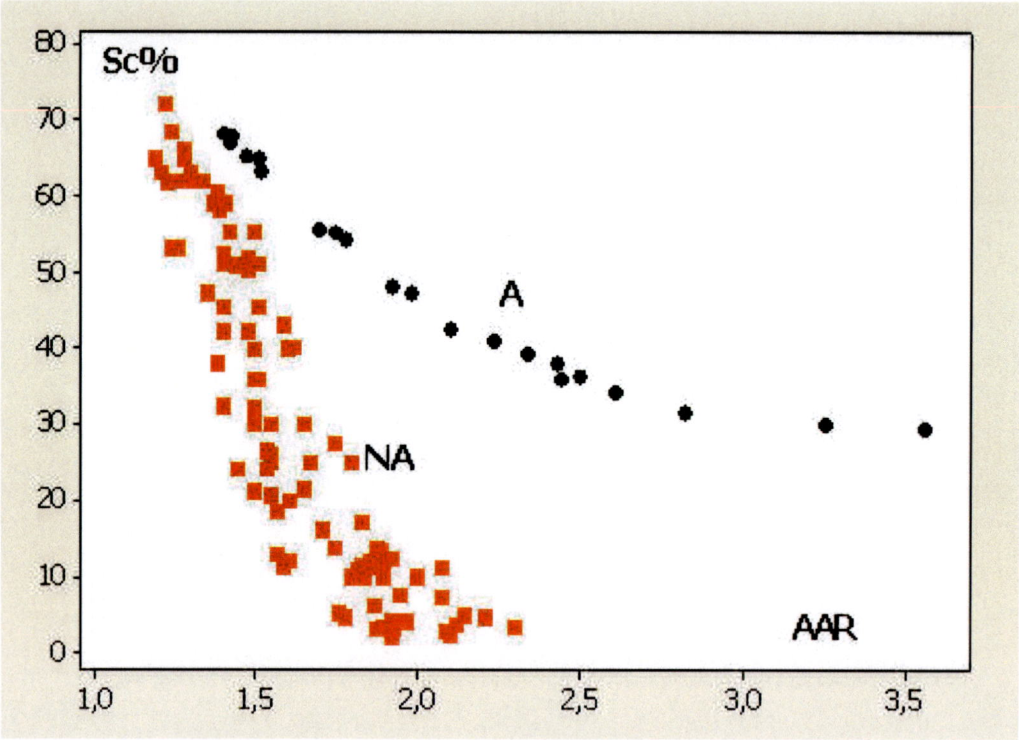

Figure 4. AAR vs S_C% scatter diagram for the Feulgen-stained BP1-T*ras* cells. Apoptotic nuclei (A, black dots) are discriminated from non-apoptotic nuclei (NA, red dots).

Present results in which BP1-T*ras* apoptotic cell nuclei were purposedly separated from non-apoptotic nuclei thus generating two patterns of point distributions in the S_C% vs. AAR diagram, would explain why in cases where this separation was not performed a few points appear as outliers in the unique whole point distribution [Mello et al., 1999]. These points certainly represent apoptotic nuclei.

In the scatter diagram for the Feulgen-stained erythrocyte nuclei most points displayed AAR values larger than 2.0, and S_C% values higher than 25.0 (Figure 6). This means that at least 25% of the nuclear area in these cells is covered with condensed chromatin, considering that a cutoff point defined as absorbance 0.200 was used for the identification of chromatin condensation. It also demonstrates that a sharp contrast between condensed and non-condensed chromatin occurs, inducing high AAR values. The pattern of distribution of the points in the S_C% vs. AAR diagram (Figure 6) was similar to that shown by the apoptotic nuclei of the BP1-T*ras* cells (Figure 4). Considering that the erythrocytes analyzed here were obtained from young adult bullfrogs for which DNA fragmentation has been demonstrated previously (Hasebe et al., 1996; Mello et al., 2000; Miyamoto et al., 2005) and confirmed in the TUNEL assay and with response to the Feulgen reaction (Figure 3a-c), no mixed cell populations were present that could acquire distinct distribution of points in the scatter diagram.

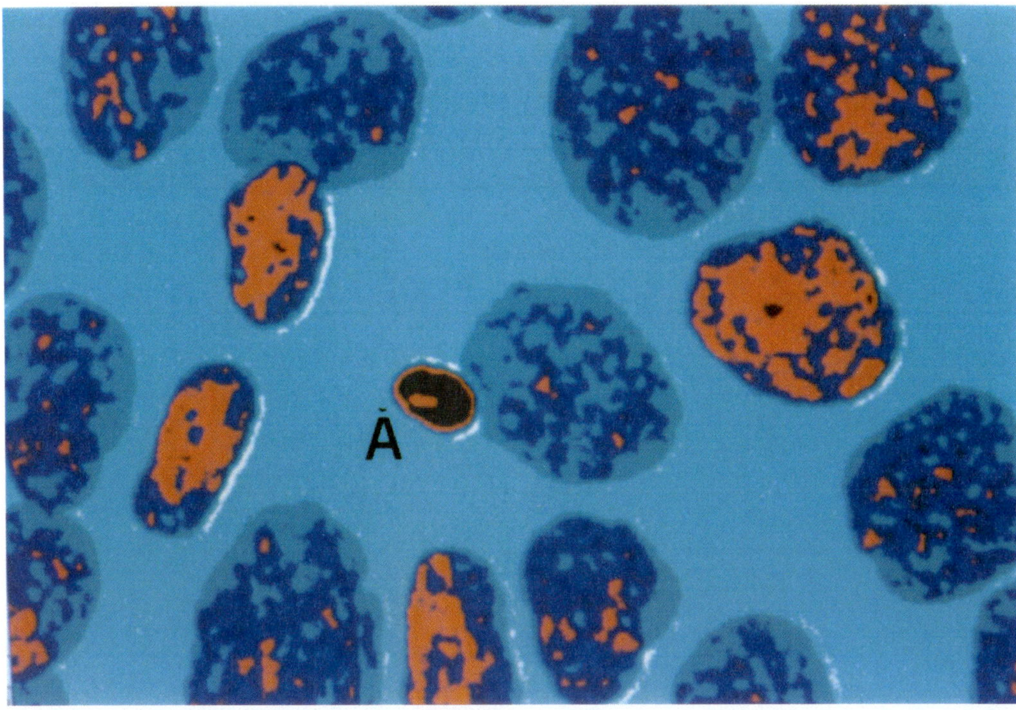

Figure 5. A sharp difference between condensed and non-condensed chromatin in an apoptotic nucleus of Feulgen-stained BP1-T*ras* cells is shown as different colors (black and red) in a pseudocolorized image photographed from the monitor of a video image analyzer (Data Translation, Inc.). A, apoptotic nucleus.

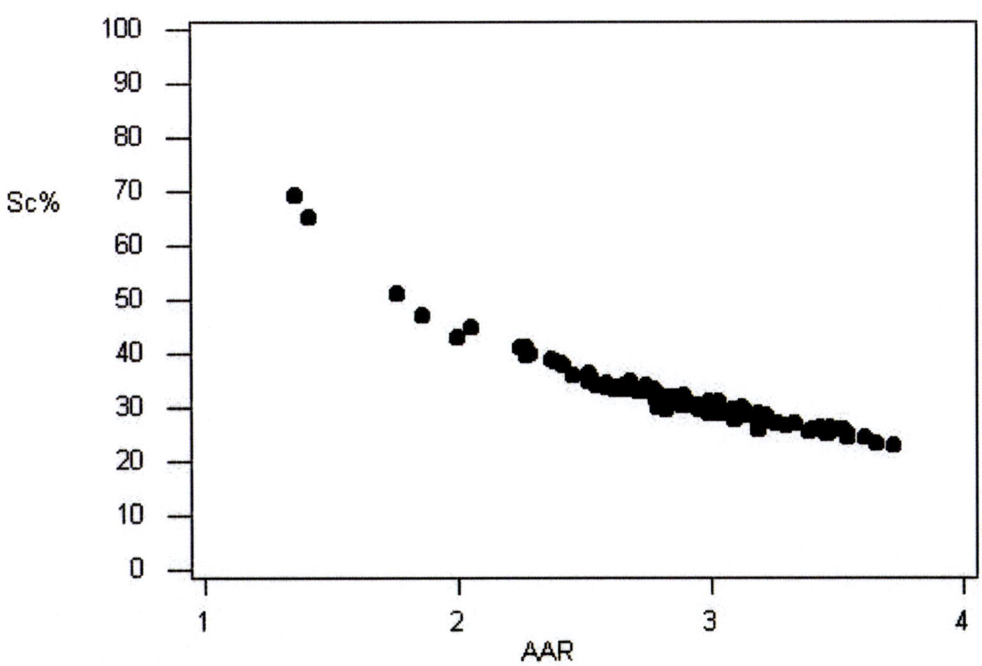

Figure 6. AAR vs S_C% scatter diagram for the Feulgen-stained bullfrog erythrocytes committed to programmed cell death.

Conclusion

As expected, the S_C% vs AAR scatter diagram was found to be a useful tool for discrimination of cells undergoing chromatin condensation in association with the cell death program in Feulgen-stained preparations. This discrimination is due in part because of a relatively high amount of condensed chromatin present but mostly because of a very sharp distinction between condensed and non-condensed chromatin states in the apoptotic nuclei and in cell nuclei under programmed cell death.

Although in the present chapter the methodological analysis relies on the use of an automated scanning microspectrophotometer, a video image analyzer would be equally pertinent to reveal the discriminatory characteristics of the apoptotic cells, provided feasible thresholds for segmentation of the nuclear and condensed chromatin images to be measured are maintained along the analysis process and values for the different parameters required are programmed to associate them to the respective nucleus. For the establishment of apoptotic ratios in automatically screened preparations, a simple examination of point distributions in the S_C% vs. AAR diagram representation would certainly be helpful.

Acknowledgments

The author is indebted to Dr. Jose Russo (Fox Chase Cancer Center, Philadelphia) for kindly providing the BP1-T*ras* cells. This study was supported by grants from the State of São Paulo Research Foundation (FAPESP), the Brazilian National Council for Research and Development (CNPq) and the State University of Campinas Research Foundation (FAEPEX/UNICAMP).

References

Alberts, B., Johnson, A., Lewis, J., Raff, M., Roberts, K. and Walter, P. *Molecular Biology of the Cell,* 4th ed. New York: Garland Science; 2002.

Calaf, G. and Russo, J. (1993). Transformation of human breast epithelial cells by chemical carcinogens. *Carcinogenesis, 14*, 483-492.

Calaf, G., Hang, P.L., Alvarado, M.V., Estrada, S. and Russo, J. (1995). c-Ha-*ras* enhances the neoplastic transformation of human breast epithelial cells treated with chemical carcinogens. *Int. J. Oncol., 6*, 5-11.

Camby, I., Salmon,I., Dauguy, A., Pasteels, J.L. and Kiss, R. (1995). The use of the digital cell image analysis of Feulgen-stained nuclei to detect apoptosis. *Histochem. Cell Biol., 104,* 407-414.

Cremer, T. and Cremer, C. (2001). Chromosome territories, nuclear architecture and gene regulation in mammalian cells. *Nature Rev. Genet., 2*, 292-301.

Creston, G.E. and Kadonga, J.T. (1993). Role of chromatin structure in the regulation of transcription by RNA polymerase II. *Curr. Opin. Cell Biol., 5*, 417-423.

Darfier, M.M. and Karaszkiewicz, J.W. (1995). *In situ* localization of apoptosis using terminal deoxynucleotidyl transference. *Focus, 17*, 81-83.

Dawson, A.E., Austin Jr., R.E. and Weinberg, D.S. (1991). Nuclear grading of breast carcinoma by image analysis. Classification by multivariate and neural network analysis. *Am. J. Clin. Path., 95 (Suppl. 1)*, 529-537.

Diaz, G., Zucca, A., Setzu, M.D. and Cappai, C. (1997). Chromatin pattern by variogram analysis. *Micr. Res. Techn., 39*, 305-311.

Gilbert, N., Gilchrist, S. and Bickmore, W.A. (2005). Chromatin organization in the mammalian nucleus. *Int. Rev. Cytol., 242*, 283-336.

Hasebe, T., Kawamura, K. and Kikuyama, S. (1996). Genomic DNA fragmentation in red blood cells of the bullfrog during metamorphosis. *Dev. Growth. Differ., 38*, 605-615.

Horvitz, H. R., Shaham, S. and Hengartner, M.O. (1994). The genetics of programmed cell death in the nematode *Caenorhabditis elegans. Cold Spring Harbor Symp. Quant. Biol., 59*, 377-385.

Jacobson, M.D., Weil, M. and Raff, M.C. (1997). Programmed cell death in animal development. *Cell, 88*, 347-354.

Kerr, J.F.R., Wyllie, A.H. and Currie, A.R. (1972). Apoptosis: a basic biological phenomenon with wide-ranging implications in tissue kinetics. *Br. J. Cancer, 26*, 239-257.

Klaude, M., Eriksson, S., Nygren, J. and Ahnstrom, G. (1996). The comet assay: mechanisms and technical considerations. *Mutat. Res., 363*, 89-96.

Knowles, D.W., Sudar, D., Bator-Kelly, C., Bissell, M.J. and Lelievre, S.A. (2006). Automated local bright feature image analysis of nuclear protein distribution identifies changes in tissue phenotype. *Proc. Natl. Acad. Sci., USA, 103*, 4445-4450.

Kong, J. and Ringer, D.P. (1995). Quantitative *in situ* image-analysis of apoptosis in well and poorly differentiated tumors from rat-liver. *Am. J. Path., 147*, 1626-1632.

Kuo, W.R., Lapidus, A. and Lambert, W.C. (1998). Identification of early apoptosis in Feulgen-stained cultured cells in situ by computerized image analysis. *Cytometry, 33*, 420-427.

Leyk, M., Nguyen, D.V., Attoor, S.N., Dougherty, E.R., Turner, N.D., Bancroft, L.K., Chapkin, R.S., Lupton, J.R., Carroll, R.J. (2005). Comparing automatic and manual image processing in FLARE assay analysis for colon carcinogenesis. *Stat. Appl. Genet. Mol. Biol., 4*, Art. No. 5.

Majno, G. and Joris, I. (1995). Apoptosis, oncosis and necrosis. An overview on cell death. *Am. J. Path., 146*, 3-15.

Mampumbu, A.R. and Mello M.L.S. (2006). DNA methylation in stingless bees with low and high heterochromatin contents as assessed by restriction enzyme digestion and image analysis. *Cytometry Part A, 69* (in press).

Maria, S.S., Vidal, B.C. and Mello, M.L.S. (2000). Image analysis of DNA fragmentation and loss in V79 cells under apoptosis. *Genet. Mol. Biol., 23*, 109-112.

Marchevsky, A.M. and Bartels, P.H. Image Analysis: a Primer for Pathologists. New York: Raven Press; 1994.

Mello, M.L.S. and Russo, J. (1990). Image analysis of Feulgen-stained c-Ha-*ras*-transformed NIH/3T3 cells. *Biochem. Cell Biol., 68*, 1026-1031.

Mello, M.L.S. and Vidal, B.C. (2000). Chromatin organization in normal and cancer cells. *J. Women's Cancer, 2*, 73-78.

Mello, M.L.S., Contente, S., Vidal, B.C., Planding, W. and Schenck, U. (1995). Modulation of *ras* transformation affecting chromatin supraorganization as assessed by image analysis. *Expt. Cell Res., 220*, 374-382.

Mello, M.L.S., Vidal, B.C. and Russo, J. (1999). Ha-*ras* oncogene effect on DNA content and chromatin supraorganization in benzo[a]pyrene-transformed human breast epithelial cells. *Anal. Cell. Path., 19*, 73-79.

Mello, M.L.S., Maria, S.S., Schildknecht, P.H.P.A. and Grazziotin, N.A. (2000). DNA fragmentation in programmed cell death in nucleate erythrocytes: a cytochemical analysis. *Acta Histochem. Cytochem., 33*, 355-359.

Mello, M.L.S., Tavares, M.C.H., Dantas, M.M., Rodrigues, V.L.C.C., Maria-Engler, S.S., Campos, S.P. and Garcia, N.L. (2001). Cell death and survival alterations in Malpighian tubules of *Triatoma infestans* following heat shock. *Biochem. Cell Biol., 79*, 709-717.

Mello, M.L.S., Barbisan, L.F., Mello, S.M., Perotto, M.C., Kato, P.N., Monte-Mor, B., Freitas, E.L., Nascimento, J.M., Vaz, T.H., Lareef, M.H., Russo, J. and Vidal, B.C. (2002). Apoptosis in benzo[a]pyrene-transformed human breast epithelial cells alter microcell-mediated transfer of chromosomes 11 and 17 [Abstract]. In: *The 28th European Congress of Cytology*, 2002, Antwerpen. Oxford: Blackwell Publ.; 2002; p. 85.

Mello, M.L.S., Vidal, B.C., Lareef, M.H., Hu, Y.F., Yang, X. and Russo, J. (2003). DNA content, chromatin texture and nuclear morphology in benzo[a]pyrene-transformed human breast epithelial cells after microcell-mediated transfer of chromosomes 11 and 17. *Cytometry Part A, 52,* 70-76.

Mello, M.L.S., Lareef, M.H., Santos, A.B., Russo, J. and Vidal, B.C. (2005). Nucleus image properties and cell death in MCF-10F cells grown on slide substrates differing in nature and size. *In Vitro Cell. Dev. Biol.-Animal, 41*, 92-96.

Miyamoto, M., Vidal, B.C. and Mello, M.L.S. (2005). Chromatin supraorganization, DNA fragmentation, and cell death in snake erythrocytes. *Biochem. Cell Biol., 83*, 15-27.

Oberholzer, M., Östreicher, M., Christen, H. and Brühlmann, M. (1996). Methods in quantitative image analysis. *Histochem. Cell Biol., 105*, 333-355.

Singh, N.P., McCoy, M.T., Tice, R.R. and Schneider, E.L. (1988). A simple technique for quantitation of low levels of DNA damage in individual cells. *Expt. Cell Res., 175*, 184-191.

Tsujimoto, Y. and Shimizu, S. (2000). Bcl-2 family: life-or-death switch. *FEBS Letters, 466*, 6-10.

van Driel, R. and Otte, A.P. Nuclear Organization, Chromatin Structure, and Gene Expression. Oxford: Oxford Univ. Press; 1997.

Vidal, B.C. (1984). Polyploidy and nuclear phenotypes in salivary glands of the rat. *Biol. Cell, 50*, 137-146.

Vidal, B.C., Russo, J. and Mello, M.L.S. (1998). DNA content and chromatin texture of benzo[a]pyrene-transformed human breast epithelial cells as assessed by image analysis. *Expt. Cell Res., 244*, 77-82.

Vidal, B.C., Schlüter, G. and Moore, G.W. (1973). Cell nucleus pattern recognition: influence of staining. *Acta Cytol., 17*, 510-521.

Wallet, F., Gérard, H., Martin, P.M. and Dussert, C. (1996). Toward a new method to *in situ* study of apoptosis and its relations with cell cycle. *Cytometry, 25*, 263-270.

Wijsman, J.H., Jonker, R.R., Keijzer, R., van de Velde, C.J.H., Cornelisse, C.J. and van Dierendonck, J.H. (1993). A new method to detect apoptosis in paraffin sections: *in situ* end-labelling of fragmented DNA. *J. Histochem. Cytochem., 41*, 7-12.

Wyllie, A.H., Beathe, G.J. and Hargreaves, A.D. (1981). Chromatin changes in apoptosis. *Histochem. J., 13*, 681-692.

Yatouji, S., Liautaud-Roger, F. and Dufer, J. (2000). Nuclear chromatin texture and sensitivity to DNase I in human leukaemic CEM cells incubated with nanomolar okadaic acid. *Cell Prolif., 33*, 51-62.

Index

A

B

C

D

E

F

I

J

K

L

M

N

O

P

Q

R

S

T

U

V

W

X

Y

Z